Crossing Numbers of Graphs

DISCRETE MATHEMATICS AND ITS APPLICATIONS

Series Editors

Miklos Bona
Patrice Ossona de Mendez
Douglas West

Titles (continued)

Abhijit Das, Computational Number Theory

Matthias Dehmer and Frank Emmert-Streib, Quantitative Graph Theory: Mathematical Foundations and Applications

Martin Erickson, Pearls of Discrete Mathematics

Martin Erickson and Anthony Vazzana, Introduction to Number Theory

Steven Furino, Ying Miao, and Jianxing Yin, Frames and Resolvable Designs: Uses, Constructions, and Existence

Mark S. Gockenbach, Finite-Dimensional Linear Algebra

Randy Goldberg and Lance Riek, A Practical Handbook of Speech Coders

Jacob E. Goodman and Joseph O'Rourke, Handbook of Discrete and Computational Geometry, Third Edition

Jonathan L. Gross, Combinatorial Methods with Computer Applications

Jonathan L. Gross and Jay Yellen, Graph Theory and Its Applications, Second Edition

Jonathan L. Gross, Jay Yellen, and Ping Zhang Handbook of Graph Theory, Second Edition

David S. Gunderson, Handbook of Mathematical Induction: Theory and Applications

Richard Hammack, Wilfried Imrich, and Sandi Klavžar, Handbook of Product Graphs, Second Edition

Darrel R. Hankerson, Greg A. Harris, and Peter D. Johnson, Introduction to Information Theory and Data Compression, Second Edition

Darel W. Hardy, Fred Richman, and Carol L. Walker, Applied Algebra: Codes, Ciphers, and Discrete Algorithms, Second Edition

Daryl D. Harms, Miroslav Kraetzl, Charles J. Colbourn, and John S. Devitt, Network Reliability: Experiments with a Symbolic Algebra Environment

Silvia Heubach and Toufik Mansour, Combinatorics of Compositions and Words

Leslie Hogben, Handbook of Linear Algebra, Second Edition

Derek F. Holt with Bettina Eick and Eamonn A. O'Brien, Handbook of Computational Group Theory

David M. Jackson and Terry I. Visentin, An Atlas of Smaller Maps in Orientable and Nonorientable Surfaces

Richard E. Klima, Neil P. Sigmon, and Ernest L. Stitzinger, Applications of Abstract Algebra with Maple™ and MATLAB®, Second Edition

Richard E. Klima and Neil P. Sigmon, Cryptology: Classical and Modern with Maplets

Patrick Knupp and Kambiz Salari, Verification of Computer Codes in Computational Science and Engineering

William L. Kocay and Donald L. Kreher, Graphs, Algorithms, and Optimization, Second Edition

Donald L. Kreher and Douglas R. Stinson, Combinatorial Algorithms: Generation Enumeration and Search

Titles (continued)

Hang T. Lau, A Java Library of Graph Algorithms and Optimization

C. C. Lindner and C. A. Rodger, Design Theory, Second Edition

San Ling, Huaxiong Wang, and Chaoping Xing, Algebraic Curves in Cryptography

Nicholas A. Loehr, Bijective Combinatorics

Nicholas A. Loehr, Combinatorics, Second Edition

Toufik Mansour, Combinatorics of Set Partitions

Toufik Mansour and Matthias Schork, Commutation Relations, Normal Ordering, and Stirling Numbers

Alasdair McAndrew, Introduction to Cryptography with Open-Source Software

Pierre-Loïc Méliot, Representation Theory of Symmetric Groups

Elliott Mendelson, Introduction to Mathematical Logic, Fifth Edition

Alfred J. Menezes, Paul C. van Oorschot, and Scott A. Vanstone, Handbook of Applied Cryptography

Stig F. Mjølsnes, A Multidisciplinary Introduction to Information Security

Jason J. Molitierno, Applications of Combinatorial Matrix Theory to Laplacian Matrices of Graphs

Richard A. Mollin, Advanced Number Theory with Applications

Richard A. Mollin, Algebraic Number Theory, Second Edition

Richard A. Mollin, Codes: The Guide to Secrecy from Ancient to Modern Times

Richard A. Mollin, Fundamental Number Theory with Applications, Second Edition

Richard A. Mollin, An Introduction to Cryptography, Second Edition

Richard A. Mollin, Quadratics

Richard A. Mollin, RSA and Public-Key Cryptography

Carlos J. Moreno and Samuel S. Wagstaff, Jr., Sums of Squares of Integers

Gary L. Mullen and Daniel Panario, Handbook of Finite Fields

Goutam Paul and Subhamoy Maitra, RC4 Stream Cipher and Its Variants

Dingyi Pei, Authentication Codes and Combinatorial Designs

Kenneth H. Rosen, Handbook of Discrete and Combinatorial Mathematics, Second Edition

Marcus Schaefer, Crossing Numbers of Graphs

Yongtang Shi, Matthias Dehmer, Xueliang Li, and Ivan Gutman, Graph Polynomials

Douglas R. Shier and K.T. Wallenius, Applied Mathematical Modeling: A Multidisciplinary Approach

Alexander Stanoyevitch, Introduction to Cryptography with Mathematical Foundations and Computer Implementations

Jörn Steuding, Diophantine Analysis

Titles (continued)

Douglas R. Stinson, Cryptography: Theory and Practice, Third Edition

Roberto Tamassia, Handbook of Graph Drawing and Visualization

Roberto Togneri and Christopher J. deSilva, Fundamentals of Information Theory and Coding Design

W. D. Wallis, Introduction to Combinatorial Designs, Second Edition

W. D. Wallis and J. C. George, Introduction to Combinatorics, Second Edition

Jiacun Wang, Handbook of Finite State Based Models and Applications

Lawrence C. Washington, Elliptic Curves: Number Theory and Cryptography, Second Edition

Crossing Numbers of Graphs

Marcus Schaefer

DePaul University

Chicago, IL, USA

CRC Press

Taylor & Francis Group

Boca Raton   London   New York

CRC Press is an imprint of the
Taylor & Francis Group, an **informa** business

A CHAPMAN & HALL BOOK

CRC Press
Taylor & Francis Group
6000 Broken Sound Parkway NW, Suite 300
Boca Raton, FL 33487-2742

First issued in paperback 2022

Version Date: 20171107

ISBN 13: 978-1-03-247644-5 (pbk)
ISBN 13: 978-1-4987-5049-3 (hbk)

DOI: 10.1201/9781315152394

Library of Congress Cataloging-in-Publication Data

Names: Schaefer, Marcus, 1969- author.
Title: Crossing numbers of graphs / Marcus Schaefer.
Description: Boca Raton : CRC Press, 2018. | Includes bibliographical references and index.
Identifiers: LCCN 2017036245 | ISBN 9781498750493
Subjects: LCSH: Topological graph theory--Textbooks. | Graph theory--Textbooks. | Geometry, Plane--Textbooks.
Classification: LCC QA166.195 .S33 2018 | DDC 511/.5--dc23
LC record available at https://lccn.loc.gov/2017036245

To Bernard.
Lovingly.

Contents

Preface

In Luca Pacioli's puzzle of the "three castles and the three fountains" Pacioli asks "whether it is possible that each castle is connected to its own gate by streets that do not cross each other in any way". The setting of the puzzle is shown in Figure 1. The three castles are on top, pictured as circles, and surrounded by a wall with three gates, one for each castle.[1]

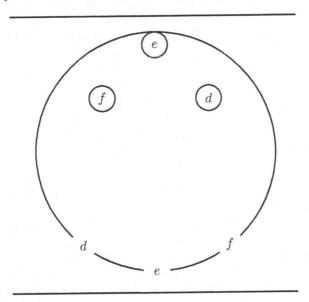

FIGURE 1: The Three Castles and the Three Fountains.

Pacioli's puzzle is remarkable for many reasons; it is one of the oldest known graph drawing puzzles, contained in a manuscript written in the first half of the 16th century, and it is has survived over the centuries in various forms.[2] It is also one of the first times the concept of crossing occurs in the context of visualization, and, already, it has a bad reputation. Crossings are to be avoided.

---

[1]Pacioli gives two versions of the puzzles, castles and gates, and monasteries and fountains; why he mixed the two in the name of the puzzle, only he knew. Thanks to my colleague Lucia Dettori for help with translating the original text.

[2]Numberlink is a current game played on a grid, based on the same idea.

And crossings were avoided; planarity and non-planarity had long made their journey from puzzles into mathematical research before crossing numbers appeared as a topic in mathematics, starting with Turán's brick factory problem in the early 1950s. Since then, the topic has steadily grown in both breadth and depth, and crossing number research has become entrenched as one of the core areas in graph drawing. Crossings will not go away, and the recent rise of information visualization has made it woefully clear that we are not really prepared yet to answer the question of how to visualize large, dense networks, which require lots of crossings (by the so-called crossing lemma).

The time then seems right for a book on the crossing number; the field has matured, there is a large body of work, with identifiable core results and techniques, and there are applications outside the field itself. In writing this book, I have tried to present a uniform view of the crossing number universe, and I hope to have accomplished this by emphasizing intuitive and visual proofs (over a more formal approach).

The book consists of three parts. The first part deals with the traditional crossing number, while the second part studies some of the more important crossing number variants that have been introduced over the years. The third part consists of two appendices.

The four chapters of Part I are as follows: Chapter 1 reviews many of the fundamental and classical results on the crossing number—including the parity and crossing lemmas—in particular as they relate to two of the two foundational conjectures of the field: Zarankiewicz' conjecture on the crossing number of complete bipartite graphs, and Hill's conjecture on the crossing number of complete graphs; both conjectures remain open. Chapter 2 develops basic tools on good drawings and studies the crossing number of families of graphs other than the complete and complete bipartite graphs. Chapter 3 compares the crossing number to other parameters, such as thickness, chromatic number and genus. Two general methods for proving crossing number lower bounds are developed based on bisection width and embedding congestion. Chapter 4 studies the computational complexity of the crossing number problem, and presents results and algorithms useful for dealing with drawings and crossing numbers.

Part II collects eight chapters on various crossing number variants. Chapters 5 and 6 deal with the crossing number of rectilinear drawings, the rectilinear crossing number, one of the oldest, and most important, alternatives to the crossing number. Chapter 5 studies rectilinear and pseudolinear drawings, and Chapter 6 looks at values of the rectilinear crossing number, specifically for the complete graph, a problem that has seen extraordinary progress recently using tools from discrete geometry. Chapter 7 concerns one of the oldest, and most neglected, crossing numbers, the local crossing number, which counts crossings along edges. Recent applications, and the rise of 1-planarity has sparked renewed interest in this variant. Chapter 8 combines material on the crossing number of monotone and book drawings of a graph. One important special case studied in that chapter is that of the crossing number of convex drawings. Chapter 9 studies the pair crossing number, which has often been mistaken for the crossing number; the pair crossing number has strong ties to string graphs, intersection graphs of curves, and we review this fruitful interaction. In Chapter 10 we look at drawing graphs in a fixed number of planes, a crossing number notion inspired by circuit layout. Chapter 11 investigates whether a theory of crossing numbers can be based on counting crossings algebraically or modulo 2, as envisioned by Tutte [340]. Chapter 12 is the only chapter in which instead of minimizing crossings, we attempt to maximize them; there are many intriguing conjectures in this area, not least of all, Conway's infamous thrackle conjecture.

Part III consists of two appendices. Appendix A sketches the foundations of topological graph theory, and developing some specialized results we need in the rest of the book. Appendix B is a very brief introduction to computational complexity.

We assume that the reader knows basic graph theory (we recommend and use Diestel [113] for reference). Many of the chapter exercises are results from research papers; to make it easier to resist temptation, I have not included names and references for exercises, but this information is available if needed for scholarly purposes.

Acknowledgments

This book was written at the invitation of Miklos Bona, and I would like to give belated thanks to Dan Archdeacon for suggesting my name to him in the first place; also thanks go to Éva Czabarka and László Székely for their gentle encouragement to take on this project. There are too many researchers who have talked to me about crossing numbers to acknowledge all of them, but I would like to single out my frequent coauthors Michael Pelsmajer and Daniel Štefankovič.

We like to think that research is cumulative, but it is sometimes hard to locate older papers; I have not included a source, unless I had a copy of the original article, and I would like to thank the DePaul librarians who made this possible. I would also like to thank DePaul University and David Miller, dean of the College of Computing and Digital Media, for a two-quarter research leave during which much of the writing for this book was done in Toronto.

Finally, I want to thank Bernard for his loving support while I was preoccupied with the writing of this book during my stay in Toronto.

List of Figures

List of Tables

Symbol Description

general

$\sim$	asymptotically equivalent

graph families

C_n	cycle on n vertices
K_n	complete graph on n vertices
$K_{m,n}$	complete bipartite graph on m and n vertices
P_n	path on n vertices
Q_d	hypercube of dimension d
W_n	wheel on n vertices

graph parameters

$\deg_G(v)$	degree of v in G
ssqd	sum of squared degrees, $\sum_v \deg^2(v)$
bt	book thickness
bw	bisection width
cw	cut width
hbw	hereditary bisection width
sk	skewness
χ	chromatic number
ϕ	congestion of embedding
σ	size of balanced separator
θ	thickness
ϑ	thrackle bound
ϑ'	subthrackle bound

graph operators

$+$	join
$\square$	Cartesian product
$\hat{\square}$	capped Cartesian product
$\odot_\sigma$	zip-product wrt σ
E_k	(k)-edges
$E_{\leq k}$	$(\leq k)$-edges
$E_{\leq\leq k}$	$(\leq\leq k)$-edges
$N_G(v)$	open neighborhood of v in G

crossing numbers

acr	algebraic crossing number
bkcr_k	k-page book crossing number
cr	crossing number
$\overline{\mathrm{cr}}$	rectilinear crossing number
$\widetilde{\mathrm{cr}}$	pseudolinear crossing number
cr_Σ	crossing number on Σ
cr_-	independent crossing number
cr_k	k-planar crossing number
$\overline{\mathrm{cr}}_k$	rectilinear k-planar crossing number
$\overline{\overline{\mathrm{cr}}}_k$	geometric k-planar crossing number
ecr	edge crossing number
iacr	independent algebraic crossing number
iocr	independent odd crossing number
lcr	local crossing number
lcr*	simple local crossing number
lpcr	local pair crossing number
max-cr	maximum crossing number
max-$\overline{\mathrm{cr}}$	maximum rectilinear crossing number
mcr	minor crossing number
mon-cr	monotone crossing number
ocr	odd crossing number
pcr	pair crossing number
$\overline{\nu}^*$	rectilinear crossing constant

functions

$ES(n)$	Erdős-Szekeres function
sgn	signum (sign) function
$X(n)$	$\lfloor\frac{n}{2}\rfloor\lfloor\frac{n-1}{2}\rfloor$
$Z(n)$	$\frac{1}{4}\lfloor\frac{n}{2}\rfloor\lfloor\frac{n-1}{2}\rfloor\lfloor\frac{n-2}{2}\rfloor\lfloor\frac{n-3}{2}\rfloor$
$Z(m,n)$	Zarankiewicz fct $X(m)X(n)$

Part I

The Crossing Number

Chapter 1

The Conjectures of Zarankiewicz and Hill

1.1 Drawings with Crossings

1.1.1 A Puzzle

In his "Perplexities" column for *The Strand Magazine*, Henry Dudeney posed a puzzle he called "Water, Gas, and Electricity", accompanied by the illustration in Figure 1.1. Is it possible, he asked, that A, B, and C each have access to (W)ater, (G)as, and (E)lectricity, without their paths crossing?

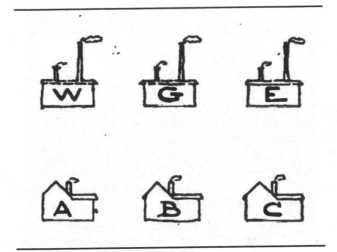

FIGURE 1.1: The Water, Gas, and Electricity puzzle.

Dudeney writes his problem is "as old as the hills", but his 1913 column remains the earliest known publication of this problem. We would now answer that a solution would require a planar drawing of a $K_{3,3}$ which we know to be impossible by Kuratowski's theorem (Appendix A reviews the basics of topological graph theory). Dudeney admits that the problem has no solution, and then presents the solution in Figure 1.2.

3

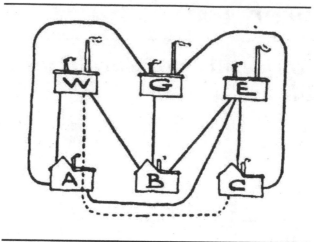

FIGURE 1.2: A solution to the Water, Gas, and Electricity puzzle?

After some initial outrage, our reaction should be constructive: it will pay off to carefully consider what we mean by a drawing of a graph, to avoid sophistries.

In a *drawing* of a graph, every vertex is assigned a different location, and edges are drawn as simple curves between the locations assigned to their endpoints. If two edges intersect finitely often, and an edge does not contain a point in its interior, then intersections between two curves are either a common endpoint, a *crossing*, or a *touching point*. The difference between a crossing and a touching point is that in every sufficiently small neighborhood of the point, crossing edges alternate as they intersect the boundary of the neighborhood, while touching edges intersect consecutively.

A drawing is *normal*, if every pair of edges has at most a finite number of intersections, there are no touching intersections, and no crossing belongs to more than two edges. We will always assume that **drawings are normal**, unless clearly stated otherwise.

Remark 1.1. Why do we not allow more than two edges crossing in a point? One possible answer is that it is not necessary: if k edges cross in the same point, this point corresponds to $\binom{k}{2}$ crossings (between pairs of edges), and the same number can be achieved by perturbing the edges slightly (as was observed by Tait [330] in 1877, also see Lemma A.23). If we want to count a point in which multiple edges cross as a single crossing, we get the concept of the *degenerate crossing number*, a topic we do not pursue [248].

A drawing is *good* if edges do not have self-crossings (they are simple curves), and every pair of edges has at most one intersection (including a shared endpoint). In particular, adjacent edges do not cross, and independent edges cross at most once.

With these clarifications, we are ready for the central definition of this chapter.

Definition 1.2 (Crossing Number). The *crossing number*, $\mathrm{cr}(D)$, of a drawing D is the number of crossings in D. The *crossing number*, $\mathrm{cr}(G)$, of a graph G is the smallest $\mathrm{cr}(D)$ of any drawing D of G. We say a drawing D of G is cr-*minimal*, or that D *realizes* $\mathrm{cr}(G)$, if $\mathrm{cr}(D) = \mathrm{cr}(G)$.

The following lemma explains why we often work with good rather than (normal) drawings.

Lemma 1.3. *If a drawing is not good, we can redraw it so it becomes good, the crossing number of the drawing decreases, and redrawn edges are routed arbitrarily close to edges of the original drawing. In particular, every* cr-*minimal drawing of a graph is good.*

Proof. Suppose D is a drawing of G which is not good. If D has a self-crossing, it can be removed by locally rerouting the edge close to the self-crossing; see Figure 1.3. (We could also short-cut the edge at the self-crossing; this is fine for the purposes of this proof, but sometimes we need to keep all pieces of an edge to satisfy some other condition.) So D must contain two edges e_1

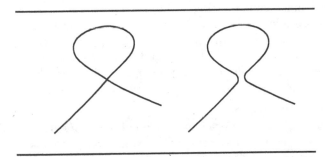

FIGURE 1.3: Removing a self-crossing.

and e_2 which intersect at least twice, and therefore cross at least once, since we do not allow multiple edges. Let p_1 and p_2 be two intersections of e_1 and e_2, consecutive along e_1, and let $\gamma_i \subseteq e_i$ be the arcs connecting p_1 and p_2. Rerouting at p_i if it is a crossing (not if it is a shared endpoint), for $i = 1, 2$, gives us a new drawing with fewer crossings, since at least one of the p_i is a crossing, and gets removed, and the resulting drawing is a drawing of the same graph (though the order of ends at a shared endpoint may have changed). This redrawing, which is often called *swapping* the arcs, may have introduced self-crossings, but these can be removed as we saw. Figure 1.4 illustrates different configurations. In all cases, we have strictly reduced the number of crossings, hence repeating this process must terminate with a good drawing. If D is cr-minimal, then it has to be good, since otherwise, it could be redrawn so as to reduce the number of crossings as we just saw. □

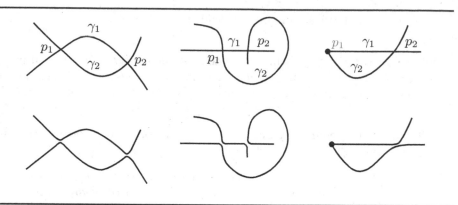

FIGURE 1.4: Removing multiple crossings by swapping arcs.

Remark 1.4 (Rules + and −). Lemma 1.3 seems to say that crossings between adjacent edges do not matter. Pach and Tóth [264] introduced two rules to study the effect of adjacent crossings (short for "crossings between adjacent edges") more closely. *Rule* + only allows good drawings of a graph, and *Rule* − counts only crossings between independent pairs of edges (adjacent edges are allowed to cross for free). Applying the rules to cr, we get two variants cr_+ and cr_-, the *independent crossing number*. By definition, $cr_-(G) \le cr(G) \le cr_+(G)$, and Lemma 1.3 implies that $cr_+ = cr$. On the other hand, $cr_- = cr$ is a major open research problem. Even showing that the two are polynomially related is not trivial. The best bound $cr(G) \le \binom{cr_-(G)}{2}$ follows from Theorem 11.13.

We saw that cr-minimal drawings are good; the reverse is not true. Nevertheless, it can be useful to study good drawings, as we will see when talking about the crossing number of complete and complete bipartite graphs.

1.1.2   Distinguishing Drawings of Graphs

Let us study some small, concrete examples. Figure 1.5 shows cr-minimal drawings of K_3 through K_7 (we will show that these drawings are cr-minimal later).

Based on these drawings one may conjecture that all cr-minimal drawings of complete graphs are *rectilinear* (edges are straight-line segments), which is false, as we will see in Chapter 6, or that there always is a crossing-free triangle in the drawing, which is false in general, see Exercise (1.9), but is true for rectilinear drawings (Theorem 6.6).

A drawing of a graph separates the plane (or the surface it is drawn on) into several *regions*, maximal connected subsets. The *size* of a region is the number of edge segments the region is incident to (an *edge segment* is a maximal connected subset of an edge not containing a vertex or a crossing), where

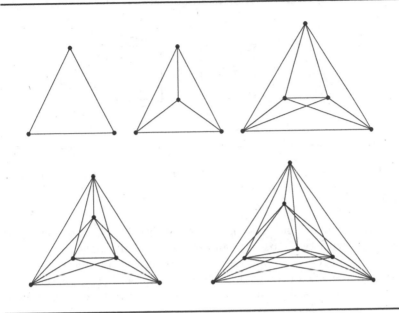

FIGURE 1.5: Crossing-minimal drawings of K_3 through K_7.

we distinguish the two sides of an edge, so every edge segment contributes two incidences, possibly to the same region. A *k-region* is a region of size k. A region which is not incident on any crossings is a *face*, or *k-face* if it is incident to k edges. For example, the drawing of K_5 in Figure 1.5 consists of eight 3-regions, four of which are 3-faces. It is sometimes convenient to treat crossings like vertices; the *planarization* of a drawing of a graph, is the graph that results from replacing each crossing with a vertex of degree 4.

We use this terminology to show that there are not too many ways to draw K_4 and K_5. We say two drawings of a graph are *isomorphic* if there is a homeomorphism of the surface that maps one drawing to the other. Figure 1.6 shows the three non-isomorphic good drawings of K_4 in the plane. The second and third are isomorphic if we work in the sphere. A drawing in the plane is a drawing in the sphere with a designated outer (unbounded) region; drawings in the plane are easier to visualize, but drawings in the sphere are fewer, so simpler to reason with.

Lemma 1.5. *There are two non-isomorphic good drawings of K_4 on the sphere.*

Proof. Figure 1.6 shows the two good drawings of K_4 on the sphere (the third drawing being isomorphic to the second drawing on the sphere). The two drawings are not isomorphic, since their crossing numbers differ.

To show there are no other non-isomorphic good drawings, we first note

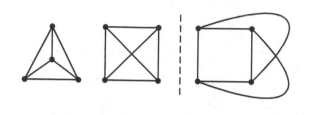

FIGURE 1.6: Non-isomorphic good drawings of K_4 in the plane. The last two are isomorphic on the sphere.

that a C_4 has at most one crossing in a good drawing: if there were a good drawing of C_4 with two crossing pairs of independent edges, let $uvwx$ be a 3-path along the cycle and suppose uv and wx have a crossing c. Then u and x are on the same side of the triangle vcw, but ux has to cross vw once. That is not possible without crossing vc or vw which is not allowed. So any good drawing of a C_4 has at most one crossing.

Since any two disjoint pairs of independent edges in a K_4 belong to a C_4, our observation on C_4 implies that K_4 can have at most one crossing in a good drawing. Since K_4 is 3-connected, its planar drawing is unique (up to a homeomorphism of the sphere) by Whitney's theorem (Theorem A.9). If the K_4 has a single crossing, the planarization of the drawing is 3-connected, and, so, it also has a unique embedding.   □

Figure 1.7 shows five non-isomorphic drawings of K_5 on the sphere.

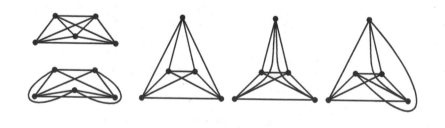

FIGURE 1.7: Non-isomorphic good drawings of K_5.

Lemma 1.6. *There are five non-isomorphic drawings of K_5 on the sphere.*

Proof. Figure 1.7 shows five good drawings of K_5. To argue that they are not isomorphic, it is sufficient to compare them by the largest number of crossings along any edge (the local crossing number of the drawing).

Since K_5 is non-planar, any drawing must contain at least one non-planar K_4. Fix the drawing of the K_4 (by Lemma 1.5 it is unique on the sphere, to visualize it in the plane, we choose the 4-region as the unbounded face). The fifth vertex can only lie in two different types of regions: a 3-region, or the 4-region. If it lies in a 3-region, its edges to the two vertices incident to that region have to lie inside the region. For each of the remaining two edges, there are three choices, resulting in two non-isomorphic drawings overall, see the top of Figure 1.8.

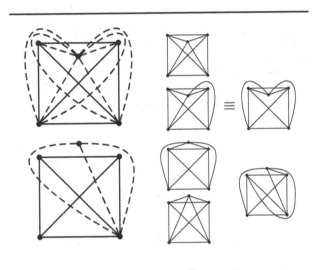

FIGURE 1.8: Building good drawings of K_5. *Top left:* Different ways a vertex in a 3-region can connect to the K_4. *Bottom left:* Different ways a vertex in the 4-region can connect to a vertex of the K_4. *Right:* Five non-isomorphic drawings of K_5 (two of the six drawings are isomorphic).

If the fifth vertex lies in the 4-region, each edge connecting it to a vertex again has three choices, one of them crossing-free, and two requiring two crossings each, see the bottom left of Figure 1.8. Since we can have at most five crossings overall—as there are only five K_4-subgraphs in the drawing—at most two of the edges incident to the fifth vertex can be involved in crossings. If both are, we get one of the two drawings with five crossings; if one is, we get one of the drawings with three crossings we already saw; if no edge incident to the fifth vertex is involved in a crossing, we obtain the unique drawing of K_5 with a single crossing. □

Another look at Figures 1.6 and 1.7 shows that the cr-minimal drawings of K_4 and K_5 are unique; this is still true for K_6, but fails for K_7 (see Exercise (1.4)).

Drawing isomorphism is a very strong notion, and often hard to handle; we introduce two weaker notions that lend themselves more easily to combinatorial analysis: rotation-system equivalence and weak isomorphism. Recall that the *rotation* at a vertex in a drawing is the cyclic permutation of edges induced by the clockwise order of their ends at the vertex, and a *rotation system* for a graph assigns a rotation to each vertex. We can consider two drawings of a graph *rotation-equivalent* if they have the same rotation system (up to a vertex-isomorphism). If a drawing of a graph has a specific rotation system, we say the drawing *realizes* the rotation system. Two drawings of a graph are *weakly isomorphic* if up to a vertex-isomorphism, the same pairs of edges cross in the two drawings. A *weak isomorphism type* is a graph together with a list of pairs of edges.[1] We say a drawing *realizes* a weak isomorphism type if exactly the pairs of edges on the list cross in the drawing. Rotation equivalence and weak isomorphism are both strictly weaker notions than drawing isomorphism; Figure 1.9 shows two rotation-equivalent drawings which are not weakly isomorphic, and hence not isomorphic.

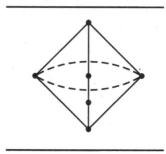

FIGURE 1.9: Two non-isomorphic realizations (indicated by dashed edges) of rotation-equivalent drawings.

On the other hand, while any two plane drawings of a planar graph are weakly isomorphic, they do not have to be isomorphic or rotation-equivalent.

1.1.3   The Crossing Lemma

By Euler's formula, a planar graph on n vertices can have at most $3(n-2)$ vertices (Corollary A.5). Therefore, any sufficiently dense graph must have many crossings. To make this precise, we start with a simple, but useful bound.

Lemma 1.7. *If G is a graph with $n \geq 3$ vertices, and m edges, then $\mathrm{cr}(G) \geq m - 3(n-2)$.*

Proof. Fix a cr-minimal drawing of G. Since at most two edges are involved in every crossing, we can turn the drawing into a plane embedding by removing

---

[1] In the literature weak isomorphism types are more commonly known as *abstract topological graph*, or *AT-graph*, for short.

at most $\mathrm{cr}(G)$ edges, so, by Euler's formula, $m - \mathrm{cr}(G) \leq 3(n-2)$, so $\mathrm{cr}(G) \geq m - 3(n-2)$. □

Amplifying this simple bound yields the powerful crossing lemma.

Theorem 1.8 (Crossing Lemma. Ajtai, Chvátal, Newborn, and Szemerédi [23]; Leighton [225]). *If G is a graph with n vertices, and m edges, and so that $m \geq \lambda n$, then*

$$\mathrm{cr}(G) \geq c\frac{m^3}{n^2},$$

with $c = (\lambda^{-2} - 3\lambda^{-3})$.

Clearly we need $\lambda > 3$ for the lower bound to be non-trivial. Two special cases of interest are $m \geq 4n$ with a factor of $c = 1/64$, and $m \geq 9/2n$ with the best lower bound (using this method) of $c = 4/243 = 1/60.75$. The probabilistic proof of the crossing lemma presented here is due to Chazelle, Sharir, and Welzl, see Aigner and Ziegler [22].

Proof. Fix a drawing of G with $\mathrm{cr}(G)$ crossings. Let H be a random subgraph of G, picking each vertex of G with probability p. Then the expected number of vertices of H is pn, the expected number of edges $p^2 m$, and the expected number of crossings, in the induced drawing, $p^4 \mathrm{cr}(G)$; here we use Lemma 1.3, which implies that any crossing occurs between two independent edges.

Then $p^4 \mathrm{cr}(G) \geq p^2 m - 3pn + 6$ by Lemma 1.7. Choosing $p = \lambda n/m$ yields $\mathrm{cr}(G) \geq (\lambda^{-2} - 3\lambda^{-3})m^3/n^2$. Since p has to be a probability, we need to require $\lambda n/m \leq 1$, that is, $m \geq \lambda n$. □

To improve the value of c in the crossing lemma, we need stronger bounds than Lemma 1.7; we will return to this topic in Chapter 7 on the local crossing number.

If we define $\kappa(n,m) := \min\{\mathrm{cr}(G) : |V(G)| = n, |E(G)| = m \geq 4n\}$, then

$$c_1 m^3/n^2 \leq \kappa(n,m) \leq c_2 m^3/n^2.$$

The existence of c_1 follows from the crossing lemma, and the existence of c_2 can be seen as follows: let $t = \lceil 4m/n \rceil + 1$. If $t > n/2$, then $m \geq n^2/8$; take a subgraph of K_n with exactly m edges. It has crossing number at most $\binom{n}{4}$, which is $O(m^3/n^2)$. If $t \leq n/2$, we take $\lfloor n/t \rfloor$ copies of K_t. This graph has at most n vertices, so we can add isolated vertices to make it an n-vertex graph. The number of edges is $\lfloor n/t \rfloor \binom{t}{2} \geq n/(2t)\binom{t}{2} = n(t-1)/4 \geq m$, where we used $t \leq n/2$ in the first inequality.

Erdős and Guy [124] conjectured that the limit of $\kappa(n,m)n^2/m^3$ exists as m/n goes to infinity. We write $f(n) \ll g(n)$ if $f(n) = o(g(n))$.

Theorem 1.9 (Pach, Spencer, and Tóth [259]). *If $n \ll m(n) \ll n^2$, then* $\lim_{n\to\infty} \kappa(n, m(n))n^2/(m(n))^3 = c > 0$ *exists.*

The limit in this theorem is known as the *midrange crossing constant*. Any lower bound on the crossing lemma is a lower bound on the midrange crossing constant; the current best bounds are $0.029 < c < 0.06$ [263, Remark 3.2].

We complete this section with one of the most celebrated applications of the crossing lemma, Székely's simplified proof of the Szemerédi-Trotter theorem [328] in incidence geometry.

Theorem 1.10 (Szemerédi-Trotter Theorem). *There are at most $4(p\ell)^{2/3} + 4p + \ell$ incidences between p points and ℓ lines in the plane.*

Proof. Fix a drawing D of p points and ℓ lines. Any lines which contain one or zero points can be removed since they contribute at most one point-line coincidence, but increase the bound by at least 1. So we can assume that every line is incident on at least two points. This allows us to define a graph G: the vertices are the p points, and there are edges between two points if they are consecutive along a line. The drawing of G contained in D is rectilinear, so $\mathrm{cr}(G) \leq \mathrm{cr}(D) \leq \ell^2$. On the other hand, the crossing lemma implies that $\ell^2 \leq \mathrm{cr}(G) \geq \frac{1}{64}m^3/n^2$, unless $m < 4n$, where $m = |E(G)|$. So in either case $m \leq 4((p\ell)^{2/3} + n)$. The number of incidences on each line is one more than the number of edges on that line, so the total number of incidences is $m + \ell \leq 4((p\ell)^{2/3} + p) + \ell$.   □

Erdős and Szemerédi showed that for a finite set A of real numbers it is not possible that both the sum-set $A + A := \{a + a' : a, a' \in A\}$ and the product set $A \cdot A := \{aa' : a, a' \in A\}$ are small. Via the Szemerédi-Trotter theorem, the crossing lemma gives an easy proof of that.

Theorem 1.11 (Elekes [122]). *For any finite set A of real numbers, $|A+A| + |A \cdot A| \geq c|A|^{5/4}W$.*

Proof. Let $P := \{(a, a') : a \in A + A, a' \in A \cdot A\}$ and $L = \{y = a(x - a') : a, a' \in A\}$. Then $|P| = |A + A||A \cdot A|$ and $|L| = |A|^2$. Since each line in L is incident to $|A|$ points (of the form $a' + a''$), the number of incidences is at least $|A|^3$. On the other hand, by the Szemerédi-Trotter theorem, that number is at most $4(|P||A|^2)^{2/3} + 4|P| + |A|^2$. Since $|P| \leq |A|^4$, we have $4|P| = 4|P|^{2/3}|P|^{1/3} \leq 4(|P||A|^2)^{2/3}$. Also, $|P| \geq |A|^2$, so $|A|^2 \leq (|P||A|^2)^{2/3}$. Using these inequalities we conclude that

$$|A|^3 \leq 9(|P||A|^2)^{2/3},$$

which gives us $|P| \geq 9^{-3/2}|A|^{5/2}$ implying the claim.   □

1.1.4   The Parity Lemma

The main goal of this chapter is the study of crossing numbers of complete and complete bipartite graphs. An elegant, and somewhat surprising, parity phenomenon occurs for good drawings of some of these graphs.

Theorem 1.12 (Kleitman [208]). *If D is a good drawing of K_n with n odd, then $\mathrm{cr}(D) \equiv \binom{n}{4} \bmod 2$. If D is a good drawing of $K_{m,n}$ with both m, n odd, then $\mathrm{cr}(D) \equiv \binom{n}{2}\binom{m}{2} \bmod 2$.*

To simplify the proof, we work with the *independent crossing number*, $\mathrm{cr}_-(G)$ which we introduced in Remark 1.4. Recall that $\mathrm{cr}_-(D)$ counts the number of crossings in D between pairs of independent edges only, and that $\mathrm{cr}_-(G)$ is the smallest $\mathrm{cr}_-(D)$ for any drawing D of G.

Proof. The proof consists of two parts: first we note that there are good drawings of K_n and $K_{m,n}$ for which the claims hold, and, secondly, we show that the parity of cr_- is invariant for drawings of K_n and $K_{m,n}$ as long as m and n are odd. For the first part, consider a convex, rectilinear drawing of K_n; every K_4-subgraph contributes a unique crossing, so the crossing number (and independent crossing number) of that drawing is $\binom{n}{4}$. For $K_{m,n}$ consider a bipartite, rectilinear drawing; again every K_4-subgraph contributes exactly one crossing, where every such graph is determined by 2 vertices in each partition, so there are $\binom{n}{2}\binom{m}{2}$ crossings in the drawing.

For the second part, let $G = K_n$ or $G = K_{m,n}$, with both n and m odd. Fix two good drawings D and D' of G. By Lemma A.22 we can assume that both drawings share the same vertex locations and all edges are poly-lines. Define D_i to be the drawing in which the first i edges are drawn as in D' and the remaining edges as in D. So $D_0 = D$ and $D_{|E(G)|} = D'$. We claim that the parity of cr_- remains unchanged as we go from D_i to D_{i+1} (the parity of cr may change). D_i and D_{i+1} differ on edge $e = uv$. We can assume that the two drawings of uv intersect at most finitely often (perturbing bend vertices slightly if necessary, this does not affect cr_-). The two curves representing uv form a closed curve C, so $\mathbb{R}^2 - C$ has a 2-coloring. This 2-coloring partitions $V(G) - \{u, v\}$ into two color classes: V_0 and V_1. Any edge whose endpoints have the same color crosses C an even number of times, so it has the same parity of crossing with both drawings of e. On the other hand, any edge whose endpoints have different colors, crosses C oddly, so the parity of crossing with e changes as we go from D_i to D_{i+1}. In other words, e changes parity of crossing with all edges in the (V_0, V_1)-cut. Edge e may also change parity of crossing with edges incident to u and v, but since we are working with cr_-, we can ignore those edges. If $G = K_n$, then $|V_0| + |V_1| = |V| - 2$ is odd, so there are an even number of edges in the cut. If $G = K_{m,n}$, then each partition contains an even number of vertices on one side of the cut, implying that there are an even number of edges in the cut. In both cases, the parity of cr_- of the drawing does not change. Therefore, $\mathrm{cr}_-(D_{i+1}) \equiv \mathrm{cr}_-(D_i) \bmod 2$. Since the original and final drawings are good, we know that $\mathrm{cr}(D) = \mathrm{cr}_-(D)$ and $\mathrm{cr}(D') = \mathrm{cr}'_-(D)$, and the result follows. □

One useful application of the theorem lies in reducing the number of cases for lower bound proofs. For example, it implies that $\mathrm{cr}(K_5) = \mathrm{cr}(K_{3,3}) = 1$—without the use of Kuratowski's theorem—since both numbers have to be odd, and both graphs can be drawn with a single crossing. Similarly, it is sufficient to show that $\operatorname{cr}(K_7) > 7$ to prove that $\operatorname{cr}(K_7) = 9$. This suggests exploring other moduli, in the hope of further simplifications, but Exercise (1.12) shows that the parity theorem is all we can expect.

We will see the parity theorem again as Theorem 11.6, with a slightly different proof. There also is a purely combinatorial approach to its proof, see [240].

1.2 Zarankiewicz and the Crossing Number of Complete Bipartite Graphs

1.2.1 Exact Values

Turán's brick factory problem, the search for $\operatorname{cr}(K_{m,n})$, spread very quickly through the graph-theoretic community in the early fifties, and by 1953 there were proofs by Urbanik and Zarankiewicz that $\operatorname{cr}(K_{m,n})$ equals

$$Z(m,n) := \left\lfloor \frac{m}{2} \right\rfloor \left\lfloor \frac{m-1}{2} \right\rfloor \left\lfloor \frac{n}{2} \right\rfloor \left\lfloor \frac{n-1}{2} \right\rfloor .$$

Both proofs of the lower bound were erroneous—a story told well in Guy [166], but to this day we call the equality $\operatorname{cr}(K_{m,n}) = Z(m,n)$ Zarankiewicz's conjecture and $Z(n,n)$ the *Zarankiewicz bound*.

To see that the Zarankiewicz bound can be achieved as an upper bound, consider the following construction: place the m vertices of the left partition on the y-axis, with $\lceil m/2 \rceil$ above, and $\lfloor m/2 \rfloor$ below the x-axis; do the same with the n vertices of the right partition, with the role of x- and y-axis exchanged. Then draw edges as straight-line segments, see Figure 1.10.

The number of crossings in each quadrant is of the form $\binom{r}{2}\binom{s}{2}$, where $r \in \{\lfloor m/2 \rfloor, \lceil m/2 \rceil\}$, $s \in \{\lfloor n/2 \rfloor, \lceil n/2 \rceil\}$. The sum of those four terms is $Z(m,n)$ (this is immediate if n and m are even, and requires some arithmetic for the other cases).

The drawing has some interesting properties: it is rectilinear, and its inner and outer face are crossing-free C_4s. Also, for every vertex, there is another vertex so that the two vertices *span*, that is, together with their neighborhoods they induce, a plane $K_{2,n}$ or $K_{m,2}$. Two vertices like this are called *antipodal*.

Does every cr-minimal drawing contain antipodal vertices? The answer is no, as we will see later, but a weaker conjecture: that every $K_{m,n}$ has a cr-minimal drawing with antipodal vertices is open. It would imply Zarankiewicz's conjecture, an observation that is implicit in Kleitman [207].

Theorem 1.13. *If every complete bipartite graph $K_{m,n}$ has a cr-minimal drawing with antipodal vertices, then Zarankiewicz's conjecture is true.*

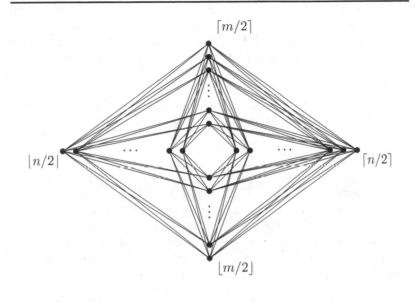

FIGURE 1.10: Drawing $K_{m,n}$ with $Z(m,n)$ crossings.

The best unconditional results for an infinite family of complete bipartite graphs only cover the cases up to $m \leq 6$.

Theorem 1.14 (Kleitman [207]). *We have* $\mathrm{cr}(K_{m,n}) = Z(m,n)$ *for* $m \leq 6$.

To prove the two theorems, we develop some basic tools for drawings of $K_{m,n}$.

Lemma 1.15. *The following are true.*

(i) *If* D *is a good drawing of* $K_{m,n}$, *then* $(m-2)\mathrm{cr}(D) \geq m\,\mathrm{cr}(K_{m-1,n})$.

(ii) *If* $\mathrm{cr}(K_{m-1,n}) = Z(m-1,n)$ *and* m *is even, then* $\mathrm{cr}(K_{m,n}) = Z(m,n)$.

(ii′) *If* $\mathrm{cr}(K_{m-1,n}) = Z(m-1,n)$ *and* m *is odd, then* $\mathrm{cr}(K_{m,n}) \geq Z(m,n) - (m-1)/(2(m-2))\lfloor n/2 \rfloor \lfloor (n-1)/2 \rfloor$. *If* $n = 5$ *and* m *is odd, then* $\mathrm{cr}(K_{m,5}) \geq Z(m,5) - 2$, *assuming that* $\mathrm{cr}(K_{m-1,5}) = Z(m-1,5)$.

(iii) *If* D *is a good drawing of* $K_{m,n}$ *containing an antipodal pair, in the right partition, say, then* $\mathrm{cr}(D) \geq \mathrm{cr}(K_{m,n-2}) + (n-2)\mathrm{cr}(K_{m,3})$.

Property (ii) explains why we always get stuck at values of $\mathrm{cr}(K_{m,n})$ with m and n odd, currently $K_{7,11}$ and $K_{9,9}$, when trying to prove Zarankiewicz's conjecture.

Proof. (*i*). Fix a good drawing D of $K_{m,n}$. It contains m good (though not necessarily cr-minimal) drawings of $K_{m-1,n}$; since each crossing is caused by a $K_{2,2}$ with two vertices in each partition, $m\,\mathrm{cr}(K_{m-1,n})$ counts each crossing of D $m-2$ times, so $(m-2)\,\mathrm{cr}(D) \geq m\,\mathrm{cr}(K_{m-1,n})$.

(*ii*) and (*ii'*). By (*i*), we have $\mathrm{cr}(K_{m,n}) \geq m/(m-2)\,\mathrm{cr}(K_{m-1,n}) = m/(m-2)Z(m-1,n)$. If m is even, then this is at least $m/(m-2)((m-2)/2)^2\lfloor n/2\rfloor\lfloor(n-1)/2\rfloor = Z(m,n)$. If m is odd, then this term is at least $\left(\frac{m}{m-2}\frac{m-3}{2}\right)\lfloor(m-1)/2\rfloor\lfloor n/2\rfloor\lfloor(n-1)/2\rfloor = Z(m,n) - \frac{m-1}{2(m-2)}\lfloor n/2\rfloor\lfloor(n-1)/2\rfloor$ which implies the claim.

(*iii*). Let u, v be the antipodal pair in the right partition. If $n = 2$, we are done, so there are vertices other than u and v in the right partition of $K_{m,n}$. Every such vertex w spans a $K_{m,3}$ together with u and v which requires $\mathrm{cr}(K_{m,3})$ crossings in D. Since none of these crossings are incident to both u and v, this forces $(n-2)\,\mathrm{cr}(K_{m,3})$ crossings with the plane $K_{m,2}$ in D. And these crossings are different from the crossings caused by the $K_{m,n-2}$ which remains after removal of u and v, forcing $\mathrm{cr}(K_{m,n-2})$ crossings. This justifies the inequality. $\square$

This is sufficient machinery to deal with the cases $m \leq 4$ of Theorem 1.14.

Proof of Theorem 1.14 for $m \leq 4$. By Lemma 1.15(*ii*), we only need to show the result for $m = 3$ ($m = 1$ being trivial). In this case, $Z(3,n) = \lfloor n/2\rfloor\lfloor(n-1)/2\rfloor$. Fix a cr-minimal drawing D of $K_{3,n}$. If there is no antipodal pair in the right partition, all $K_{3,2}$-subgraphs contain at least one crossing, so there are at least $\binom{n}{2}$ crossings, which is larger than $Z(3,n)$. Hence, there must be an antipodal pair in the right partition. By Lemma 1.15(*iii*), we know that $\mathrm{cr}(D) \geq \mathrm{cr}(K_{3,n-2}) + (n-2)\,\mathrm{cr}(K_{3,3}) = Z(3,n-2) + (n-2) = Z(3,n)$, applying induction, and pen and paper. $\square$

We can now complete the proof that Zarankiewicz's conjecture is true if there are always cr-minimal drawings with antipodal pairs.

Proof of Theorem 1.13. We use induction on $\min(m,n)$. By Theorem 1.14 Zarankiewicz's conjecture is true for $\min(m,n) \leq 3$, so we can assume that $m, n > 3$. Lemma 1.15(*iii*) implies $\mathrm{cr}(K_{m,n}) \geq \mathrm{cr}(K_{m-2,n}) + (m-2)\,\mathrm{cr}(K_{3,n}) \geq Z(m-2,n) + (m-2)Z(3,n) = Z(m,n)$, using induction. $\square$

The *responsibility*, $\mathrm{rsp}(v)$, of a vertex v in a drawing, is the number of crossings on edges incident to v. A *twin* in a graph is a vertex with the same open neighborhood as another vertex.

Lemma 1.16. *Let D be a good drawing of a graph with n vertices.*

(*i*) *There is a vertex u with $\mathrm{rsp}(u) \leq 4\,\mathrm{cr}(D)/n$ and a vertex v with $\mathrm{rsp}(v) \geq 4\,\mathrm{cr}(D)/n$.*

(*ii*) *For any vertex u we can add a twin vertex v and a crossing-free edge uv, causing at most $\mathrm{rsp}(u) + Z(3, \deg u)$ crossings.*

Proof. (*i*) is immediate, since every crossing is counted four times, once for each endpoint of the two independent edges it lies on. (*ii*) Draw v close to u and connect v to the neighbors of u following the edges between u and its neighbors, causing $\mathrm{rsp}(u)$ crossings. We can arrange it, so the edges incident to u and v cross each other at most $Z(3, \deg(u))$ many times. Finally, we can add a crossing-free edge between u and v. $\square$

To prove the second half of Theorem 1.14, Zarankiewicz's conjecture for $m = 5, 6$, we make some more observations on good drawings. A *self-crossing* of a subgraph of another graph is a crossing between two edges of the subgraph.

Lemma 1.17. *Let D be a good drawing of $K_{m,n}$ with m and n odd.*

(*i*) *Every $K_{m,2}$-subgraph is involved in an even number of crossings if $m \equiv 1 \bmod 4$, and an odd number otherwise.*

(*ii*) *If D is cr-minimal, then no $K_{m,2}$-subgraph contains more than $Z(m, 3)$ self-crossings.*

Proof. (*i*) The number of crossings a $K_{m,2}$ is involved in is the difference between the number of crossings in a good drawing of $K_{m,n}$ and $K_{m,n-2}$. By Theorem 1.12 this is the same as $\binom{m}{2}$ modulo 2. (*ii*) The $K_{m,2}$ consists of two stars, $K_{m,1}$ with centers u and v, say. Without loss of generality, let $\mathrm{rsp}(u) \leq \mathrm{rsp}(v)$. Suppressing v and reintroducing it as a twin of u using Lemma 1.16 would reduce the number of crossings, contradicting the cr-minimality of D. $\square$

Proof of Theorem 1.14 for $m \leq 6$ assuming $K_{5,5}$ and $K_{5,7}$. We already covered the cases $m \leq 4$, and Lemma 1.15(*ii*) implies that we only need to establish the bound for $m = 5$.

Fix a cr-minimal drawing D of $K_{5,n}$. If D contains two antipodal vertices in the right partition, we apply Lemma 1.15(*iii*) and induction to show that $\mathrm{cr}(D) \geq \mathrm{cr}(K_{5,n-2}) + (n-2)\mathrm{cr}(K_{5,3}) = Z(5, n-2) + 4(n-2) = Z(5, n)$, where we get $\mathrm{cr}(K_{5,3})$ by symmetry from the previous case. We conclude that every $K_{5,2}$-subgraph has at least one self-crossing. This gives us $\binom{n}{2}$ crossings which is not sufficient.

Let us try a different attack: suppose that $\mathrm{cr}(K_{5,n}) < Z(5, n)$ for some n, and let n be minimal. We will show that $n \leq 7$, so the problem reduces to checking a finite number of graphs.

By Lemma 1.15(*ii*), the number n has to be odd. In that case, part (*ii'*) of the lemma implies that $\mathrm{cr}(K_{5,n}) \geq Z(5, n) - 2$. Since $K_{5,n}$ has a good drawing with $Z(5, n)$ crossings, Theorem 1.12 implies that $\mathrm{cr}(K_{5,n}) = Z(5, n) - 2$.

Each of the vertices in the right partition has responsibility at most $2(n - 2)$: removing such a vertex results in a $K_{5,n-1}$, meaning we have removed at most $Z(5, n) - 2 - Z(5, n-1) = 2(n-2)$ crossings. Since there are $Z(5, n) - 2 = n(n-2) - 1$ crossings overall, all but two of the vertices have to be incident to $2(n-2)$ crossings. The remaining two vertices either both have responsibility

$2(n-2)-1$, or one has $2(n-2)$ and the other $2(n-2)-2$. The second case would imply that each $K_{5,2}$-subgraph has at least two self-crossings: the number of self-crossings of a $K_{5,2}$ is the number of crossings it is involved in, which is even, by Lemma 1.17, less the even responsibility of each of its degree-5 vertices, so an even number larger than 1, so at least 2. This would force at least $2\binom{n}{2} > Z(5,n)$ crossings in the drawing, a contradiction.

We conclude that all vertices have responsibility $2(n-2)$, except for two of them, call them v_1 and v_2, which have responsibility $2(n-2)-1$ each. Let u and u' be two other vertices from the right partition.

As we argued before, edges incident to u and u' cross at least twice. So u is involved in at least $2(n-3)$ crossings with edges incident to vertices u' from the right partition, other than v_1, v_2. This leaves two crossings which involve edges incident to v_1 and v_2. Since the parity of each has to be odd, edges incident to u cross exactly one edge incident to v_1 and v_2, each. This accounts for $n-2$ crossings on the edges incident to v_1 and v_2, each. Since both have responsibility $2(n-2)-1$, this means edges incident to v_1 and v_2 cross each other at least $n-3$ times. By Lemma 1.17, this number may be at most $Z(5,3)=4$, so $n \leq 7$.

So we only need to show that $\mathrm{cr}(K_{5,5}) = Z(5,5)$ and $\mathrm{cr}(K_{5,7}) = Z(5,7)$, see Theorem 4.16.   □

Kleitman's proof of Zarankiewicz's conjecture for $m = 3$ and $m = 5$ reduces the conjecture to a finite set of cases which need to be checked: $K_{3,3}$ for $m = 3$ and $K_{5,5}$ and $K_{5,7}$ for $m = 5$. Christian, Richter, and Salazar [93] were able to show that this pattern persists: Zarankiewicz's conjecture is finite in the sense that for each m, there are only finitely many n for which the conjecture has to be established.

Theorem 1.18 (Christian, Richter, and Salazar [93]). *There is a function f so that if Zarankiewicz's conjecture fails for $K_{m,n}$, then it fails for $K_{m,n'}$ for some $n' \leq f(m)$.*

While f is an explicit function, it grows too fast to be useful in practice.

Table 1.1 summarizes all known results. It is based on Kleitman's results together with Woodall's work determining $\mathrm{cr}(K_{7,7})$ and $\mathrm{cr}(K_{7,9})$. At the time, in 1993, Woodall wrote that "the smallest unsettled cases are now $K_{7,11}$ and $K_{9,9}$", a challenge that still stands.

1.2.2 Minimal Drawings

Are there cr-minimal drawings of $K_{m,n}$ other than the Zarankiewicz drawings? For $m = 3$ we can show that, in a weak sense, this is not the case.

Theorem 1.19. *In any two cr-minimal drawings of $K_{3,n}$, the rotations of the degree-3 vertices are determined (up to a permutation of the vertices).*

Proof. Fix a cr-minimal drawing of $K_{3,n}$. There are two possible choices for

	3	4	5	6	7	8	9	10	11
3	1	2	4	6	9	12	16	20	...
4		4	8	12	18	24	32	40	...
5			16	24	36	48	64	80	...
6				36	·54	72	96	120	...
7					81	108	144	180	?
8						155	192	240	?
9							?	?	?

TABLE 1.1: The known values of $\mathrm{cr}(K_{m,n})$.

the rotations at the degree 3 vertices: (123) and (321), since these are the only cyclic permutations of three elements. Suppose there are n_{123} vertices of type (123), and n_{321} vertices of type (321). Then any two vertices of the same type force at least one crossing, so there are at least $\binom{n_{123}}{2} + \binom{n_{321}}{2}$ crossings. Since $n_{123} + n_{321} = n$, the minimum is achieved for $\lfloor n/2 \rfloor$ and $\lceil n/2 \rceil$, and only for this pair of values. Since in this case the sum is exactly $Z(3,n)$, we conclude that the rotations of the degree-3 vertices are determined. $\qquad\square$

The situation for $K_{5,n}$ is already significantly more complex. Figure 1.11 shows a good drawing of $K_{5,6}$ without antipodal vertices. The drawing is not cr-minimal (since a cr-minimal drawing of $K_{5,6}$ always contains antipodal

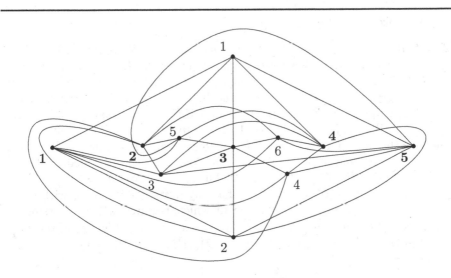

FIGURE 1.11: A drawing of $K_{5,6}$ without antipodal vertices and 25 crossings, adapted from [185].

vertices, as we will see), but it can be used to create cr-minimal drawings of $K_{5,n}$, with $n \equiv 0 \bmod 4$ and $n \geq 8$. For example, use Lemma 1.16 to create twin vertices of vertices b_0 and b_3 in the drawing of $K_{5,6}$. Both these vertices have degree 5 and responsibility 7, and their incident edges share one common crossing, so adding those two twins increases the number of crossings by $2(Z(3,5) + 7) - 1 = 23$ to $25 + 23 = 48$ which we know is optimal; since creating twins cannot create antipodal vertices, we have a cr-minimal drawing of $K_{5,8}$ without antipodal vertices. Since the rotation system determines whether a graph has antipodal vertices, this means that there are cr-minimal drawings of $K_{5,8}$ which are not rotation-equivalent.

The following theorem summarizes the known results.

Theorem 1.20 (Hernández-Vélez, Medina, and Salazar [185]). *Any cr-minimal drawing of $K_{5,n}$ contains antipodal vertices if $n \equiv 2 \bmod 4$. If $n \equiv 0 \bmod 4$ and $n \geq 8$, then there are cr-minimal drawings of $K_{5,n}$ without antipodal vertices, and all such drawings can be obtained from the drawing of $K_{5,6}$ in Figure 1.11 by creating twins of vertices.*

1.2.3 Asymptotic Behavior

While we do not know how to prove Zarankiewicz's conjecture, it is asymptotically true.

Lemma 1.21 (Guy [166]). *The limit $\lim_{m,n\to\infty} \mathrm{cr}(K_{m,n})/Z(m,n)$ exists and is at least $4/5$.*

As Szekeley would say, Zarankiewicz's conjecture is 80% true.

Proof. If we let $a_{m,n} := \mathrm{cr}(K_{m,n})/Z(m,n)$, then $a_{m,n} \leq 1$, since we have drawings realizing $Z(m,n)$. We rewrite $a_{m,n} = \mathrm{cr}(K_{m,n})/(mn)^2 \; (mn)^2/Z(m,n)$. First note that $\lim_{m,n\to\infty}(mn)^2/Z(m,n) = 1/16$. Let $a'_{m,n} := \mathrm{cr}(K_{m,n})/(mn)^2$. Then $a'_{m,n} \leq 16$, and $a'_{m,n}$ is monotonically increasing in both variables: consider $\mathrm{cr}(K_{m+1,n})/\mathrm{cr}(K_{m,n})$. By Lemma 1.15(i), this ratio is at least $(m + 1)/(m-1)$. So $a'_{m+1,n}/a'_{m,n} \geq (m+1)/(m-1)m^2/(m+1)^2 = m^2/(m^2-1) > 1$, showing that $a'_{m,n}$ increases as m increases; since $a'_{m,n}$ is symmetric, it is monotonically increasing in both variables. Hence $a'_{m,n}$—as a bounded, monotonically increasing sequence—has a limit, but then, so does $a_{m,n}$.

To prove the lower bound, note that any drawing of a $K_{m,n}$ with $m \geq 5$ contains $\binom{m}{5}$ drawings of $K_{5,n}$ each of which requires $Z(5,n)$ crossings by Theorem 1.14. Since each particular crossing is contained in $\binom{m}{3}$ of the $K_{5,m}$, this implies that $\mathrm{cr}(K_{m,n})/Z(m,n) \geq 4\binom{m}{5}/(\binom{m}{3}\lfloor m/2\rfloor\lfloor(m-1)/2\rfloor) \to 4/5$ as $m, n \to \infty$. □

The proof of the lemma actually shows that $\mathrm{cr}(K_{m,n}) \geq \frac{1}{5}m(m-1)\lfloor n/2\rfloor\lfloor(n-1)/2\rfloor$; as Kleitman already pointed out, this implies that $\mathrm{cr}(K_{7,7}) \in \{77, 79, 81\}$. Woodall showed that 81 is the correct answer, confirming Zarankiewicz's conjecture for $K_{7,7}$. The crossing number of $K_{9,9}$ is not known at this point, though the lower bound above implies that it is an even number between 232 and 256.

Using semidefinite programming, de Klerk, Maharry, Pasechnik, Richter, and Salazar [101] were able to show that $\mathrm{cr}(K_{7,n}) \geq 2.1796n^2 - 4.5n$. De Klerk, Pasechnik, and Schrijver [105] further developed the method to show that $\mathrm{cr}(K_{8,n}) \geq 2.9299n^2 - 6n$ and $\mathrm{cr}(K_{9,n}) \geq 3.8676n^2 - 8n$. The counting argument of Lemma 1.21 can then be used with $K_{9,n}$ in place of $K_{5,n}$ to give an improved bound on the asymptotic limit.

Theorem 1.22 (de Klerk, Pasechnik, and Schrijver [105]). *We have*

$$\lim_{m,n\to\infty} \mathrm{cr}(K_{m,n})/Z(m,n) \geq 0.8594.$$

1.3 Hill and the Crossing Number of Complete Graphs

1.3.1 Exact Values

How to draw the complete graph? There are two main families of drawings, both going back to the earliest crossing number literature [67, 165, 172]; for history, see [52]: the soda-can (cylindrical) drawings, and the Harary-Hill (2-page) drawings.

In the soda-can drawings, we place half the vertices on the top rim of a soda-can, and the other half on the bottom rim. Vertices on the same rim are connected using only the top and bottom of the can, while edges between the two rims are routed as shortest curves. Figure 1.12 illustrates the construction for K_{14}.

The top (as well as the bottom) part of the drawing amounts to the drawing of a complete graph in which the vertices are in convex position, and edges lie within the convex hull of the vertices. These drawings are called *convex*, or, 1-*page* (see Chapter 8), and their crossing number is $\binom{r}{4}$, if there are r vertices on the rim, since every four vertices determine a unique crossing in a convex drawing.[2] This leaves us with the cylinder part of the can. Let us analyze the case that we have r vertices each on the top and bottom rim of the cylinder, where r is odd. Space the vertices evenly, and align top and bottom vertices horizontally. Connect every pair of vertices by a shortest curve (since r is odd, there is a unique shortest curve).[3] To count the crossings, let us consider an edge whose endpoints are at (absolute) distance a (counting vertices), so a

---

[2]Determining this number occurs as a problem in an elementary algebra textbook from 1889, see [94].

[3]Not a geodesic curve. The word geodesic is often abused in the crossing number literature; it simply means a locally shortest curve, it is not the same as a shortest curve: there are infinitely many geodesic curves connecting a vertex on the top to a vertex on the bottom.

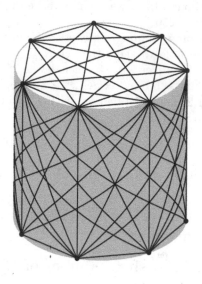

FIGURE 1.12: Soda-can drawing of K_{14}. Degenerate crossings are removed by local perturbation.

vertical edge, the steepest one, is distance 0, and the flattest edge is at distance $\lfloor r/2 \rfloor$. We count how often such an edge crosses an edge whose endpoints are at distance $b \leq a$, so the steeper edges. As Figure 1.13 shows, the number of crossings is $a - 1$ if $b = 0$, $2a - 2 = (a - 1 + b) + (a - 1 - b)$ if $0 < b < a$, and $2a - 1$ if $b = a$.

To count all crossings, we count crossings of edges of distance a with edges that are steeper (have endpoints of smaller distance), and all crossings between edges that have the same distance of endpoints. This gives us

$$2r \sum_{a=2}^{\lfloor r/2 \rfloor} \left((a - 1) + \sum_{b=1}^{a-1} (2a - 2) \right) + r \sum_{a=1}^{\lfloor r/2 \rfloor} (2a - 1),$$

since there are $2r$ edges at every given distance; in the last term, we only multiply by r so we do not double-count. Now this expression simplifies to

$$2r \sum_{a=2}^{\lfloor \frac{r}{2} \rfloor} (a - 1)(2a - 1) + r \left\lfloor \frac{r}{2} \right\rfloor^2 = \frac{1}{3} r \left\lfloor \frac{r}{2} \right\rfloor \left(4 \left\lfloor \frac{r}{2} \right\rfloor^2 - 1 \right),$$

which can be verified easily.[4] Add to this the $2\binom{r}{4}$ crossings on the top and

---

[4]For example, using https://www.wolframalpha.com.

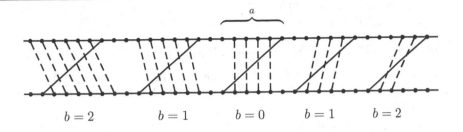

FIGURE 1.13: An edge of slope a crossing with steeper edges of slope b. For $b = 0$, there are clearly $a - 1$ edges of slope b which cross an edge of slope a. As b increases, that number increases or decreases by b depending on whether the edges lean in the opposite (on the left), or in the same (on the right) direction. For $b = a$ one of the numbers is -1, and we drop it from the count.

bottom, and we get

$$\frac{1}{3}r\left\lfloor\frac{r}{2}\right\rfloor\left(4\left\lfloor\frac{r}{2}\right\rfloor^2 - 1\right) + 2\binom{r}{4} = \frac{1}{4}r(r-1)^2(r-2),$$

using that r is odd. This gives us a drawing of K_n, for $n = 2r$, so the case that $n \equiv 2 \bmod 4$. We will write the number of crossings as $Z(n) := \frac{1}{4}\lfloor n/2\rfloor\lfloor(n-1)/2\rfloor\lfloor(n-2)/2\rfloor\lfloor(n-3)/2\rfloor$, for reasons that will become obvious. We can modify the construction for even r: if we followed the instructions as they are, we would draw two arcs between two vertices of distance $\lfloor r/2\rfloor$ (for odd r, those vertices are distinct). Whenever we have this choice, let us always pick the arc that winds around the cylinder clockwise. Reviewing our counts for odd r, we see that we are not overcounting crossings for edges of distance r: in the expression $2r\sum_{a=2}^{\lfloor r/2\rfloor}(a-1)(2a-1)$, the term for $a = \lfloor r/2\rfloor$ should have been multiplied with r, not $2r$, and in the expression $r\sum_{a=1}^{\lfloor r/2\rfloor}(2a-1)$, we should have excluded the term for $a = \lfloor r/2\rfloor$, since no two edges of distance $\lfloor r/2\rfloor$ cross for even r—they are all oriented clockwise. Therefore we only have

$$\frac{1}{3}r\left\lfloor\frac{r}{2}\right\rfloor\left(4\left\lfloor\frac{r}{2}\right\rfloor^2 - 1\right) + 2\binom{r}{4} - r\left(\left\lfloor\frac{r}{2}\right\rfloor - 1\right)\left(2\left\lfloor\frac{r}{2}\right\rfloor - 1\right) - r\left(2\left\lfloor\frac{r}{2}\right\rfloor - 1\right)$$

$$= \frac{1}{4}r(r-1)(r-1)(r-2)$$

crossings (using that r is even). This number equals $Z(n)$, for $n = 2r$, so in either case we have a drawing of K_n with $Z(n)$ crossings, assuming n is even. The responsibility of every vertex in the drawing is the same, since the drawing is vertex transitive, so all vertices must have responsibility $4Z(2r)/(2r) = \frac{1}{2}(r-1)^2(r-2)$. Removing a single vertex, from the top layer, say, then results in a drawing of a K_n, $n = 2r - 1$, with $\frac{1}{4}r(r-1)^2(r-2) -$

$\frac{1}{2}(r-1)^2(r-2) = \frac{1}{4}(r-1)^2(r-2)^2 = Z(n-1)$ crossings. So we can achieve a bound of $Z(n)$ crossings for K_n for both even and odd n.

Theorem 1.23 (Guy [165]). *We have* $\mathrm{cr}(K_n) \leq Z(n)$ *for all* n.

Again, it is interesting to observe some properties of the soda-can drawings. They contain disjoint convex subdrawings of $K_{\lfloor n/2 \rfloor}$ and $K_{\lceil n/2 \rceil}$ (see Exercise (1.15)), and there is a plane Hamiltonian cycle (with all but two edges crossing-free). There are n crossing-free edges, and all triangles are empty (do not contain a vertex).

There is a second family of drawings, mentioned by Harary and Hill [172] and analyzed by Blažek and Koman [67], which look very different. Take a regular convex n-gon on vertices $v_0, \ldots, v_{n-1}$, and separate edges into two categories, one for inside the polygon, the other for outside: draw $v_0 v_i$ inside the polygon for $1 \leq i \leq \lceil n/2 \rceil$ as a straight-line segment, and draw all other straight-line segments between vertices $v_i v_j$ inside if they are parallel to one of the initial set of edges (that is, draw $v_i v_j$ inside if $0 \leq i + j \leq \lceil n/2 \rceil$ mod 2. All other edges draw outside (e.g., draw them inside, and then project them to the outside face of the polygon). Figure 1.14 illustrates the construction for K_8.

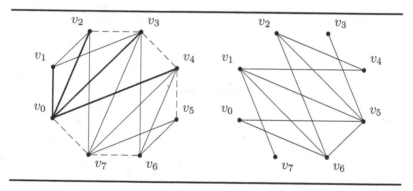

FIGURE 1.14: Harary-Hill drawing of K_8. Edges inside the polygon on left, edges outside of polygon on right.

The Harary-Hill construction is analyzed in [67]. These drawings are sometimes called cycle drawings, or, more commonly, 2-page book drawings. We will study this class of drawings in Chapter 8. Note that in this family of drawings, there is a crossing-free Hamiltonian cycle. All triangles are empty again (do not contain vertices).

With this evidence in hand, we may conjecture that equality holds in Theorem 1.23.

Conjecture 1.24 (Hill's Conjecture [165]). $\text{cr}(K_n) = Z(n)$, where

$$
\begin{aligned}
Z(n) &= \frac{1}{4} \left\lfloor \frac{n}{2} \right\rfloor \left\lfloor \frac{n-1}{2} \right\rfloor \left\lfloor \frac{n-2}{2} \right\rfloor \left\lfloor \frac{n-3}{2} \right\rfloor \\
&= \begin{cases} \frac{1}{64} n(n-2)^2(n-4) & \text{if } n \text{ is even} \\ \frac{1}{64}(n-1)^2(n-3)^2 & \text{otherwise.} \end{cases}
\end{aligned}
$$

The conjecture is named after British artist Anthony Hill, who started investigating it in the late 1950s. It is also known as Guy's conjecture (it is hinted at in Guy's paper [165]), and the Harary-Hill conjecture, since the first explicit statement in print seems to be in a paper by Harary and Hill [172].

Hill's conjecture is known to be true for $n \leq 12$; this was shown by Guy for $n \leq 10$, and Pan and Richter for $n = 11, 12$ [272]; $\text{cr}(K_{13})$ is known to be an odd number between 219 and 225 [239]. Table 1.2 lists all known exact values for complete graphs.

K_5	K_6	K_7	K_8	K_9	K_{10}	K_{11}	K_{12}	K_{13}
1	3	9	18	36	60	100	150	219, 221, 223 or 225

TABLE 1.2: Values and bounds for $\text{cr}(K_n)$, $5 \leq n \leq 13$.

Theorem 1.25 (Guy [168]). *We have* $\text{cr}(K_n) = Z(n)$ *for* $n \leq 10$.

We prove this theorem, for $n \leq 8$ only, in Lemma 1.27 *(iii)* below.

Lemma 1.26. *The following are true.*

(i) *If D is a good drawing of K_n, then $\text{cr}(D) \geq n/(n-4)\,\text{cr}(K_{n-1})$.*

(ii) *If $\text{cr}(K_{n-1}) = Z(n)$ and n is even, then $\text{cr}(K_n) = Z(n)$.*

Property *(ii)* explains why, as was the case for complete bipartite graphs, Hill's conjecture is always stuck at odd values. For example, since we already know that $\text{cr}(K_5) = 1$, property *(ii)* tells us that $\text{cr}(K_6) = 3$.

Proof. (i) By Lemma 1.16(i), there is a vertex v with $\text{rsp}(v) \geq 4\,\text{cr}(D)/n$. Removing that vertex leaves a drawing of a K_{n-1} with at most $(n-4)/n\,\text{cr}(D)$ crossings, so $\text{cr}(K_{n-1}) \leq (n-4)/n \; \text{cr}(D)$. (ii) By (i), $\text{cr}(K_n) \geq n/(n-4) \; \text{cr}(K_{n-1}) = n/(n-4)Z(n-1)$, by assumption, but for n even, $n/(n-4)Z(n-1) = \frac{1}{4}n/(n-4)((n-1)/2)^2((n-3)/2)^2 = \frac{1}{4}(n/2)((n-1)/2)^2(n-4)/2 = Z(n)$. $\square$

We can now prove that Hill's conjecture is true for $n \leq 8$.

Lemma 1.27. *The following are true.*

(i) *K_n has a unique cr-minimal drawing up to (weak) isomorphism for $n \leq 6$. For $n = 7$ this property fails.*

(ii) *Any* cr-*minimal drawing of* K_n *contains a* cr-*minimal drawing of a* K_{n-1} *for* $n \leq 8$. *For* $n = 9$, *this property fails.*

(iii) $\mathrm{cr}(K_n) = Z(n)$ *for* $n \leq 8$.

Proof. (i) This is obvious for $n \leq 3$, and for $n = 4, 5$ it follows from Lemmas 1.5 and 1.6. So $n = 6$ is the first interesting, and the last true, case of this claim. We already know that there are exactly three crossings in a cr-minimal drawing of K_6. Removing any vertex leaves us with a K_5 which we already know to have a unique drawing up to (weak) isomorphism (see Figure 1.5). In this drawing all regions are triangular, and are either incident to three vertices, or two vertices and a single crossing. If we add a vertex to any region and calculate the smallest responsibility it can have by connecting it to the other five vertices, we see that that number is 2 in the first case (three vertices on the region boundary), and 3 in the second case, which is too much. In the first case, there is only one way to realize the two crossings, so the embedding is unique. We will see later, that K_7 has five non-isomorphic, cr-minimal drawings.

(ii) and (iii) We already know $\mathrm{cr}(K_n) = Z(n)$ for $n \leq 6$, and $\lceil 4 \mathrm{cr}(K_n)/n \rceil = Z(n) - Z(n-1)$ in these cases, so removing a vertex with highest responsibility implies the result. Let D be a drawing of K_7. We know that $\mathrm{cr}(K_7) \leq Z(7) = 9$, so D contains a vertex of responsibility at least 6; removing it leaves a drawing of K_6 with 3 crossings, which we know to be minimal. This drawing has all triangular regions of three different types: there are two empty triangles on vertices of the graph, six triangles which share one edge with the first type of triangle, and six triangles which do not share an edge with the first type of triangle. As before, we can place the new vertex into each of the three types of triangles and calculate the smallest possible responsibility for the new vertex by connecting it to all other vertices. This number is 6 for all three types, and it can be realized in an actual drawing. So $\mathrm{cr}(K_7) = 6 + 3 = 9$. Lemma 1.26(i) then implies that $\mathrm{cr}(K_8) = 18$, completing the proof of (iii). To finish (ii) we can now make the same argument as before: a cr-minimal drawing of K_8 contains a vertex of responsibility 9, so removing it leaves a drawing of K_7 with 9 crossings, which we know to be minimal.

Finally, consider the drawing of K_9 in Figure 1.15; all vertices have responsibility at most 17, so no K_8-subgraph has a cr-minimal drawing. $\square$

For $n = 9$, Lemma 1.26 together with $\mathrm{cr}(K_7) = 18$ from Lemma 1.27 implies that $\mathrm{cr}(K_9) \geq 9/518 > 32$. Since $\mathrm{cr}(K_9$ is even by Theorem 1.12, and $\mathrm{cr}(K_9) \leq 36$, we know that $\mathrm{cr}(K_9)$ is 34 or 36. Excluding the possibility of 34 crossings requires some work; we recommend McQuillan and Richter [241].

Lower bound proofs for $\mathrm{cr}(K_n)$ have traditionally followed Guy's paper, building non-isomorphic good drawings of K_n incrementally, and, for larger n have used computers to do so. If one is interested in the number of non-isomorphic cr-minimal drawings, there seems to be no way around this approach at this point. McQuillan and Richter [241] have succeeded in removing computer assistance for $n \leq 12$, but their proofs are still lengthy.

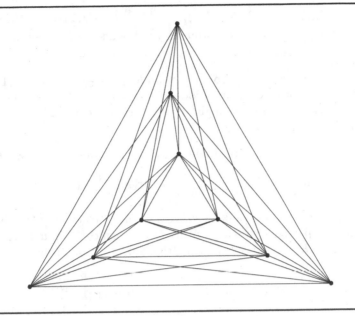

FIGURE 1.15: A cr-minimal drawing of K_9 not containing a cr-minimal K_8 (based on [168, Figure 12]).

Archdeacon [32] conjectured that any rotation system of a K_n, whether realizable or not, induces at least $Z(n)$ non-planar K_4s; this, of course, would imply Hill's conjecture, stripping it of all topological overhead.

Another conjecture going back to an early (and flawed) paper by Saaty [296] asks whether a cr-minimal drawing of a K_n always contains a cr-minimal drawing of a K_{n-2}. As we saw above that conjecture fails for K_{n-1} in place of K_{n-2} for $n = 9$. Pan and Richter showed that the conjecture is true for $n = 9$. McQuillan and Richter mention that the conjecture is not known to hold for $n = 13$ and is, with current methods, beyond the reach of computers.

1.3.2   Asymptotic Behavior

Since we cannot currently prove that $cr(K_n) = Z(n)$, we next try to establish that $Z(n)$ is the right order of magnitude for $cr(K_n)$, which is implied (if not explicitly stated) in a paper by Guy [165].

Lemma 1.28 (Guy [165]). *The limit $\lim_{n \to \infty} cr(K_n)/Z(n)$ exists.*

Proof. Let $a_n := cr(K_n)/Z(n) = cr(K_n)/n^4 \; n^4/Z(n)$. Since the second term approaches 64 as $n \to \infty$, it is sufficient to show that $a'_n := cr(K_n)/n^4$ converges. Now $a'_n \le Z(n)/n^4 \le 1/64$, so the sequence is bounded. Moreover,

$a'_{n-1} = \mathrm{cr}(K_{n-1})/(n-1)^4 \leq \mathrm{cr}(K_n)/n^4 \frac{n^4(n-4)}{n(n-1)^4} < \mathrm{cr}(K_n)/n^4 = a'_n$, implying that a'_n is strictly increasing, and thus, being bounded, must have a limit, and therefore, so must a_n. $\qquad\square$

By the crossing lemma, Theorem 1.8, $\mathrm{cr}(K_n) \geq \frac{1}{60.75}\binom{n}{2}^3/n^2 = (1/121.5 - o(1))n^4$; we could use Kleitman's lower bounds on $\mathrm{cr}(K_{5,n})$ to get a better bound. This is not necessary though; the many complete bipartite subgraphs of K_n connect the conjectures of Zarankiewicz and Hill, at least asymptotically.

Theorem 1.29 (Kainen [200], Richter, and Thomassen [286]). *We have*

$$\lim_{n\to\infty} \mathrm{cr}(K_n)/Z(n) \geq \lim_{n,m\to\infty} \mathrm{cr}(K_{m,n})/Z(m,n).$$

In Székely's words: if Zarankiewicz's conjecture is c percent true, then Hill's conjecture is c percent true, asymptotically, of course. Lemma 1.21 implies that $\lim_{n\to\infty} \mathrm{cr}(K_n)/Z(n) \geq 4/5$; using the improved bounds in Theorem 1.22 yields a better bound.

Corollary 1.30 (de Klerk, Pasechnik, and Schrijver [105]). *We have*

$$\lim_{n\to\infty} \mathrm{cr}(K_n)/Z(n) \geq 0.8594.$$

Proof of Theorem 1.29. The statement of the theorem is meaningful since both limits exist: the first by Lemma 1.28, the second by Lemma 1.21.

Fix a drawing D of a K_{2n}. There are $\binom{2n}{n}/2$ complete bipartite subgraphs $K_{n,n}$, each of which requires at least $\mathrm{cr}(K_{n,n})$ crossings. Any crossing in D occurs in at most $2\binom{2n-4}{n-2}$ of the $K_{n,n}$-subgraphs: the endpoints of the two crossing edges must belong to different partitions in the $K_{n,n}$ (so the crossing occurs in the drawing of the $K_{n,n}$), and there are two ways to do this. For each of these, we can fill up one of the partitions with $n-2$ of the $2n-4$ available vertices.

We conclude that $\mathrm{cr}(D) \geq \frac{1}{4}\binom{2n}{n}/\binom{2n-4}{n-2}\mathrm{cr}(K_{n,n})$. Choosing D to be cr-minimal gives us

$$\frac{\mathrm{cr}(K_{2n})}{Z(2n)} \geq \frac{1}{4}\frac{\binom{2n}{n}}{\binom{2n-4}{n-2}}\frac{\mathrm{cr}(K_{n,n})}{Z(2n)}.$$

Let us rewrite the right-hand side a bit:

$$\frac{1}{4}\frac{\binom{2n}{n}}{\binom{2n-4}{n-2}}\frac{\mathrm{cr}(K_{n,n})}{Z(n,n)}\frac{(Z(n,n))}{Z(2n)}$$

As we let n go to infinity, the first term converges to 4 cancelling out the last term, which converges to $1/4$, proving the result. $\qquad\square$

For some application an actual, rather than an asymptotic, lower bound can be useful.

Lemma 1.31. *We have* $\mathrm{cr}(K_n) \geq 0.8Z(n)$.

Proof. Consider a cr-minimal drawing of K_n. By Theorem 1.14, every $K_{5,n-5}$-subgraph has at least $Z(5, n-5)$ crossings, and there are $\binom{n}{5}$ such subgraphs. Since each crossing occurs in $4\binom{n-4}{3}$ of these subgraphs, the number of crossings in the drawing is at least

$$\frac{Z(5, n-5)\binom{n}{5}}{4\binom{n-4}{3}} = 3.2 \frac{\lfloor\frac{n-5}{2}\rfloor}{n-5} \frac{\lfloor\frac{n-6}{2}\rfloor}{n-6} \left(\frac{1}{4} \frac{n}{2} \frac{n-1}{2} \frac{n-2}{2} \frac{n-3}{2} \right).$$

Since we know that $\mathrm{cr}(K_n) = Z(n)$ for $n < 12$, we can assume $n \geq 12$. (Alternatively, Theorem 1.25 together with the observation that $\mathrm{cr}(K_{11}) \geq 11/7 \, \mathrm{cr}(K_{10}) = 95$, by Lemma 1.26, is sufficient for this conclusion.) Now

$$\frac{\lfloor\frac{n-5}{2}\rfloor}{n-5} \frac{n-1}{2} \frac{n-3}{2} \geq \frac{1}{2} \left\lfloor \frac{n-1}{2} \right\rfloor \left\lfloor \frac{n-3}{2} \right\rfloor$$

for $n \geq 9$: for odd n we have equality, and for even n, the inequality is equivalent to $(n-6)/(2(n-5))(n-1)/2 \, (n-3)/2 \geq \frac{1}{2}(n-2)/2 \, (n-4)/2$ which is true for real $n \geq 9$. Similarly,

$$\frac{\lfloor\frac{n-6}{2}\rfloor}{n-6} \frac{n}{2} \frac{n-2}{2} \geq \frac{1}{2} \left\lfloor \frac{n-2}{2} \right\rfloor \left\lfloor \frac{n-4}{2} \right\rfloor$$

is true for $n \geq 12$. Combining these inequalities, we conclude that

$$\frac{Z(5, n-5)\binom{n}{5}}{4\binom{n-4}{3}} \geq 0.8Z(n).$$

$\square$

1.3.3 Random Drawings

We conclude this section with a beautiful result on random geodesic drawings of complete graphs in the sphere.

Theorem 1.32 (Moon [251]). *If we place n vertices randomly on a sphere (uniformly and independently), and connect any two vertices by a shortest arc, the expected number of crossings is* $1/16\binom{n}{2}\binom{n-2}{2}$.

Moon's paper shows that the crossing number of a random drawing of a complete graph has a normal distribution. He applies the theorem to routing aircrafts; from an economic point of view, he claims, we will try to minimize the length of routes, what the random model does; from a safety point of view, we will try to minimize crossings. The theorem shows that they are asymptotically the same (assuming the truth of Hill's conjecture).

Proof. We first note that there is no ambiguity in the description: the probability that two vertices coincide or that there is more than one shortest arc to choose from is 0, so we can ignore those cases. We need to determine the probability that two edges cross, so consider two pairs of points on the sphere, chosen uniformly and independently at random. Each pair of points lies on a great circle, and the two great circles cross in two points c and c'. The probability that one of the edges passes through c or c' is $1/2$, since the probability that it does not is the probability that both endpoints lie in the same half of the circle, which is $1/2$. So the probability that both edges pass through c or c' is $1/4$, and, since they cannot pass through both, the probability that they pass through the same vertex, and therefore cross, is $1/8$.

By linearity of expectation, the expected number of crossings is the number of pairs of independent edges, $\frac{1}{2}\binom{n}{2}\binom{n-2}{2}$ times $1/8$, which is the result we claimed. □

1.4 Notes

It is somewhat surprising that the study of $\mathrm{cr}(K_{m,n})$ preceded $\mathrm{cr}(K_n)$, and that the latter was not suggested by the mathematical community, but by a non-mathematician, Anthony Hill, albeit with a strong interest in geometry. The early history of Zarankiewicz's and Hill's conjectures is told in detail by Beineke and Wilson [52]. For a report on the status quo, see Székely's recent survey [329].

The crossing lemma has probably received more attention than any other result on crossing numbers; this is in no little part due to its connection to discrete geometry, discovered by Székely [328], a paper still worth reading (we only included one of his applications, the Szemerédi-Trotter Theorem). There has been much work on improving the crossing lemma (some of which we will see in later chapters), and proving it for variants of the crossing number. Pach, Spencer, and Tóth [259] showed that

$$\mathrm{cr}(G) \geq c_g \frac{m^{g/2+2}}{n^{g/2+1}},$$

for graphs G of girth g and $m \geq 4n$. Surveys on crossing numbers include Richter and Salazar's "Crossing numbers" [284], and Buchheim, Chimani, Gutwenger, Jünger, and Mutzel's "Crossings and Planarization" [76].

1.5 Exercises

(1.1) What is wrong with the following proof that $\text{cr}_- = \text{cr}$? Take a cr_- minimal drawing of a graph. Use Lemma 1.3 to remove any crossings between adjacent edges. The resulting drawing shows that $\text{cr} = \text{cr}_-$.

(1.2) Show that $\text{cr}_-(K_5) = 1$. *Hint:* The proof of Theorem 1.12 will be helpful.

(1.3) Show that $\text{cr}_-(K_6) = 3$. *Hint:* Exercise (1.2) will come in handy.

(1.4) Show that K_6 has a unique cr-minimal drawing on the sphere.

(1.5) Show that K_7 has five non-isomorphic cr-minimal drawings on the sphere.

(1.6) Show that $K_{3,3}$ has a unique cr-minimal drawing on the sphere.

(1.7) Find a vertex-minimal graph with crossing number 2.

(1.8) Determine the crossing number of the Petersen graph.

(1.9) Find a cr-minimal drawing of K_7 which does not contain a crossing-free triangle, and therefore is not isomorphic (on the sphere) to a drawing with a triangle as an outer face.

(1.10) Use Euler's formula to show that if $K_{m,n}$ is planar, then $\min(m, n) \leq 2$. *Hint:* What is the largest number of edges a planar $K_{m,n}$ can have?

(1.11) Show that $K_{n,n}$ has a cr-minimal drawing with a plane Hamiltonian cycle.

(1.12) Call a graph *n-constant*, if all good drawings of the graph have the same crossing number modulo n. Show that if G is n-constant for $n \geq 3$ if and only if it is a star or a triangle, and G is 2-constant if and only if it is a K_n or $K_{n,m}$ with n and m odd.

(1.13) The authors of [185] write "An elegant argument using $\text{cr}(K_{3,3}) = 1$ plus purely combinatorial arguments (namely, Turán's theorem on the maximum number of edges in a triangle-free graph) shows that $\text{cr}(K_{3,n}) = Z(3, n)$." Find the elegant argument. *Hint:* The special case of Turán's theorem used here is known as Mantel's theorem. What could it bound (from above)?

(1.14) Give a purely combinatorial proof of Theorem 1.12. *Hint:* count K_5- and $K_{3,3}$-subgraphs. The base cases are trickier.

(1.15) Show that if D is a soda-can drawing of K_n, that is, there are disjoint convex drawings of $K_{\lceil n/2 \rceil}$ and $K_{\lfloor n/2 \rfloor}$, then $\mathrm{cr}(D) \geq Z(n)$. *Note:* We will improve this result in Exercise (6.7). *Hint:* For every vertex x on one of the rims, there is an edge e on the other rim, so that the triangle consisting of x and e separates the drawings of the complete disjoint graphs on the top and bottom disks. Starting with two vertices x and y, pairs of vertices on the other rim can be divided into two groups based on whether the K_4 induced by them and x and y has a crossing or not.

(1.16) Show that any cr-minimal drawing of a K_9 contains a cr-minimal drawing of a K_7.

(1.17) Consider the following way of creating a drawing of K_n: start with a single vertex, and do the following $n - 1$ times: pick a vertex of smallest responsibility, and duplicate it. (*a*) Prove or disprove: the resulting drawing achieves the $Z(n)$ bound of Hill's conjecture. (*b*) Does the resulting drawing have any interesting properties? (E.g., any plane substructures?)

(1.18) Show that the following is true: if we place n black and m white vertices randomly on a sphere (uniformly and independently), and connect any two vertices of different color by a shortest arc, the expected number of crossings is $1/4 \binom{n}{2} \binom{m}{2}$.

Chapter 2

Drawings and Values

2.1 Good Drawings of Complete Graphs

We continue our investigation of good drawings of complete graphs begun in Section 1.1.2. We investigate the relationship between isomorphism, weak isomorphism and rotation systems, study the enumeration problem, and finally look for unavoidable substructures in good drawings.

2.1.1 Basic Properties

For good drawings of the complete graph, rotation systems and weak isomorphism types are in a 2:1 relationship as explained by the following lemma. We say that two rotation systems of a graph are *equivalent up to flipping* if they are the same or the two cyclic permutations at each vertex are inverses of each other; visually, the second case corresponds to flipping the plane (or turning the sphere inside out).

Lemma 2.1 (Kynčl [221]). *Any two good drawings of the complete graph with the same rotation system up to flipping are weakly isomorphic, and any two weakly isomorphic drawings have the same rotation system up to flipping.*

In other words, the rotation system in a good drawing of the complete graph determines which pairs of edges cross, and knowing which pairs of edges cross determines the rotation system up to flipping the plane.

Proof. Suppose we are given the rotation system of a complete graph in a good drawing. No two adjacent edges cross, so we only need to determine whether a pair xy, uv of independent edges crosses. Consider the triangle xuv. Flipping the rotation of all vertices if necessary, we can assume that at least two of the three edges incident to y start inside the triangle at x, u and v. Then y has to lie inside the triangle: if y were outside the triangle, then the (unique) C_4 on $\{x, y, u, v\}$ containing the two inside edges would require two crossings, which is impossible by Theorem 12.2. The two inside edges then have to lie entirely inside the triangle xuv, since otherwise they would have to leave and reenter the triangle, which they cannot. The third edge incident to y then crosses the opposite edge in the triangle xuv (the one it is not adjacent to) if and only if

33

it starts outside the triangle. If this edge is xy, then xy crosses uv, otherwise it does not.

The other direction is immediate for $n = 3$. For $n = 4$ and $n = 5$, the claim follows by inspection of the non-isomorphic drawings of K_n, see Figures 1.6 and 1.7. The weak isomorphism type determines the number of crossings on each edge, and, as we observed earlier, any two non-isomorphic drawings can be distinguished by the largest number of crossings along any edge. If $n >$ 5, pick a vertex v and consider all K_5-subgraphs containing v. Each such subgraph gives us two possible (inverse of each other) orderings of four edges incident to v, ignoring all other edges. If two partial orderings overlap in at least three edges, and order the overlapping edges in the same cyclic order, the two orderings can be combined into a single order, and the same is true for their inverses. We conclude that there are at most two possible rotations at v as determined by the weak isomorphism type, and these two are inverses of each other. Choose one of the two rotations. For any vertex $u \neq v$, we can now pick a K_5-subgraph containing both u and v, which allows us to reconstruct the rotation at u uniquely. Since u was arbitrary, and we only had one choice (inverting the rotation at v), the result follows. □

The proof showed that weak isomorphism and isomorphism are the same for good drawings of K_n up to $n = 5$. For $n = 6$, this is no longer true, see Exercise (2.5). What is not determined by the rotation system, or by the information which edges cross, is what *order* edges cross in. Recall that in a planarization of a drawing, every crossing is replaced by a vertex of degree 4.

Lemma 2.2 (Pan and Richter [272]). *The planarization of a good drawing of K_n, $n \geq 4$, is 3-connected.*

The lemma implies that the order of crossings along an edge is sufficient to determine a good drawing of K_n up to isomorphism. We used the special case of K_4 earlier.

Proof. Fix a drawing D of K_n, and suppose it can be disconnected by removing a set S of two vertices or crossings. By inspection, we see that this is not possible for $n = 4$, so we can assume $n \geq 5$. Let u and v be any two vertices in $V(K_n) - S$. If uv contains no crossings in S, then u and v are still connected. If uv contains a crossing in S, there is at most one other crossing or vertex in S. However, there are at least three paths of length 2 in K_n between u and v. Since a crossing can belong to at most two edges and a vertex to at most one of the paths, at least one of the paths is disjoint from S, so u and v are still connected. Any crossing is incident to four disjoint paths (since the original drawing was good) to vertices in $V(K_n)$, so every crossing that survives in $V(K_n) - S$ is connected to at least one vertex in $V(K_n) - S$. Since, as we just argued, any two vertices of $V(K_n) - S$ are still connected, this means that the planarization of K_n remains connected, even after the removal of S. In other words, the planarized K_n is 3-connected. □

2.1.2 Enumeration

In Figure 1.15 we saw a cr-minimal drawing of K_9 which does not contain a cr-minimal drawing of K_8. If we want to establish lower bounds on $cr(K_n)$ by building cr-minimal drawings inductively, we need to work with a larger class of drawings, and good drawings are a natural choice; if we can exhaustively generate all good drawings of a K_n, then we can easily determine $cr(K_n)$. To understand how far such an approach can possibly take us, we should figure out how many different good drawings of K_n there are. We showed in Section 1.1.2 that there are two non-isomorphic drawings of K_4 and five non-isomorphic drawings of K_5. In a tour de force, Gronau and Harborth [161] determined all 121 non-isomorphic drawings of K_6 (displayed in their paper). This was later corroborated computationally by Ábrego et al. [1], who also determined the values in Table 2.1.

graph	K_3	K_4	K_5	K_6	K_7	K_8
#D	1	2	5	121	46,999	502,090,394
#rot	1	2	5	102	11,556	5,370,725

TABLE 2.1: Number of non-isomorphic good drawings, $\#D$, and realizable rotation systems, #rot, of K_n.

The numbers clearly show what we already knew, that the rotation system of a good drawing, even of a complete graph, cannot determine the drawing up to isomorphism. So how many good drawings of K_n are there? There are two answers, depending on whether we count up to weak isomorphism or isomorphism.

Theorem 2.3 (Pach and Tóth [268]). *The number of good drawings of K_n up to weak isomorphism is $2^{\Omega(n^2)}$ and $2^{O(n^2 \log n)}$.*

Proof. We start with the lower bound. Let $n \equiv 3 \bmod 4$ and place $(n+1)/4$ vertices u_i at $(i, 1)$, $1 \le i \le (n+1)/4$, and $(n+1)/4$ vertices v_j at $(j, -1)$, $1 \le j \le (n+1)/4$. Any two straight-line segments of the form uv cross $y = 0$ in the $(n-1)/2$ points w_k, $1 \le k \le (n-1)/2$, at $((k+1)/2, 0)$, which we make the remaining $(n-1)/2$ vertices. Draw all edges $u_i u_{i+1}$, $v_i v_{i+1}$, and $w_i w_{i+1}$ as straight-line segments. For each edge $u_i v_j$ we can draw it as either passing slightly to the left or to the right of w_k, where $k = i - j$. See Figure 2.1 for an illustration of this stage, for $n = 27$ (with edges passing to the left of each w_k). This gives us $2^{(n/4)^2}$ choices. We add the remaining edges as follows: edges from w to u or v vertices we draw as straight-line segments. Edges between w vertices as one-bend polygonal arcs close to $y = 0$. The convex hull of the drawing is a crossing-free Hamiltonian cycle. This limits the number of weak isomorphisms between any two of the drawings we created to be at most a linear in n, since the Hamiltonian cycles have to align, so the number of realizable weak isomorphism types is at least $2^{\Omega(n^2)}$.

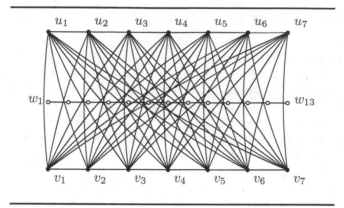

FIGURE 2.1: A partial drawing of K_{27}.

By Lemma 2.1, the rotation system of a complete graph determines which pairs of edges cross. Since there are at most $((n-1)!)^n = 2^{O(n^2 \log n)}$ this proves the upper bound for weakly isomorphic drawings. □

Edges in the lower bound construction can be made x-monotone (crossing every vertical line at most once), 1-bend polygonal arcs; while this is close to a rectilinear drawing, the asymptotic behavior in that case is quite different, see Exercise (5.8).

Theorem 2.4 (Kynčl [221]). *The number of good drawings of K_n up to isomorphism is $2^{\Theta(n^4)}$.*

Since a good drawing of a complete graph is determined by the order of crossings along each edge together with the rotation at each vertex and every crossing, we can easily get an upper bound of $n!4^{\binom{n}{4}}(n^2)!$ which is $2^{O(n^4 \log n)}$, the theorem shows that we can remove the $\log(n)$ factor in the exponent.

Proof. Consider a rectilinear drawing of K_n with all vertices placed on a circle. The planarization of this drawing has $n' = n + \binom{n}{4}$ vertices and $m' = m + 2\binom{n}{4}$ edges. We want to show that there are many triangular faces, so let $f' = f'_3 + f'_{\geq 4}$, the number of faces split into triangular faces, and faces of size at least 4. Then $2m' \geq 3f'_3 + 4f'_{\geq 4}$, so $f'_3 \geq 4f' - 2m' \geq 2m' - 4n'$, using $f' \geq m' - n'$. This gives us $f'_3 \geq 2\binom{n}{4} + 3\binom{n}{2} - 4n$. Every vertex of K_n can be incident to at most $n - 1$ of these triangles, so there are at least $2\binom{n}{4} - 4n$ triangles with crossings as vertices. Each crossing vertex can occur in at most 4 triangles, so there are at least $(2\binom{n}{4} - 4n)/10$ disjoint triangles. For each of these triangles, we can choose one of the sides, and slide it over the crossing of the other two sides in the triangle. We call this a *slide (move)*. More officially, it is called a Reidemeister move of type 3, see Figure 2.2, which changes the isomorphism type of the drawing (independent of what we do with the other triangles).

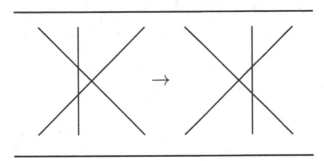

FIGURE 2.2: A slide (Reidemeister move of type 3).

This gives us $2^{\Omega(n^4)}$ non-isomorphic drawings of K_n as long as the order of the vertices around the circle is considered fixed. However, there are at most $n!$ such orders, and dividing by that number still leaves us with $2^{\Omega(n^4)}$ non-isomorphic drawings.

For the upper bound, consider a star $K_{1,n-1}$ in a drawing of K_n. The star does not have any self-crossings, since adjacent edges cannot cross. Remove the star from the plane, and fill in the resulting hole with a disk; turn crossings into vertices on the boundary of the disk. In this fashion we obtain a drawing of a matching (independent edges) with all vertices on a disk. Theorem 8.11 tells us that the number of non-isomorphic drawings of this type only depends on the number of crossings, which is $\binom{n}{4}$, and is at most $2^{\binom{n}{4}}$. There are n ways to choose the star, and at most $4^n n^{2n}$ ways edges of the K_n can cross the edges of the star. This gives us a bound on the number of non-isomorphic drawings of K_n of at most $2^{O(\binom{n}{4})}$. $\qquad\square$

We will later see similar results for rectilinear and pseudolinear drawings.

2.1.3 Extremal Properties

Does every good drawing of a K_n contain a crossing-free edge? It turns out that the answer depends. Let $h(n)$ and $H(n)$ denote the smallest and largest number of crossing-free edges in a good drawing of K_n.

Theorem 2.5 (Ringel [290], Harborth, and Mengersen [181]). *The following equalities hold.*

(i) $H(n) = 2n - 2$ *for* $n \geq 4$, *and*

(ii) $h(n) = 0$ *for* $n \geq 8$.

Harborth and Mengersen also determine the values of $h(n)$ for $n \leq 7$.

Proof. To see that $H(n) = 2n - 2$ take a convex drawing of K_{n-1} and add a vertex connecting it to every vertex of K_{n-1} in the outer face. This drawing has $2(n - 1)$ crossing-free edges. Also, $H(n) \leq 2n - 2$. If not, let n be the smallest counterexample, so $H(n) \geq 2n - 1$, but $H(n - 1) \leq 2(n - 2)$, and fix a drawing with $H(n)$ crossing-free edges. Removing any vertex from this drawing leaves us with at most $2(n - 2)$ crossing-free edges. Since each edge is counted $n - 2$ times, this implies $H(n) \leq 2(n - 2)/(n - 2)n = 2n$, so there must be a vertex v incident to at most 4 crossing-free edges. This vertex must be incident to at least 3 crossing-free edges, since removing it would otherwise create a drawing of K_{n-1} with more than $2(n - 2)$ crossing-free edges. Let C be the 3- or 4-cycle passing through the endpoints of the crossing-free edges at v, without v, in the order determined by the rotation of the edges at v. Then C consists of crossing-free edges: if any of its edges were involved in a crossing, it could be rerouted along the two crossing-free edges connecting its endpoints to v, increasing the number of crossing-free edges, which is not possible by choice of the drawing. So v is at the center of a crossing-free wheel graph with 3 or 4 spokes. There can be no other vertices, since any such vertex cannot connect to both v and all edges of C without causing a crossing. So we either have $n = 4$, and a planar drawing of K_4 showing $H(4) = 6$, or a drawing of K_5 with eight crossing-free edges, and two crossing edges, which we know cannot be reduced, so $H(5) = 8$.

To draw a K_n for even $n \geq 8$ without crossing-free edges, we proceed as follows: take a convex drawing of a K_{n-2} on vertex set $v_1, \ldots, v_{n-2}$, and swap inner and outer region, so the convex polygon is empty, except we keep the diagonal edge $v_1 v_{n/2+1}$ on the inside. Add two vertices u_i, $i \in \{1, 2\}$, one on each side of the diagonal and close to v_1. For the even-numbered vertices $v_k \in \{v_2, v_4, \ldots v_{n-2}$ do the following: add straight-line segments $u_1 v_k$, $u_2 v_k$ and edges $u_1 v_{k-1}$, $u_2 v_{k+1}$ as follows: draw a straight-line segment u_1 so it crosses $v_k v_{k+1}$ close to v_k, then turn left outside the polygon and connect to v_{k-1}; similarly draw a straight-line segment u_2 so it crosses $v_k v_{k-1}$ close to v_k, and then turn right and connect it to v_{k+1}. By staying close to the polygon, we can ensure that we are not creating crossings between adjacent edges. See Figure 2.3.

Finally, connect $u_1 u_2$ by a straight-line segment, crossing the diagonal. The way we constructed the drawing, every edge is involved in a crossing, and there are no crossings between adjacent edges. Moreover, inside the polygon all edges are drawn as straight-line segments, and there are cells which are not incident to vertices, for example the cell formed by $u_1 v_2$, $u_2 v_2$, $u_1 v_{n-3}$, $u_2 v_{n-3}$ (and possibly the diagonal $v_1 v_{n/2+1}$ if it crosses that cell). We can place a vertex into that cell and connect it to every other vertex using a straight-line segment, giving us a good drawing of K_{n+1} without any crossing-free edges. □

Crossing-free edges have also been studied for complete bipartite graphs.

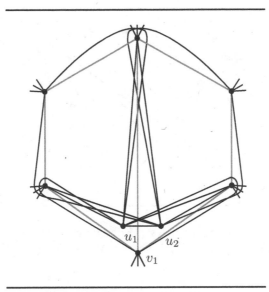

FIGURE 2.3: A drawing of K_8 without any crossing-free edges. Edges with both ends outside the C_6 are only shown partially.

Theorem 2.6. *Every good drawing of a $K_{2,n}$, $n \geq 2$, contains at least two disjoint, crossing-free edges.*

For $n = 1$ there are two crossing-free edges, albeit not disjoint. The result is optimal in that the usual bipartite rectilinear drawing of a $K_{m,n}$ has exactly two crossing-free edges. We suspect this result may have been proved in [242].

Proof. Fix a good drawing of a $K_{2,n}$, $n \geq 1$, and let u and v be the two vertices of the left partition. We claim that u is incident to a crossing-free edge. By symmetry, v will be as well. If the two crossing-free edges are adjacent, we can remove them from the drawing. We apply induction to the resulting drawing of $K_{2,n-1}$ to obtain a crossing-free edge incident to u. Together with the crossing-free edge incident to v we removed, this gives us two disjoint, crossing-free edges.

We prove the claim by induction. For $n = 2$, the result follow by inspection of the two good drawings of C_4. So we can assume that $n > 2$. Pick any vertex $w \neq u, v$ and remove it from the drawing of $K_{2,n}$. By induction, the resulting drawing has a crossing-free edge ux. If ux is crossing-free in the original drawing, we are done. Otherwise, ux must cross vw in some point c. Consider the drawing of a complete bipartite graph induced by u, v and all vertices inside the triangle ucw, including w. This graph is smaller than the original graph, since it does not contain ux: that edge cannot cross uw, so x is outside the triangle, and it contains at least uwv, so, by induction, it contains a crossing-free edge uy, where $y = w$ is allowed. Then uy is crossing-free in

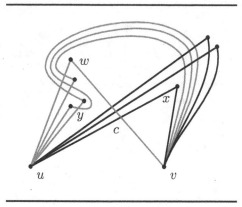

FIGURE 2.4: Good drawing of $K_{2,n}$. Edge ux crosses vw in c but is crossing-free in $K_{2,n} - \{w\}$. Subgraph with vertices in ucw in gray.

the original drawing: if it were involved in a crossing, it would have to be with an edge vz, $z \neq w$, incident to v which is not part of the subgraph. This edge cannot cross the boundary of the triangle ucw: uc only crosses vw, vw is adjacent to vz, so this only leaves uw, but iv vz crossed uw, then z would have to lie inside uwc, and vz would have been part of the subgraph we considered, which was not the case. On the other hand, uy lies entirely inside ucw, so vz and uy cannot cross. □

We include two applications of Theorem 2.6, that significantly improve previously known results. The first concerns empty triangles, the second, pairwise disjoint edges. In a good drawing of a graph, an *empty triangle* is a K_3-subgraph whose interior does not contain any vertices, though there may be edges intersecting the triangle. Harborth [179] introduced $t(n)$, the smallest number of empty triangles in a good drawing of K_n. Figure 2.5 shows that K_n has a good drawing with $2n - 4$ empty triangles.

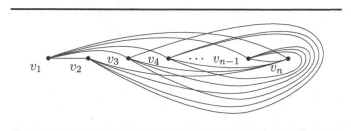

FIGURE 2.5: A good drawing of K_n with $2n - 4$ empty triangles formed by vertices (v_1, v_2, v_i) and (v_i, v_{n-1}, v_n) (adapted from [179, Fig. 1]).

Every vertex in the drawing is incident to two empty triangles, and it turns out this is true for all good drawings of K_n.

Theorem 2.7 (Ruiz-Vargas [295]). *Every vertex in a good drawing of K_n is incident to two empty triangles. In particular, $t(n) \geq 2/3n$.*

A close analysis of the proof performed by Aichholzer et al. [20] shows that even $t(n) \geq n$ is true. Using computer search they verified that $t(n) = 2n - 4$ for $3 \leq n \leq 8$.

Proof. Fix a good drawing D of K_n. We will show that every vertex is incident to an empty interior triangle. This is sufficient, since we can move the point at infinity into that triangle, and use the proof to find a second empty interior triangle.

For every vertex v, the star S_v centered at v is free of self-crossings, since its edges are adjacent. Let $u \neq v$ be another vertex of the graph. Then by Theorem 2.6, there is an edge uw so that $D[S_v \cup \{uw\}]$ is free of crossings. If $\Delta = vuw$ is empty, we are done. Otherwise, Δ contains a vertex u'. Then vu' lies in the interior of Δ, but $u'u$ and $u'w$ may cross the boundary of Δ: $u'u$ in vw, and $u'w$ in vu. Theorem 2.6 applied to the complete bipartite graph between $\{u', v\}$ and $\{u, w\}$ guarantees that one of these two cases, say $u'u$ crossing vw, does not happen, but then $\Delta' := vu'u$ is properly contained in Δ and contains fewer vertices than Δ. So if we repeat this process, it will eventually stop with an empty interior triangle incident to v. $\square$

In our second application of Theorem 2.6 we are looking for large, plane subgraphs of good drawings of the complete graph.

Corollary 2.8 (Fulek [140]). *Every good drawing of a complete graph on n vertices contains a plane subgraph of minimum degree at least 2 and size at least $3(n - 1)/2$.*

Proof. We build a plane subgraph H inductively. We start with $H := S_v$, a star centered at an arbitrary vertex v. Since the drawing is good, S_v is a plane subgraph of our drawing. As long as there is a vertex of degree 1 in H, we proceed as follows: let u be the vertex of degree 1, so uv is the edge it is incident to in H. Let S be the set of vertices on the boundary of the face of H to which u belongs. By Theorem 2.6, the complete bipartite subgraph between $\{u, v\}$ and $S - \{u, v\}$ contains a crossing-free edge e incident to u. If e crossed an edge f of H, then f must be incident to v. If it were not, say $f = xy$, then the triangle vxy belongs to H, and the endpoints of e lie on opposite sides of the triangle, which is not possible, since all vertices in S lie in the same face of H as u. However, if f is incident to v, say $f = vx$, then x cannot have belonged to S, so e must also cross some edge not incident to v to reach x, and that case we already excluded. We can now add e to H and repeat the process until all vertices have degree at least 2.

Removing v from H gives us a subgraph with minimum degree 1, so H contains at least $n - 1 + \lceil (n - 1)/2 \rceil = 3\lceil (n - 1)/2 \rceil$ edges. $\square$

While Corollary 2.8 is a good start, we really want to find a large plane subgraph consisting of independent edges, sometimes called a *disjoint matching*, or *pairwise disjoint edges*.

Theorem 2.9 (Suk [324]). *Every good drawing of a complete graph on n vertices contains $\Omega(n^{1/3})$ pairwise disjoint edges.*

Theorem 2.9 dramatically improved a long line of previous logarithmic lower bounds; our proof here will follow the elegant approach of Fulek [140]. We start with an intermediate result. Let us call two vertices in a good drawing of a graph *antipodal*, if the drawing contains a spanning complete bipartite subgraph in which the two vertices are antipodal.

Theorem 2.10 (Fulek [140]). *If a good drawing of a complete graph has two antipodal vertices, then the drawing contains $\Omega(n^{1/2})$ pairwise disjoint edges.*

For the proof we use a result by Tóth [334]; we call a drawing of a graph in the plane *x-monotone*, if every edge crosses every vertical line at most once. Tóth showed that a good, *x*-monotone drawing of a graph with at most k pairwise disjoint edges can have at most $O(k^2 n)$ edges. We will prove a weaker bound of $O(k^4 n)$ in Theorem 5.35.

Proof. Let u and v be the antipodal vertices; since the drawing of the complete bipartite graph for which u and v are antipodal is planar, we can find a homeomorphism of the plane which turns it into a rectilinear drawing with u and v on the y-axis, and the remaining vertices along the x-axis. Moreover, we can assume that there are $\lfloor n/2 \rfloor - 1$ vertices on the left, and $\lceil n/2 \rceil - 1$ vertices on the right side of uv; we label these vertices $w_1, \ldots, w_{n-2}$. The drawing of the complete bipartite subgraph looks as shown in Figure 2.6. Consider an edge $w_i w_j$, $i < j$. This edge cannot cross the four edges it is

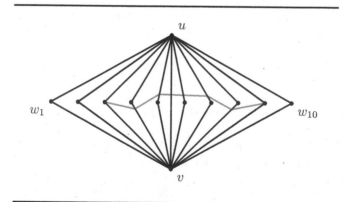

FIGURE 2.6: Antipodal vertices u, v, and their neighbors. Edge $w_3 w_9$ (in gray) is drawn as an *x*-monotone curve.

adjacent to, so $w_i w_j$ either lies entirely in the 4-cycle $uw_i v w_j$ or entirely outside of it. Similarly, $w_i w_j$ can cross each $uw_k v$ at most once: if it crossed it more than once, it would have to cross it at least three times, but then it would have to cross one of the edges twice, which is not possible. Hence $w_i w_j$ enters and leaves each 4-cycle $uw_k v w_{k+1}$ at most once. Within each 4-cycle—except the outer one: $uw_1 v w_{n-2}$—we can then replace the drawing of each edge by a straight-line segment connecting its crossings intersections with the 4-cycle; this does not change whether two edges cross or not in a particular 4-cycle, and the drawing remains good. Moreover, we can move the intersections close to the x-axis, so that the drawings of all edges which do not pass through the outer 4-cycle are x-monotone (intersect every vertical line at most once); in Figure 2.6 this is shown for edge $w_3 w_9$. There are at least $(n/2-1)^2$ edges between $\{w_1, \ldots, w_{\lfloor n/2 \rfloor -1}\}$ and $\{w_{\lfloor n/2 \rfloor -1}\}, \ldots, w_{n-2}\}$. Each of these edges either crosses uv or passes through the outer 4-cycle $uw_1 v w_{n-2}$. Suppose that at least half of the edges intersect uv. Since they do not pass through the outer 4-cycle, they induce a good, x-monotone drawing of a graph with $\Omega(n^2)$ edges. Using the aforementioned theorem of Tóth, this implies that there must be at least $\Omega(n^{1/2})$ pairwise disjoint edges. If, on the other hand, fewer than half of the edges between $\{w_1, \ldots, w_{\lfloor n/2 \rfloor -1}\}$ and $\{w_{\lfloor n/2 \rfloor -1}\}, \ldots, w_{n-2}\}$ intersect uv, then at least half of the edges must pass through the outer 4-cycle; we then redraw so the vertices along the x-axis occur in order $w_{\lfloor n/2 \rfloor -1}\}, \ldots, w_{n-2}, w_1, \ldots, w_{\lfloor n/2 \rfloor -1}$. So the former outer face $uw_1 v w_{n-2}$ lies where uv used to be. We can then repeat the same argument to again find $\Omega(n^{1/2})$ pairwise disjoint edges. $\square$

With Theorem 2.10 we can now complete our proof that every good drawing of a complete graph on n vertices contains $\Omega(n^{1/4})$ pairwise disjoint edges.

Proof of Theorem 2.9. Fix a good drawing of K_n. We saw in Corollary 2.8 that the drawing contains a plane subgraph H of minimum degree at least 2, with one vertex v of degree $n-1$. Let Δ be the largest degree in $H - \{v\}$. Since every vertex in $H - \{v\}$ has degree at least 1, there must be a matching of size $(n-1)/(2\Delta)$. On the other hand, if we let u be the vertex in $H - \{v\}$ of degree Δ, then our drawing of K_n contains a drawing of a $K_{2+\Delta}$ with two antipodal vertices u and v, which, by Theorem 2.10, contains $\Omega(\Delta^{1/2})$ pairwise disjoint edges. Since the maximum of $(n-1)/(2\Delta)$ and $\Omega(\Delta^{1/2})$ is always at least $\Omega(n^{1/3})$, this completes the proof. $\square$

The weaker bound of $O(k^4 n)$ we prove in Theorem 5.35 yields a lower bound of $\Omega(n^{1/5})$ for Theorem 2.9. On the other hand, if the bound in Theorem 5.35 can be improved to $O(kn)$, as is conjectured, then the lower bound for Theorem 2.9 improves to $\Omega(n^{1/2})$. We are not aware of any reasonable upper bounds on the number of pairwise disjoint edges.

We will return to extremal questions later; for example, Theorem 7.20 deals with forcing k pairwise crossing edges, and Corollary 11.32 with k pairwise disjoint edges.

2.2　Crossing Numbers of Graphs

It is often claimed that we do not know the crossing numbers of many infinite families of graphs, but we do know several, and for many we have asymptotic results. We cannot review all of these results in either depth of breadth (a survey on this topic would be very welcome); instead, we highlight some representative results in this area. We will discuss the crossing number of random graphs in Section 3.1.

2.2.1　Cartesian Products of Graphs

The *Cartesian product*, $G \square H$ of two graphs G and H consists of $|V(G)|$ copies H_v of H, $v \in V(G)$. We add an edge between $x \in H_u$ and $y \in H_v$ if uv is an edge of G and x and y are copies of the same vertex in H. Note that $G \square H$ and $H \square G$ are isomorphic. Both the $m \times n$ square grid, $P_m \square P_n$, and

$$Q_n := \overbrace{K_2 \square \cdots \square K_2}^{n}$$ is the *n-dimensional hypercube*, or *n-cube*, for short, are examples of Cartesian products of graphs. We use P_n to denote the path on n vertices, so P_n has length $n - 1$.

Since square grids are planar, the simplest case of non-planar Cartesian products occurs for stars and paths; after many partial results, the exact answer was found recently.

Theorem 2.11 (Bokal [69]). *If $m, n \geq 1$, then*

$$\mathrm{cr}(K_{1,m} \square P_n) = (n - 2) \left\lfloor \frac{m}{2} \right\rfloor \left\lfloor \frac{m-1}{2} \right\rfloor.$$

The proof makes use of the *zip-product* of two graphs, introduced by Bokal [71]. Let G_1, G_2 be two graphs with vertices $v_i \in V(G_i)$ of the same degree. Let σ be a bijection between the neighborhoods of v_1 and v_2, the *zip-function*. Then $G_1 \odot_\sigma G_2$ is the graph obtained from taking $G_1 \cup G_2$, removing v_1 and v_2 and connecting their former neighbors by a matching according to σ.

Lemma 2.12 (Bokal [69]). *If there are cr-minimal drawings of G_1 and G_2 so that the neighbors of $v_i \in V(G_i)$ have the same order in the rotation system when identified according to a zip-function σ, then*

$$\mathrm{cr}(G_1 \odot_\sigma G_2) \leq \mathrm{cr}(D_1) + \mathrm{cr}(D_2).$$

Proof. Take a mirror copy of D_2 in which v_2 is in the outer face. Remove v_2 from D_2, and add it to the region containing v_1 in D_1. Remove v_1. Since the zip-function matched the neighbors of v_1 and v_2 in the same order, and one of these orders has now been reversed, we can connect the neighbors according to σ without introducing any crossings, yielding a drawing of $\operatorname{cr}(G_1 \odot_\sigma G_2)$ with at most $\operatorname{cr}(D_1) + \operatorname{cr}(D_2)$ crossings. $\qquad\square$

To make the lemma useful, we need equality. If the graphs are sufficiently connected, this can be guaranteed. We say that a subgraph F of G is a *star-set* for $S \subseteq V(G)$ if F is either a subdivision of $K_{1,|S|}$ with S the leaves of the subdivided star, or F is a subdivision of $K_{1,|S|-1}$ with S being the leaves and the center of $K_{1,|S|-1}$; we say that a set of vertices S in G is *k-star-connected* if there are k disjoint star-sets for S.

Lemma 2.13 (Bokal [69]). *If G_1 and G_2 each contain a vertex, $v_i \in V(G_i)$, of the same degree so that the neighborhoods of v_i are 2-star connected in $G_i - \{v_i\}$, then*

$$\operatorname{cr}(G_1 \odot_\sigma G_2) \geq \operatorname{cr}(G_1) + \operatorname{cr}(G_2),$$

for any zip-function σ between the neighborhoods of the v_i.

In the proof we will use the notation $\operatorname{cr}_D(A, B)$ to denote the number of crossings in D that occur between an edge in A and an edge in B.

Proof. Fix a cr-minimal drawing D of $G_1 \odot_\sigma G_2$ for an arbitrary zip-function σ between the neighborhoods $N_i := N_{G_i}(v_i)$ of v_i, $i = 1, 2$. D contains two drawings of a subdivision of G_1: restrict D to $E_1 := E(G_1 - \{v_1\})$, add the matching F between N_1 and N_2 and one of the two star-sets F_1^2 and F_2^2 in $G_2 - \{v_2\}$ connecting N_2. Suppressing the subdivision vertices in $F \cup F_1^2$ and $F \cup F_2^2$ yields two drawings of G_1. If there are crossings between edges adjacent to v_1 (which can happen if the edges of F cross each other, or F_1^2 or F_2^2), then remove those using Lemma 1.3. We conclude that $\operatorname{cr}(G_1) \leq \operatorname{cr}_D(E_1, E_1) + \operatorname{cr}(E_1, F) + \operatorname{cr}(E_1, F_j^2)$, for $j = 1, 2$, so

$$\operatorname{cr}(G_1) \leq \operatorname{cr}_D(E_1, E_1) + \operatorname{cr}_D(E_1, F) + \operatorname{cr}_D(E_1, F_1^2 \cup F_2^2)/2.$$

A symmetric argument for G_2 gives us

$$\operatorname{cr}(G_2) \leq \operatorname{cr}_D(E_2, E_2) + \operatorname{cr}_D(E_2, F) + \operatorname{cr}_D(E_2, F_1^1 \cup F_2^1)/2,$$

with $E_2 := E(G_2 - \{v_2\})$ and F_j^1, $j = 1, 2$ being the two star-sets in $G_1 - \{v_1\}$. Then

$$
\begin{aligned}
\operatorname{cr}(G_1) + \operatorname{cr}(G_2) \;\leq\;& \operatorname{cr}_D(E_1, E_1) + \operatorname{cr}_D(E_2, E_2) + \operatorname{cr}_D(E_1 \cup E_2, F) + \\
& \frac{1}{2}(\operatorname{cr}_D(E_1, F_1^2 \cup F_2^2) + \operatorname{cr}_D(E_2, F_1^1 \cup F_2^1)) \\
\leq\;& \operatorname{cr}_D(E_1, E_1) + \operatorname{cr}_D(E_2, E_2) + \operatorname{cr}_D(E_1 \cup E_2, F) + \\
& \frac{1}{2}(\operatorname{cr}_D(E_1, E_2) + \operatorname{cr}_D(E_2, E_1)) \\
\leq\;& \operatorname{cr}_D(E_1, E_1) + \operatorname{cr}_D(E_2, E_2) + \operatorname{cr}_D(E_1 \cup E_2, F) + \\
& \operatorname{cr}_D(E_1, E_2) \\
\leq\;& \operatorname{cr}(D).
\end{aligned}
$$

since each crossing in D is counted at most once. $\qquad\square$

Recall that the join $G+H$ of two graphs G and H on disjoint vertex sets is the union of the two graphs together with all edges between $V(G)$ and $V(H)$ and that a *dominating vertex* in a graph is a vertex adjacent to every vertex of the graph (except itself). We introduce $G\,\hat{\square}\,H$, the *capped Cartesian product* of G and H just like $G\square H$ with the copies of G replacing vertices of degree 1 in H get contracted to a single vertex, called a *cap*. (The capped Cartesian product, different from the Cartesian product, is no longer commutative. For example, $P_2\,\hat{\square}\,P_3$ is isomorphic to $K_4 - e$, while $P_3\,\hat{\square}\,P_2$ is simply P_2.)

Theorem 2.14 (Bokal [69]). *If G has a dominating vertex, then*

$$\mathrm{cr}(G\,\hat{\square}\,P_{n+2}) = n\,\mathrm{cr}(G + 2K_1),$$

for $n \geq 0$.

Proof. Let $G_n := G\,\hat{\square}\,P_{n+2}$, for $n \geq 1$. Then G_1 is isomorphic to $G + 2K_1$, settling the base case. For $n > 1$ note that $G_{n+1} = G_n \odot_\sigma (G + 2K_1)$, where σ matches the neighborhood of a cap of G_n to one of the $2K_1$-vertices of $G+2K_1$. Both $G+2K_1$ and G_n are 2-star connected with respect to the vertices identified by σ: $G + 2K_1$ by construction, and G_n by using the dominating vertex of G. So Lemma 2.13 implies that $\mathrm{cr}(G_{n+1}) \geq \mathrm{cr}(G_n) + \mathrm{cr}(G + 2K_1)$, implying the lower bound.

The upper bound is an immediate consequence of Lemma 2.12 and choosing σ so that the two rotations match. $\square$

Proof of Theorem 2.11. First note that $K_{1,m} + 2K_1$ is $K_{3,m}$ with two edges between the three degree m vertices. Hence, the crossing number is at least $\lfloor m/2 \rfloor \lfloor (m-1)/2 \rfloor$ by Theorem 1.14, and this bound can easily be achieved.

Since $K_{1,m}$ has a dominating vertex, Theorem 2.14 implies that $\mathrm{cr}(K_{1,m}\,\hat{\square}\,P_n) = (n-2)\,\mathrm{cr}(K_{1,m}+2K_1) = (n-2)\lfloor m/2 \rfloor \lfloor (m-1)/2 \rfloor$, for $n \geq 2$. This completes the proof, since $K_{1,m} \cap P_n$ is a subdivision of $K_{1,m}\,\hat{\square}\,P_n$. $\square$

The zip-product method can be extended to work for stars and trees [70], and there is also extensive work on Cartesian products of paths and small graphs [209].

2.2.2 Cycles and the Harary-Kainen-Schwenk Conjecture

The graphs $C_m\square C_n$ are known as *toroidal grids (meshes)*, since they embed on the torus. If we assume that $3 \leq m \leq n$, the drawing in Figure 2.7 shows that $C_m\square C_n$ can be drawn with at most $(m-2)n$ crossings, even in a rectilinear drawing.

Conjecture 2.15 (Harary, Kainen, and Schwenk [173]). *For $m \leq n$,*

$$\mathrm{cr}(C_m\square C_n) = (m-2)n.$$

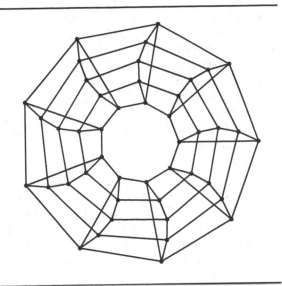

FIGURE 2.7: How to draw $C_m \square C_n$ with $(m-2)n$ crossings. Illustrated for $(m,n) = (5,9)$.

Glebsky and Salazar [152] showed that the conjecture is true for $n \geq m(m+1)$, and, by a long series of papers, for $m \leq 7$, see [152] for references. Here we will show that the conjecture is true for $3 \leq m \leq 4$ and that it is true to within a factor of $1/2$.

Theorem 2.16 (Beineke and Ringeisen [55,287]). *For* $3 \leq m \leq 4$ *and* $m \leq n$,

$$\mathrm{cr}(C_m \square C_n) = (m-2)n.$$

We color the edges of $C_m \square C_n$ red if they belong to a C_m and blue if they belong to a C_n. The red and blue cycles are the *principal* cycles. We say a cycle C *separates* two cycles C' and C'', if C' and C'' lie in different components of $\mathbb{R}^2 - C$. One simple observation: if all principal cycles of the same color, say red, are free of crossings, then there is no principal red cycle that separates two other principal red cycles, since those two cycles are connected by a blue path which is vertex-disjoint from the supposedly separating cycle, and thus would have to cross it.

Two principal cycles (of the same color) are *neighbors* if they are connected by an edge (of the opposite color).

Proof of Theorem 2.16 for $m = 3$. Let $m = 3$, and fix a cr-minimal drawing of $C_3 \square C_n$, $n \geq 3$. Consider the case that all red triangles are free of crossings. Let H be a subgraph induced by two neighboring (red) triangles; we call such a subgraph a *slice*. If H is plane (as part of the chosen drawing), then every other red triangle must lie in one of the 4-faces of H (by the observation on separating cycles preceding the proof). This implies that H is involved in at least two crossings with blue edges (to connect to the two vertices of H which are not incident to the 4-face to the vertices in the 4-face). If H is not plane, then two of its blue edges must cross each other (since we assumed that the red triangles are crossing-free). If we define the *force* of a subgraph as the number of times an edge of that subgraph is involved in a crossing, we have shown that H has force at least 2. Hence, the sum of the forces of all slices H is $2n$, which is twice the number of crossings in the drawing. Therefore, the number of crossings in the drawing is at least n.

We are now ready to prove the result by induction on n. In the base case $n = 3$, and $\mathrm{cr}(C_3 \square C_3) = 3$, see Exercise (2.7). So we can assume $n > 3$. If all red triangles are free of crossings, we are done by the earlier argument, so there must be a red triangle which is involved in a crossing. Removing the edges of the triangle and suppressing its vertices yields a drawing of $C_3 \square C_{n-1}$ with one fewer crossing (the edges of the red triangle cannot cross each other, since the initial drawing was good). Using Lemma 1.3, we can ensure that the drawing is good. By induction, it contains at least $n - 1$ crossings, which then implies that the original drawing had n crossings. $\square$

The proof of Theorem 2.16 for $m = 3$ contains a germ of a more general idea, which is extracted in the following lemma. Call a crossing *ordinary* if it occurs between two edges not belonging to the same principal cycle.

Lemma 2.17 (Juarez and Salazar [198]). *Any drawing of $C_m \square C_n$, $3 \le m \le n$, in which all principal cycles of one of the two colors are pairwise disjoint, has at least $(m - 2)n$ ordinary crossings.*

Proof. For the proof we will use the following claim:

> Any drawing of $C_p \square C_q$, with $p, q \ge 3$, in which the principal p-cycles are pairwise disjoint, and no principal p-cycle separates two other principal p-cycles, has at least $(p - 2)q$ ordinary crossings.

Let us show how the claim implies the lemma. Let D be the drawing described in the lemma. We prove the result by induction on $m + n$. For $m + n = 6$, the crossing number of $C_3 \square C_3$ is 3, see Exercise (2.7), and all crossings are ordinary, since a non-ordinary crossing in this case would have to occur between two adjacent edges; removing that crossing using Lemma 1.3 would yield a drawing of $C_3 \square C_3$ with fewer than 3 crossings, a contradiction. This establishes the base case.

Assume that the result has been shown for all $m' + n' < m + n$, and we want to establish it for $C_m \square C_n$. We start with the case that all red cycles are pairwise disjoint. If no red cycle separates two other red cycles, we can apply the claim with $p = m$ and $q = n$ to obtain the result. Hence, there must be a red cycle separating two other red cycles. The separating cycle must then be intersected twice by each of the blue cycles, with one intersection being a shared vertex, and the other intersection being a crossing. We conclude that the red, separating cycle has at least m ordinary crossings. If we remove the edges of that cycle, and suppress its vertices, we obtain a drawing of $C_m \square C_{n-1}$ which, by induction, has at least $(m-2)(n-1)$ ordinary crossings if $n-1 \geq m$, so the original drawing has at least $m + (m-2)(n-1) > (m-2)n$ ordinary crossings; or at least $(n-3)m$ ordinary crossings if $n-1 < m$, in which case the original drawing has at least $m + (n-3)m = (n-2)m = (m-2)n$ ordinary crossings (since $m \leq n$, $n-1 < m$ implies $m = n$). In both cases we are done.

If all blue cycles are pairwise disjoint, we proceed similarly: if no blue cycle separates two other blue cycles, we apply the claim with $p = n$ and $q = m$, to see that the drawing has $(n-2)m \geq (m-2)n$ ordinary crossings. Hence, there is a blue cycle involved in n ordinary crossings. If $m = 3$, this completes the proof. For $m > 4$, we again remove the edges of the blue cycle and suppress its vertices to obtain a drawing of $C_{m-1} \square C_n$ which, by induction, has $(m-3)n$ ordinary crossings, implying that the original drawing has $m + (m-3)n = (m-2)n$ ordinary crossings, and we are done with this case.

We are left with the proof of the claim. Color the edges in the principal p-cycles red and the remaining edges, belonging to the principal q-cycles, blue. By orienting one of the C_q, we induce a natural order on the red cycles. We consider *slices* H induced by two neighboring red cycles, so H consists of a *left* and a *right* red cycle (as determined by the orientation of C_q) as well as the blue edges between them. Let H' be the next slice (again determined by the orientation of the C_q). We define the *force* of H as the number of crossings of (i) a blue edge in H with a red edge in H', (ii) any edge of H with a blue edge in H', and (iii) a blue edge in H with another blue edge in H. We note that each crossing in a drawing of $C_p \square C_q$ contributes to the force of at most one such H as long as $q > 3$. If $q = 3$ a blue edge in H crossing a red edge belonging to the right red cycle of H' counts as a type (i) crossing in H as well as a type (ii) crossing in H'', the slice after H', since H is preceded by H''. Since the red cycles are pairwise disjoint, any blue edge of H must cross the right red cycle in H' an even number of times: the edge is not incident to the cycle, and its endpoints are on the same side of the cycle. We can therefore assign half the crossings of this type to H and the other half to H'', ensuring that even in the case $q = 3$ every crossing counts towards the force of at most one of the slices.

We are done, if we can show that every slice has force at least $p-2$, since all three types of crossings are ordinary, and there are q slices, giving us $(p-2)q$ ordinary crossings. So let H be an arbitrary slice, and let B be a maximal subset of blue edges of H which do not cross a red edge in H', and do not cross each other. Any blue edge of H not in B must either cross a red edge of H', so type (i), or an edge in B, type (iii), showing that H has at least $p - |B|$ crossings of types (i) and (iii). Note that the left and right red cycles of H together with the edges of B form a graph whose only crossings are between red edges of the same cycle. The right red cycle of H' must lie in one of the regions of this graph, and the boundary of that region does not consist of red edges only, since otherwise one of the red cycles would separate two other red cycles. But at least $|B| - 2$ vertices of the right red cycle of H are not incident to this region, so there are at least $|B| - 2$ blue edges of H' which have to cross edges on the boundary of that region, and those edges belong to H. Therefore these are crossings of type (ii) in H, and we conclude that the total force of H is at least $p - |B| + |B| - 2 = p - 2$. $\qquad\square$

We use Lemma 2.17 to complete the proof of Theorem 2.16; a direct proof is also possible. Another consequence of the lemma for $C_3 \square C_n$ can be found in Exercise (2.10).

Proof of Theorem 2.16 for $m = 4$. Fix a cr-minimal drawing of $C_4 \square C_n$, $n \geq 4$. If all the red cycles are pairwise disjoint, then Lemma 2.17 implies the result. Hence, there are two red cycles which cross. Since the cycles are vertex-disjoint, they must have at least two crossings. If $n \geq 5$, we can remove the edges of one of the 4-cycles and suppress its vertices to obtain a drawing of $C_4 \square C_{n-1}$ which, by induction, has at least $2(n - 1)$ crossings; together with the 2 crossings we removed, this gives $2n$ crossings in the original drawing. The case $n = 4$ is Exercise (2.8). $\qquad\square$

Lemma 2.17 is a powerful tool, and it allows us a relatively short proof of a general lower bound on $\mathrm{cr}(C_m \square C_n)$.

Theorem 2.18 (Shahrokhi, Sýkora, Székely, and Vrt'o [310]). *If $m, n \geq 3$, then*

$$\mathrm{cr}(C_m \square C_n) \geq c(m - 2)n.$$

This was first shown, for $c = 1/9$ in [310], the best current bound is $c = 0.8 - o(1)$ [297], we follow [198], with $c = 1/2$.

Proof. Fix a cr-minimal drawing of $C_m \square C_n$. We consider *slices* consisting of three neighboring red cycles $R_\leftarrow$, $R_\uparrow$, and $R_\rightarrow$ (left, middle, right), and the blue edges between left and middle, and middle and right cycle. Let S be a maximal set of vertices of $R_\uparrow$ so that (i) the color rotation at each vertex is $rrbb$ (recording the colors of incident edges), (ii) the blue edges incident to S do not cross edges of $R_\leftarrow$, $R_\uparrow$, and $R_\rightarrow$, and (iii) none of the blue edges incident to S cross each other. We claim that there are $|S| - 2$ ordinary crossings among edges of $R_\leftarrow$, $R_\uparrow$ and $R_\rightarrow$. Consider the $|S|$ blue paths of length two with a midpoint in S. These paths are pairwise disjoint, by (iii), they are disjoint from the three red cycles, by (ii), and both edges of each path lie on the same side of $R_\uparrow$, by (i). If we add an edge to each path connecting its endpoints close to the two blue edges, we obtain a drawing of $C_3 \square C_{|S|}$ in which the blue triangles are pairwise disjoint, so by Lemma 2.17, there are at least $|S| - 2$ ordinary crossings which must be between edges of the three red cycles (because the blue triangles do not cross the red cycles, by (ii)).

Define the *force* of the slice to be the number of crossings (i) between edges of the slice, and (ii) between $R_\uparrow$ and blue edges not belonging to the slice.

Note that each crossing counts towards at most two slices (assuming $n \geq 4$, which we can, because of Theorem 2.16). Consider a path P of length two with midpoint v in $V(R_\uparrow) - S$. Then either P crosses a blue edge incident to S or one of the red edges of $R_\leftarrow$, $R_\uparrow$, $R_\rightarrow$, or the rotation at v is *rbrb*. In the first two cases the edges of P clearly contribute at least one to the force of the slice. In the third case, the principal blue cycle containing P must intersect $R_\uparrow$ an even number of times, and since the two cycles have exactly one vertex in common, namely v, there must be at least one crossing of a blue edge with $R_\uparrow$, which contributes one to the force. In summary, the force of a slice is at least $|V(R_\uparrow)| - |S| + |S| - 2 = m - 2$. Since there are n slices, this completes the proof for $c = 1/2$. $\qquad\square$

2.2.2.1 The Hypercube

Theorem 2.19 (Faria, de Figueiredo, Sýkora, and Vrt'o [131]). *For $n \geq 0$ we have*

$$\operatorname{cr}(Q_n) \leq \frac{5}{32} 4^n - \left\lfloor \frac{n^2 + 1}{2} \right\rfloor 2^{n-2}.$$

This bound had been conjectured as the correct value for $\operatorname{cr}(Q_n)$ by Eggleton and Guy [121], and again by Erdős and Guy [124], after the earlier attempt to show it is an upper bound failed. The construction is rather intricate, so instead we present a construction due to Madej [234] which achieves the slightly weaker upper bound of $\frac{1}{6} 4^n - 2^{n-4}(2n^2 + 3 - (-1)^n/3)$. The construction will produce a 2-page drawing (to be studied in more detail in Chapter 8); that is, all vertices will lie on a horizontal line, the *spine*, and edges are drawn as circular arcs either above or below the spine. We start with Q_0, an isolated vertex. Suppose we have a drawing of Q_{n-1}. Mirror that drawing at a vertical line to the right of itself, and connect corresponding vertices by circular arcs. Draw arcs in the upper page if n is even, and the lower page otherwise. See Figure 2.8 for an illustration.

To count the crossings v_n in Madej's drawing of Q_n we first count u_n,

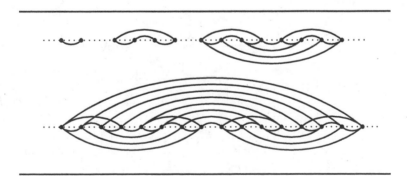

FIGURE 2.8: 2-page drawings of Q_1 through Q_4.

the number of times edges in Madej's drawing of Q_n cross any vertical line through a vertex. So $u_0 = u_1 = 0$, $u_2 = 2$, $u_3 = 16$, and, more generally, $u_n = 2u_{n-1} + \sum_{i=1}^{2^{n-1}} 2(i-1)$, which solves as $u_n = 2^{n-1}(2^n - n - 1)$. Madej's drawing of Q_n consists of four copies of Q_{n-2}; the additional edges of the last two steps can be viewed as horizontal edges of the Q_{n-2}, so $v_n = 4v_{n-2} + 4u_{n-2} = 4v_{n-2} + 2^{n-1}(2^{n-2} - n - 3)$ and $v_0 = v_1 = v_2 = v_3 = 0$, which satisfies $v_n \leq \frac{1}{6}4^n$ (a slightly better upper bound can be derived, but it fails to be sharp for $n = 5$ already).

Table 2.2 collects the upper bounds implied by Theorem 2.19 for small n.

G	bounds
Q_4	8 [108]
Q_5	≤ 56 [234]
Q_6	≤ 352 [131]

TABLE 2.2: Upper bounds on $\mathrm{cr}(Q_n)$ for small n.

The crossing number of Q_4 is 8 (this is the last case for which Madej's bound is sharp), a result due to Dean and Richter [108], see Exercise (2.8), and note that $C_4 \square C_4$ is isomorphic to Q_4. The bound of $\mathrm{cr}(Q_5) \leq 56$ is expected to be sharp, also see Exercise (2.9).

Asymptotically, it is conjectured that $\lim_n \mathrm{cr}(Q_n)/4^n = 5/32$. In the next section we will use Leighton's embedding method to obtain a lower bound on $\mathrm{cr}(Q_n)$ and show that $\lim_n \mathrm{cr}(Q_n)/4^n > 1/19$ (Corollary 3.9).

2.2.3 More Complete Graphs

After our careful investigation of the crossing number of $K_{m,n}$, it appears natural to ask what we know about k-partite graphs. The answer is: not much. Early on, Harborth [174] was able to establish a general upper bound using the function

$$
\begin{aligned}
Z(x_1, \ldots, x_n) \; := \; \frac{1}{8} \Bigg(& \sum_{1 \leq a < b < c < d \leq n} 3 x_a x_b x_c x_d + 3 \binom{\lfloor c/2 \rfloor}{2} \\
& - \sum_{1 \leq a < b \leq n} \left\lfloor \frac{c - ((x_a \bmod 2) + (x_b \bmod 2))}{2} \right\rfloor x_a x_b \Bigg) \\
& + \sum_{a=1}^{n} \left\lfloor \frac{x_a}{2} \right\rfloor \left\lfloor \frac{x_a - 1}{2} \right\rfloor \left\lfloor \frac{m - x_a}{2} \right\rfloor \left\lfloor \frac{m - x_a - 1}{2} \right\rfloor \\
& - \sum_{1 \leq a < b \leq n} \left\lfloor \frac{x_a}{2} \right\rfloor \left\lfloor \frac{x_a - 1}{2} \right\rfloor \left\lfloor \frac{x_b}{2} \right\rfloor \left\lfloor \frac{x_b - 1}{2} \right\rfloor,
\end{aligned}
$$

where m is the sum of the x_i and c is the number of odd x_i. For $n = 2$, the formula agrees with the usual Zarankiewicz function $Z(x_1, x_2)$.

Theorem 2.20 (Harborth [174]).

$$\text{cr}(K_{x_1,\ldots,x_n}) \leq Z(x_1, \ldots, x_n).$$

A closer look at the function Z shows that is has three parts depending on how many different partitions the endpoints of a crossing belong to (4, 3, or 2). The drawing achieving the bound Z is a generalization of the Zarankiewicz drawing, Figure 1.10, for the bipartite case, where instead of 2 axes, we use n evenly spaced axes. See Harborth's paper for the details and an analysis (in German).

Harborth conjectures that the function Z is the correct value of $\text{cr}(K_{x_1,\ldots,x_n})$, so it is of interest to consider special cases, like tripartite graphs. To shorten the formulas, we write $H(x) := \lfloor x/2 \rfloor \lfloor (x - 1)/2 \rfloor$. Note that $H(x + 1) - H(x) = \lfloor x/2 \rfloor$.

Corollary 2.21.

$$\begin{aligned}
\text{cr}(K_{x_1,x_2,x_3}) \leq\ & H(x_1)(H(x_2 + x_3) - H(x_2)) \\
& + H(x_2)(H(x_1 + x_3) - H(x_3)) \\
& + H(x_3)(H(x_1 + x_2) - H(x_1)).
\end{aligned}$$

For tripartite graphs, it is known that the upper bound can be achieved by a rectilinear drawing [149]. Harborth's conjecture is consistent with all the particular values we have so far. For $x_1 = 1$, it gives us $H(x_2)(H(x_3 + 1) - H(x_3)) + H(x_3)H(x_2 + 1) = \lfloor x_3/2 \rfloor H(x_2) + H(x_3)H(x_2 + 1)$. So $Z(1, 3, n) = \lfloor n/2 \rfloor + Z(4, n)$ and $Z(1, 4, n) = 2\lfloor n/2 \rfloor + Z(5, n) = n(n - 1)$, which are the correct values for these cases [39, 189]. There are more results for tripartite crossing number, but we will show only one.

Theorem 2.22 (Asano [39]). *For $n \geq 1$, we have*

$$\text{cr}(K_{1,3,n}) = Z(4, n) + \left\lfloor \frac{n}{2} \right\rfloor.$$

We follow a proof idea from [255]. The *maximum crossing number*, $\text{max-cr}(G)$, of a graph G is the largest number of crossings in any good drawing of G (Chapter 12 will deal with max-cr in detail).

Lemma 2.23 (Ouyang, Wang, and Huang [255]). *Let $m = |V(G)|$, then for $n \geq 1$ we have*

$$(n - 1)\,\text{cr}(G + nK_1) \geq n\,\text{cr}(G + (n - 1)K_1) + \text{cr}(K_{m,n}) - \text{max-cr}(G).$$

Here $G + nK_1$ is the join of G with n isolated vertices; in other words, we add a complete bipartite $K_{m,n}$ between the m vertices of G and n new vertices to G.

Proof. Fix a cr-minimal drawing D of $G+nK_1$. It has n subgraphs of the form $G+(n-1)K_1$ and one subgraph of the form $K_{m,n}$. Counting the crossings in those subgraphs gives us at least $n\operatorname{cr}_D(G+(n-1)K_1)+\operatorname{cr}_D(K_{m,n})$ crossings. Any crossing between two edges of G is counted n times; a crossing between an edge of G and an edge of $K_{m,n}$ is counted $n-1$ times, and a crossing between edges of $K_{m,n}$ is counted $(n-2)+1=n-1$ times. Therefore, $(n-1)\operatorname{cr}(D)+\operatorname{cr}_D(G)$ is an upper bound on $n\operatorname{cr}_D(G+(n-1)K_1)+\operatorname{cr}_D(K_{m,n})$. It follows that $n\operatorname{cr}(G+(n-1)K_1)+\operatorname{cr}(K_{m,n})\leq(n-1)\operatorname{cr}(G+nK_1)+\text{max-cr}(G)$, which is what we had to show. $\qquad\square$

Proof of Theorem 2.22. It is easy to modify an optimal drawing of $K_{4,n}$, with $Z(4,n)$ crossings, to add a $K_{1,3}$ between the four vertices of the left partition incurring at most $\lfloor n/2 \rfloor$ additional crossings. For the lower bound, let $G = K_{1,3}$, so $K_{1,3,n} = G+nK_1$. By Lemma 2.23, we have

$$\operatorname{cr}(K_{1,3,n}) \geq \frac{n}{n-1}\operatorname{cr}(K_{1,3,n-1}) + Z(4,n)\frac{1}{n-1} - \text{max-cr}(G).$$

Since G is a star, $\text{max-cr}(G) = 0$, because adjacent edges do not cross in a good drawing. We conclude that

$$\operatorname{cr}(K_{1,3,n}) \geq \frac{n}{n-1}\operatorname{cr}(K_{1,3,n-1}) + 2\left\lfloor\frac{n}{2}\right\rfloor\left\lfloor\frac{n-1}{2}\right\rfloor\frac{1}{n-1}.$$

We can now complete the proof by induction. For $n < 3$ the result is immediate: $K_{1,3,1}$ is planar, and $K_{1,3,2}$ can be drawn with one crossing, which, since the graph contains a $K_{3,3}$-subdivision, is optimal. So assume $n \geq 3$. Then

$$\begin{aligned}
\operatorname{cr}(K_{1,3,n}) &\geq \frac{n}{n-1}\operatorname{cr}(K_{1,3,n-1}) + 2\left\lfloor\frac{n}{2}\right\rfloor\left\lfloor\frac{n-1}{2}\right\rfloor\frac{1}{n-1} \\
&\geq \frac{n}{n-1}\left(Z(4,n-1) + \left\lfloor\frac{n-1}{2}\right\rfloor\right) + 2\left\lfloor\frac{n}{2}\right\rfloor\left\lfloor\frac{n-1}{2}\right\rfloor\frac{1}{n-1} \\
&\geq Z(4,n) + \left\lceil\frac{n}{2}\right\rceil - 2/3,
\end{aligned}$$

which, since $\operatorname{cr}(K_{1,3,n})$ is an integer, completes the proof. $\qquad\square$

Another special case already emphasized by Harborth, occurs when all x_i are even, and, in particular, if all $x_i = 2$. In that case, Harborth's formula implies $\operatorname{cr}(K_{2,\ldots,2}) \leq 6\binom{n}{4}$; as far as we know, this is only known up to $n = 4$, see Exercise (2.19). $K_{2,\ldots,2}$ is a K_n in which every vertex has been duplicated, a smoothed version of K_n.

We can show that asymptotically, the bound suggested by Harborth is right in the tripartite, diagonal case, that is, for the crossing number of $K_{n,n,n}$.

Theorem 2.24 (Gethner, Hogben, Lidický, Pfender, Ruiz, and Young [149]). *For sufficiently large n we have*

$$\operatorname{cr}(K_{n,n,n}) \geq 2/3 Z(n,n,n).$$

Proof. We use the standard counting method, applied to $K_{2,3,n}$-subgraphs. As Exercise (2.18) shows, $\text{cr}(K_{2,3,n}) \sim n^2$ (the exact number is $Z(5,n)+n$, we do not need that information for this proof). Clearly, $K_{2,3,n}$ occurs $6\binom{n}{2}\binom{n}{3}\binom{n}{n} \sim n^5/2$ times as a subgraph of n. So a cr-minimal drawing of $K_{n,n,n}$ has $n^7/2$ crossings, asymptotically, with double-counting. How often is each crossing counted? This depends on how its four endpoints are distributed among the 3 partitions. If the endpoints avoid one of the partitions, a crossing is counted

$$\left(\binom{n-2}{0}\binom{n-2}{1}\binom{n}{n} + \binom{n-2}{0}\binom{n}{3}\binom{n-2}{n-2} + \binom{n}{2}\binom{n-2}{1}\binom{n-2}{n-2}\right)$$

corresponding to endpoints of the crossing being distributed as 220, 202, and 022 (each of these occurs in two ways). This number is asymptotically equivalent to $2(n + n^3/6 + n^3/2) \sim 4/3n^3$. If the four endpoints are distributed over all 3 partitions, the options are 211, 121, and 112, and the corresponding count is

$$2\left(\binom{n-2}{0}\binom{n-1}{2}\binom{n-1}{n-1} + \binom{n-1}{1}\binom{n-2}{1}\binom{n-1}{n-1} + \binom{n-1}{1}\binom{n-1}{2}\binom{n-2}{n-2}\right),$$

which is asymptotically equivalent to $2(n/2 + n^2 + n^2/2) \sim 3n^2$. So a crossing occurs in at most $4/3n^3$ subgraphs of type $K_{2,3,n}$, which implies that there are at least $n^7/2/(4/3n^3) = 3/8n^4$ crossings, asymptotically.

Working with the expression in Corollary 2.21, we see that $Z(n,n,n) = 3H(n)(H(2n) - H(n)) \sim 3n^2/4 \left((3n)^2/4 - n^2/4\right) = 9/16n^4 = 3/2(3/8n^4)$, concluding the proof. $\qquad \square$

As Gethner, Hogben, Lidický, Pfender, Ruiz, and Young [149] point out, the counting approach is limited: we count crossings which avoid a partition, whereas in a cr-minimal drawing, crossings are mostly of the other type. They manage to prove a much stronger bound for rectilinear drawings using Razborov's flag algebra.

2.2.4 Crossing-Critical Graphs

A graph G is *crossing critical* if $\text{cr}(H) < \text{cr}(G)$ for every proper subgraph H of G. We call G *k-crossing critical* if $\text{cr}(G) \geq k$, and $\text{cr}(H) < k$ for every proper subgraph H of G.

For example, $K_{3,3}$ and K_5 are 1-crossing critical. Since subdividing an edge, or suppressing a degree-2 vertex, or removing vertices of degree one or zero does not change the crossing number of a graph, we can restrict ourselves to graphs of minimum degree at least 3. In that class, $K_{3,3}$ and K_5 are the only 1-crossing-critical graphs; the original proof of this fact was part of a proof of Kuratowski's theorem by Dirac and Schuter, now we see it as an easy consequence of that result, see Exercise (2.13). There are k-crossing critical graphs for every k, for example, kK_5. Kochol [211], improving an earlier result due to Širáň [346], showed that there are infinitely many k-crossing critical graphs.

Theorem 2.25 (Širáň [346]; Kochol [211]). *There are infinitely many 3-connected, k-crossing critical graphs for every $k \geq 2$.*

The main idea behind the construction is contained in Figure 2.9. The graph in the figure is 3-connected, and it has crossing number 2. Removing any of its edges reduces the crossing number, since part of the graph can then be twisted to make the two crossings at the bottom redundant. The repeating upside/downside pentagon can be repeated arbitrarily often, leading to an infinite family. See Exercise (2.14).

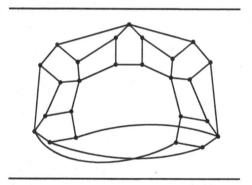

FIGURE 2.9: A 2-crossing critical graph.

A k-crossing critical graph need not have crossing number k. For example, $C_3 \square C_3$ has crossing number 3 as we saw (Exercise (2.7)), but removing any of its edges (since the graph is edge-transitive, it does not matter which edge) leaves a graph which can be drawn with a single crossing. Richter and Thomassen [285] showed that the gap cannot be too large, however.

Theorem 2.26 (Richter and Thomassen [285]). *If G is a k-crossing critical graph, then $\mathrm{cr}(G) \leq 2.5k$.*

It is not known whether the bound is asymptotically correct; the best lower bound is furnished by $G = K_{3,n}$, for which $\mathrm{cr}(G) = \lfloor n/2 \rfloor \lfloor (n-1)/2 \rfloor$, while $G-e$ has crossing number at most $\lfloor n/2 \rfloor \lfloor (n-1)/2 \rfloor - (\lceil n/2 \rceil - 1)$: the standard drawing of $K_{3,n}$ contains an edge with $\lceil n/2 \rceil - 1$ crossings; since $K_{3,n}$ is edge-transitive, we can assume that edge is e. Therefore, removing any edge from $K_{3,n}$ drops the crossing number by the square root of the crossing number.

Call a cycle C in a graph d-*light*, if there is an edge $e \in E(C)$ so that $\sum_{v \in V(C), v \notin e} (\deg(v) - 2) \leq d$. We also say that C is d-light *with respect to* e.

Lemma 2.27. *A planar graph of minimum degree at least* 3 *contains a* 4*-light (facial) cycle.*

This type of argument goes back to a paper by Lebesgue, and was developed by Richter and Thomassen [285].

Proof. We can assume that G is connected. We let $n = |V(G)|$, $m = |E(G)|$, and f the number of faces in the embedding. Define the weight of a face of the embedding as $w(F) := \sum_{v \text{ on } F} 1/\deg(v)$. (Note that a vertex may occur multiple times along a facial boundary, we count it for each such occurrence.) Then $\sum_F w(F) = n := |V(G)|$. The *length* of a face, $\ell(F)$, is the number of edges in F's boundary walk, so $2m = \sum_F \ell(F)$.

By the Euler-Poincaré Formula, Theorem A.15, we have $2 \leq n - m + f = \sum_F w(F) - \ell(F)/2 + 1$. Since $w(F) \leq \ell(F)/3$, by the assumption on minimum degree, there must be a face with $w(F) - \ell(F)/2 + 1 > 0$. For such a face, $\ell(F) < 6$, since otherwise, $w(F) - \ell(F)/2 + 1 \leq 0$. Let C be the cycle bounding F; note that C is indeed a cycle, and not a walk, since otherwise, F would be incident to a cut-vertex, which would force a vertex of degree 1.

If $\ell(F) = 3$, then $w(F) > 1/2$; this means that the minimum degree of vertices on F has to be at most 5, so C is 3-light. If $\ell(F) = 4$, then $w(F) > 1$, and the minimum degree of vertices of F has to be 3. Let v be the vertex of degree 3 in C. If both its neighbors in C have degree at least 6, then $w(F) \leq 1/3 + 2/6 + 1/3 = 1$, a contradiction, so at least one of the neighbors, u, has degree at most 5. Choosing e to be the edge on C not incident to u and v shows that C is 4-light. If $\ell(F) = 5$, then $w(F) > 3/2$. There cannot be two vertices of degree at least 4, since otherwise $w(F) \leq 2\frac{1}{4} + 3\frac{1}{3} = 3/2$. So there are at least four vertices of degree 3, and we can choose e so that C is 3-light. $\qquad\square$

Lemma 2.28. *If G has minimum degree at least* 3 *and there are k edges so that removing these edges makes the graph planar, then G contains a $(4+k)$-light cycle.*

The smallest number of edges whose removal makes a graph planar is the *skewness* of the graph; we will study this notion more closely in Section 3.3.

Proof. Let E' be the set of at most k edges so that $G - E'$ is planar. We show the result by induction on k. If $t = 0$, then the result follows from Lemma 2.27. So assume that $k > 0$, and let $f \in E'$. If $G - f$ has minimum degree at least 3, then, inductively, there is a $(4+k)$-light cycle C in $G - f$; that is, C contains an edge e so that $\sum_{u \in V(C), u \notin e}(\deg(u) - 2) \leq 4 + k - 1$. If at most one end of f is incident to C, then C and e are as needed. If both ends of f are incident to C, then f is a chord of C. Let C' be the cycle in $C \cup \{f\}$ avoiding e. Then C' is $(4 + k - 1)$-light with respect to f.

We conclude that one or both endpoints of f have degree 2 in $G - f$. If either (or both) of the endpoints of f is part of a triangle in G, then that triangle is 1-light, and we are done. Suppressing degree-2 vertices in $G - f$ cannot lead to duplicate edges in that case, so we can apply induction to the resulting graph G' to find a $(4 + k - 1)$-light cycle C' in G'. If C' contains only one edge on which an endpoint of f has been suppressed, then the corresponding cycle C in G is $(4 + k)$-light. If C' contains two such edges, then the corresponding cycle C in G is only $(4 + k + 1)$-light, but f is a chord of C, and we can choose a cycle in $C \cup \{f\}$ using f which is at most $(4 + k)$-light. $\qquad \square$

Proof of Theorem 2.26. We can assume that G has minimum degree at least 3, since suppressing vertices of degree at most 2 does not change the crossing number. Since G is k-crossing critical, there is a set E' of at most k edges whose removal leaves G planar: removing an arbitrary edge e leaves a graph $G - e$ with crossing number at most $k - 1$; starting with $E' = \{e\}$, and adding one edge from each crossing, gives us a planar graph. We can then apply Lemma 2.28 to find a $(4 + k)$-light cycle C in G, that is, there is an edge $e = xy \in E(C)$ so that $\sum_{u \in V(C), u \notin e} (\deg(u) - 2) \leq 4 + k$.

Since G is k-critical, $\mathrm{cr}(G - e) \leq k - 1$, and we can fix a drawing of $G - e$ with at most $k - 1$ crossings. In this drawing we can reconnect the endpoints x and y of e by following $C - e$ closely, and we can do so in two ways, on either side of $C - e$. The two potential drawings of e pass by $\sum_{u \in V(C), u \neq x, y} (\deg(u) - 2)$ edges incident to vertices in $V(C) - \{x, y\}$. Each crossing along $C - e$ is picked up by the two drawings, resulting in at most $2(k - 1)$ crossings (if edges of $C - e$ cross each other, such a crossing is picked up twice by the drawing of e, however, it results in a self-crossing of e which we can remove, so we can ignore such crossings). This implies that one of the two drawings is involved in at most $(2(k - 1) + 4 + k)/2 = 1.5k + 1$ crossings, and this is the drawing of e we choose. Together with the $k - 1$ of the drawing of $G - e$, this gives us the bound we claimed. $\qquad \square$

We can rephrase Theorem 2.26 as a result on the *decay* of the crossing number: every graph contains an edge whose removal reduces the crossing number by at most sixty percent, up to a small constant.

Corollary 2.29 (Richter and Thomassen [285]). *Every non-empty graph G contains an edge e so that $\mathrm{cr}(G - e) \geq 2/5 \, \mathrm{cr}(G) - 1$.*

Proof. If $\mathrm{cr}(G - e) = \mathrm{cr}(G)$ for some edge e in G, then there is nothing to prove, so we can assume that $\mathrm{cr}(G - e) < \mathrm{cr}(G)$ for all edges in G, so G is crossing-critical. Let e be the edge of G which maximizes $k = \mathrm{cr}(G - e) + 1$. Then $\mathrm{cr}(G - f) < k \leq \mathrm{cr}(G)$ for every edge f in G, and G is k-crossing critical, and, by Theorem 2.26, we have $\mathrm{cr}(G) \leq 2.5k$ which implies the claim. $\qquad \square$

On the other hand, Kochol [212] constructed 3-connected graphs G with an edge e so that $\mathrm{cr}(G - e) \leq 3/4 \, \mathrm{cr}(G)$, and $\mathrm{cr}(G)$ arbitrarily large.

Corollary 2.30 (Richter and Thomassen [285]). *If G is k-crossing critical and has minimum degree at least 7, then $n \leq 2(k - 6)$, where $n = |V(G)|$.*

The contrapositive of the corollary is of interest as well: if a graph has more than $2(\text{cr}(G) - 6)$ vertices, it contains an edge whose removal does not change the crossing number.

Proof. Theorem 2.26 implies that G has a drawing with at most $2.5k$ crossings. The planarization of this drawing has $n + k$ vertices, and at least $7/2n + 2k$ edges. By Euler's theorem, Corollary A.5, $7/2n + 2k \leq 3(n + k - 2)$ which implies $n \leq 2(k - 6)$. $\qquad\square$

The proof fails for $k = 6$, but a similar bound is still true [285]. This implies that there are only finitely many k-crossing critical graphs of minimum degree 6. For smaller minimum degree, even smaller average degree, infinite families of k-crossing critical graphs exist for sufficiently large k, see [71].

While the project of classifying all 2-crossing critical graphs is ongoing, if mostly complete, we do have one structural result on k-crossing critical graphs in general, due to Hliněný [187] who showed that there is a function $f(k)$ so that every k-crossing critical graph has pathwidth bounded by $f(k)$.

2.3 Notes

Good drawings of complete graphs have become a well-studied topic, but there is little literature on other types of graphs. There are few exceptions, including Mengersen's work on the number of crossing-free edges in drawings of the complete k-partite graphs [242][1], as well as a paper by Cardinal and Felsner [85] who have started looking at good drawings of complete bipartite graphs. This is in stark contrast to the large number of papers studying the crossing number of specific types of graphs; we have only scratched the surface on this topic in this chapter. Unfortunately, there currently is no survey which focusses on the crossing numbers of different graph families.

2.4 Exercises

(2.1) Recall that $h(n)$ is the largest number of crossing-free edges in any good drawing of a K_n. Determine $h(n)$ for $3 \leq n \leq 7$.

(2.2) Show that if $n \geq 4k + 8$, then there is a good drawing of K_n in which every edge has more than k crossings.

---

[1]Only some of that work has appeared in English, including [243].

(2.3) Show that every K_n has a drawing in which every pair of edges crosses exactly once or twice.

(2.4) Show that for a graph with rotation system, a cr-minimal drawing realizing the rotation system may not be good. *Hint:* K_5 will work.

(2.5) Show that there is a rotation system for K_6 which has three non-isomorphic realizations.

(2.6) Show that the expected number of crossings in a random drawing of $K_{m,n}$ on the sphere (vertices chosen uniformly and independently at random, edges drawn as shortest path), is $\frac{1}{4}\binom{m}{2}\binom{n}{2}$.

(2.7) Show that $\mathrm{cr}(C_3 \square C_3) = 3$.

(2.8) Show that $\mathrm{cr}(C_4 \square C_4) = 8$. *Hint:* Show that there is a principal cycle involved in four crossings.

(2.9) Find a drawing showing that $\mathrm{cr}(Q_5) \leq 56$. *Hint:* Start with the optimal drawing of Q_3 we saw.

(2.10) Show that in a cr-minimal drawing of $C_3 \square C_n$ the triangles are pairwise disjoint.

(2.11) Let G^k be the graph on $V(G)$ in which there is an edge between two vertices u and v, if u and v have distance at most k in G. Show that P_n^3 is planar for all n, and $\mathrm{cr}(P_n^4) = n - 4$ for $n \geq 5$. *Note:* One reason this is interesting is that $K_n = P_n^{n-1}$, so if we could calculate $\mathrm{cr}(P_n^k)$ for arbitrary k we could find $\mathrm{cr}(K_n)$.

(2.12) Determine $\mathrm{cr}(K_{1,m} \square C_n)$. *Hint:* The cases $n < 6$ will be special.

(2.13) Show that $K_{3,3}$ and K_5 are the unique 1-crossing critical graphs of minimum degree 3.

(2.14) Show that there are infinitely many 3-connected, k-crossing critical graphs for every $k \geq 2$. *Hint:* Generalize Figure 2.9.

(2.15) Show that Theorem 2.26 remains true for multigraphs.

(2.16) Show that the factor in Theorem 2.26 can be improved to 2 if the graph has minimum degree at least 4.

(2.17) Show that $\mathrm{cr}(K_{4,n} - e) = Z(4,n) - \lceil n/2 \rceil + 1$. *Hint:* Use Lemma 2.23.

(2.18) Show that $\mathrm{cr}(K_{2,3,n}) \sim n^2$, that is, $\mathrm{cr}(K_{2,3,n})/n^2 \to 1$ as $n \to \infty$. *Note/Hint:* The exact crossing number of $K_{2,3,n}$ is known to be $Z(5,n) + n$, but we have an easier proof for the asymptotic behavior using Lemma 2.23.

(2.19) Show that $\mathrm{cr}(K_{2,2,2,}) = 6$.

(2.20) Show that $\operatorname{cr}(K_n - C_n) = \Theta(n^4)$. *Hint:* There are no complete subgraphs. So what else is there?

Chapter 3

The Crossing Number and Other Parameters

3.1 Bisection Width and Graph Layout

The *bisection width*, $\text{bw}(G)$, of a graph $G = (V, E)$ is the smallest $|E(V_1, V_2)|$ for any partition (V_1, V_2) of V with $|V_1|, |V_2| \geq \frac{1}{3}|V|$, where $E(V_1, V_2)$ is the set of all edges between V_1 and V_2. Bisection width is closely related to the crossing number. This connection was discovered by Leighton [225] in his work on VLSI layout, while looking for tools to bound the area necessary to draw a network. We present a slightly stronger version of his original result. For this we need the parameter $\text{ssqd}(G) := \sum_{v \in V(G)} \deg_G^2(v)$, the *sum of squared degrees*.

Theorem 3.1 (Pach, Shahrokhi, and Szegedy [258]). *We have*

$$16 \operatorname{cr}(G) + \text{ssqd}(G) \geq \left(\frac{\text{bw}(G)}{1.58} \right)^2.$$

While the term $\text{ssqd}(G)$ is annoying, it is necessary: an $n \times n$-grid has bisection width $\Omega(n)$, but is planar, see Exercise (3.2).

Proof. Let D be a drawing of G realizing $\operatorname{cr}(G)$, and let G' be the graph obtained by replacing each crossing in D with a new vertex (of degree 4). Then G' has $\operatorname{cr}(G) + n$ vertices. Assign a weight of 0 to each of the new vertices, and a weight of $1/n$ to the original vertices from G. By the planar edge separator theorem, Theorem A.12, there is a set of edges $T \subseteq E(G')$ of size at most $1.58(\text{ssqd}(G'))^{1/2} = 1.58(16 \operatorname{cr}(G) + \text{ssqd}(G))^{1/2}$ so that the vertices of G' can be partitioned into two sets V_1 and V_2 of weight at most $2/3$ so that any edges between V_1 and V_2 belong to T. Since crossings have weight 0, $V_1 \cap V(G)$ and $V_2 \cap V(G)$ each contain at least $1/3$ of the vertices of G, and, since they are separated by T, form a bisection, showing that $\text{bw}(G) \leq 1.58(16 \operatorname{cr}(G) + \text{ssqd}(G))^{1/2}$, from which the result follows. $\square$

Theorem 3.1 is a very useful tool, opening up crossing number problems to recursive attacks; the next theorem is a fine illustration of this idea. In a *convex* drawing, all vertices are on the outer face. The following result traces its history back to Bhatt and Leighton [61]; it was improved by Even, Guha, and Schieber [129]; our presentation combines elements from Kolman and Matoušek [213] and Shahrokhi, Sýkora, Székely, and Vrt'o [311], which give the currently best bound.

Theorem 3.2 (Bhatt and Leighton [61]). *Every graph has a convex, rectilinear drawing with at most $c \log(n)(\mathrm{cr}(G) + \mathrm{ssqd}(G))$ crossings.*

We derive the theorem from a result on planar graphs.

Lemma 3.3. *Every planar graph has a convex, rectilinear drawing with at most $c \log(n) \, \mathrm{ssqd}(G)$ crossings, and such a drawing can be found in polynomial time.*

Applying Lemma 3.3 to the planarization of a graph G yields Theorem 3.2, since crossings contribute as $16 \, \mathrm{cr}(G)$ to ssqd of the planarization.

Proof. We describe a recursive procedure which places vertices of a planar graph G on a convex arc, and draws edges as straight-lines, so that there are at most $c \log(n)(\mathrm{cr}(G) + \mathrm{ssqd}(G))$ crossings. For such a drawing D, we measure $\ell(D)$, the largest number of edges whose endpoints are separated on the arc by some vertex.

Assign weight $w(v) := \deg(v)^2 / \mathrm{ssqd}(G)$ to every vertex $v \in V(G)$. If $w(v) > 2/3$ for some vertex v, we let $V_1 := \{v\}$ and $V_2 := V - V_1$; in this case, $w(V_2) < 1/3$. Otherwise, all vertices have weight at most $2/3$, and we use the planar edge separator theorem, Theorem A.12, to partition $V = V_1 \cup V_2$, so that $b := |E(V_1, V_2)| \le 1.58 \, \mathrm{ssqd}(G)^{1/2}$. Both $w(V_1)$ and $w(V_2)$ are at least $1/3$, so we must have $w(V_1), w(V_2) < 2/3$.

Let $G_i := G[V_i]$, for $i = 1, 2$. In all cases, we have $\mathrm{ssqd}(G_i) < 2/3 \, \mathrm{ssqd}(G)$ (in the first case, G_1 consists of an isolated vertex, so $\mathrm{ssqd}(G_1) = 0$). Recursively find rectilinear drawings D_i of G_i on a convex arc with at most $c \log(n_i) \, \mathrm{ssqd}(G_i)$ crossings, where $n_i := |V_i|$, for $i = 1, 2$. When we combine the two drawings, and add the b edges in the separator, we obtain a drawing D of G with $\ell(D) \le b + \max(\ell(D_1), \ell(D_2))$. By induction, $\ell(D) \le c' \, \mathrm{ssqd}(G)^{1/2}$, for some constant c'.

What about the number of crossings? Since each edge in the separator can cross at most $\ell(D_i)$ edges in D_i, we have

$$
\begin{aligned}
\mathrm{cr}(D) \ &\le \ \binom{b}{2} + \mathrm{cr}(D_1) + \mathrm{cr}(D_2) + b(\ell(D_1) + \ell(D_2)) \\
&\le \ \mathrm{cr}(D_1) + \mathrm{cr}(D_2) + 2b(\ell(D) \\
&\le \ \mathrm{cr}(D_1) + \mathrm{cr}(D_2) + c'' \, \mathrm{ssqd}(G)
\end{aligned}
$$

since each edge in the separator can cross at most $\ell(D_i)$ edges in D_i. Let us bound the terms in this expression. Inductively, we can then show that $\mathrm{cr}(D) \le c \, \mathrm{ssqd}(G)$ for sufficiently large c. Since the separator in Theorem A.12 can be found in polynomial time, the whole construction can be performed in polynomial time. □

The *cutwidth*, cw(G), of a graph G is the smallest k so that there is an ordering $v_1, \ldots, v_n$ of the vertices of G with $|E(\{v_1, \ldots, v_i\}, \{v_{i+1}, \ldots, v_n\})| \leq k$ for all $1 \leq i < n$. (If we imagine the vertices placed along a horizontal line for a fixed ordering, this is the largest number of crossings with any vertical line.) By definition, cw(G) $\geq$ bw(G), and the two parameters can be arbitrarily far apart: two copies of K_n connected by an edge have bisection width 1, but cutwidth $\Omega(n^2)$. This begs the question whether the bisection width lower bound can be extended to cutwidth, and the answer, found by Djidjev and Vrt'o [116], shows that this is possible, with a slightly worse constant.

Theorem 3.4 (Djidjev and Vrt'o [116]). *We have*

$$16\,\mathrm{cr}(G) + \mathrm{ssqd}(G) \geq \left(\frac{\mathrm{cw}(G)}{4.74}\right)^2.$$

The proof relies on the following lemma, which is of interest in its own right. It uses ideas similar to the ones in Lemma 3.3.

Lemma 3.5. *Any planar graph has cutwidth at most* $4.74\,\mathrm{ssqd}(G)^{1/2}$.

Proof. Let $G = (V, E)$ be a plane graph; to each vertex $v \in V$, assign weight $w(v) := \deg(v)^2 / \mathrm{ssqd}(G)$. If there is a vertex v with weight $w(v) > 2/3$, the rest of the graph $G - v$ has weight at most $1/3$, so $\mathrm{ssqd}(G - v) \leq 1/3\,\mathrm{ssqd}(G)$. We then have

$$\mathrm{cw}(G) \leq \mathrm{cw}(G - v) + \deg(v) \leq \mathrm{cw}(G - v) + \mathrm{ssqd}(G)^{1/2}.$$

If all vertices have weight at most $2/3$, then we use the planar edge separator theorem, Theorem A.12, to partition $V = V_1 \cup V_2$, so that $|E(V_1, V_2)| \leq 1.58\,\mathrm{ssqd}(G)^{1/2}$, and each of $w(V_1)$ and $w(V_2)$ are at least $1/3$, so both $\mathrm{ssqd}(G_i) \leq 2/3\,\mathrm{ssqd}(G)$ for $i = 1, 2$. By placing G_1 to the left of G_2, we can achieve

$$\mathrm{cw}(G) \leq \max\{\mathrm{cw}(G_1), \mathrm{cw}(G_2)\} + 1.58\,\mathrm{ssqd}(G)^{1/2}.$$

So in both cases, there is a G' so that

$$\mathrm{cw}(G) \leq \mathrm{cw}(G') + 1.58\,\mathrm{ssqd}(G)^{1/2},$$

G' has weight at most $2/3$, and $\mathrm{ssqd}(G') \leq 2/3\,\mathrm{ssqd}(G)$. Hence,

$$\mathrm{cw}(G) \leq 4.74 \sum_{i=0}^{\infty} \left(\frac{2}{3}\right)^i \mathrm{ssqd}(G)^{1/2}.$$

$\square$

Proof of Theorem 3.4. Fix a cr-minimal drawing of G, and let G' be the planarization of G, so $\mathrm{ssqd}(G') = 16\,\mathrm{cr}(G) + \mathrm{ssqd}(G)$. By Lemma 3.5, $\mathrm{cw}(G') \leq 4.74\,\mathrm{ssqd}(G')^{1/2}$, which implies the theorem. $\square$

We conclude this section with a nice application of bisection width by determining the crossing number of a random graph, in which every edge is present with probability p; this is known as the *Erdős-Rényi model* of a random graph, introduced in [126]. We write $G \sim G(n, p)$ if G is a subgraph of K_n in which each edge gets chosen independently with probability p. We say a graph $G \sim G(n, p)$ has a particular property *almost surely* if the probability of G having this property goes to 1 as n goes to infinity. The original paper by Erdős and Rényi already contained a relevant result. If $G \sim G(n, c/n)$, then G is almost surely planar if $c < 1/2$ and non-planar almost surely if $c > 1/2$.

Let us say two disjoint sets of vertices A and B are *ε-dense*, if there are at least $\varepsilon|A||B|$ edges between them. In a random graph, we would expect two sets to be $p|A||B|$-dense; we make this precise below.

Lemma 3.6. *Suppose $G \sim G(n, p)$ with $p = c/n$, and let $0 < \alpha < 1/2$ be so that $c \geq 8\ln(4)/\alpha^2$. Almost surely, any two disjoint sets of vertices of size at least αn in G are $p/2$-dense.*

Proof. Let A and B be the two disjoint sets. The number of edges between A and B has a binomial distribution, so the expected number of edges is $p|A||B|$. The probability that a binomially distributed variable is less than half its expected value μ is at most $e^{-\mu/8}$ using Chernoff bounds; in our case, this probability is at most $e^{-p|A||B|/8} \leq e^{-c\alpha^2/n}$. Since there are at most 4^n pairs of subsets of n vertices, the probability that any two disjoint sets of size at least αn are not $p/2$ dense is at most $4^n e^{-p|A||B|/8} \leq e^{-c\alpha^2/8n} = e^{-n(c\alpha^2/8-\ln(4))}$ which goes to 0 as n goes to infinity as long as $c \geq 8\ln(4)/\alpha^2$. So almost surely, all pairs of disjoint sets of size at least αn are $p/2$-dense. $\square$

The expected number of edges in a random graph $G \sim G(n, p)$ is $p\binom{n}{2}$. For $p = c/n$, the case we are interested in, this is a linear number of edges, so the crossing lemma does not yield a good lower bound. The bisection method, however, works.

Theorem 3.7 (Pach and Tóth [264]). *If $G \sim G(n, p)$ with $p = c/n$ and $c \geq 100$, then almost surely G has crossing number at least $\Omega(p^2 n^4)$.*

For stronger results on crossing numbers of random graphs, see [319].

Proof. Lemma 3.6 implies that a random graph almost surely has bisection width $p/2(n/3)(2n/3) = pn^2/9$ as long as $c \geq 100 \geq 72\ln(4)$. By Theorem 3.1, the crossing number of the random graph G is at least

$$\mathrm{cr}(G) \geq \frac{1}{16}\left(\left(\frac{\mathrm{bw}(G)}{1.58}\right)^2 - \mathrm{ssqd}(G)\right) \geq \frac{p^2 n^4}{3236} - \frac{\mathrm{ssqd}(G)}{16}.$$

Now $\mathrm{ssqd}(G) \leq \Delta(G)n$, where $\Delta(G)$, the maximum degree, is binomially distributed with expected value $p(n-1) \leq c$. The probability that $\Delta(G)$ is at least $n^{1/2}$ is at most $c/n^{1/2}$ by Markov's bound, a probability which goes to 0. So, almost surely, $\mathrm{ssqd}(G)/16 \leq c/16\, n^{3/2}$ which is $o(p^2 n^4)$. $\square$

3.2 Embeddings and Congestion

An *embedding*, φ, of a graph H in a graph G maps vertices of H to distinct vertices of G, and edges of H to paths of G so that $\varphi(uv)$ connects $\varphi(u)$ to $\varphi(v)$ for all edges $uv \in E(H)$. There are various parameters associated with embeddings, the most notable one being its *(edge) congestion*. For an embedding φ, the *congestion* of an edge e, $\phi_\varphi(e)$ is the number of paths $\varphi(f)$ containing e, for $f \in E(H)$. The *congestion* of φ is the largest congestion $\phi_\varphi(e)$ of any edge $e \in E(G)$. Finally, the *congestion of embedding H in G*, written $\phi(H \hookrightarrow G)$ is the smallest congestion of any embedding of H in G.

Lemma 3.8 (Leighton [225]). *For any two graphs G and H we have*

$$\mathrm{cr}(H) \leq \phi(H \hookrightarrow G)^2(\mathrm{ssqd}(G) + \mathrm{cr}(G)).$$

Proof. Let φ be the embedding of H in G with congestion $\phi(H \hookrightarrow G)$, and let D be a drawing of G with $\mathrm{cr}(G)$ crossings. We use D as a blueprint for drawing G: place $v \in V(H)$ close to $\varphi(v)$. Draw $uv \in E(G)$ as a curve by following the path $\varphi(uv)$, from $\varphi(u)$ to $\varphi(v)$. We can draw these curves so they do not overlap, and only cross close to crossings in D and vertices of G. This gives us a drawing of H; let us count the crossings: close to a crossing in D we have at most $\phi(H \hookrightarrow G)^2$ crossings, close to a vertex v of D we have at most $(\phi(H \hookrightarrow G) \deg_G(v))^2$ crossings, which sums up to $\phi(H \hookrightarrow G)^2 \mathrm{ssqd}(G)$. Combining the two counts yields the result. □

Lemma 3.8 allows us to derive a crossing number lower bound for the hypercube.

Corollary 3.9 (Sýkora and Vrt'o [327]). *For $n \geq 0$, we have*

$$\mathrm{cr}(Q_n) \geq \frac{1}{20}4^n - (n^2 + 1)2^n.$$

Proof. For the lower bound, we rewrite Lemma 3.8 as

$$\mathrm{cr}(G) \geq \mathrm{cr}(H)/\phi(H \hookrightarrow G)^2 - \mathrm{ssqd}(G).$$

Let $G = Q_n$, and $H = K_{2^n}$. We claim $\phi(H \hookrightarrow G) \leq 2^{n-1}$. Together with $\mathrm{cr}(K_{2^n}) \geq 0.8Z(2^n)$, from Lemma 1.31, we obtain

$$
\begin{aligned}
\mathrm{cr}(Q_n) &\geq \frac{2^{n-1}(2^{n-1} - 1)(2^{n-1} - 1)(2^{n-1} - 2)}{52^{2n-2}} - 2^n n^2 \\
&\geq \frac{1}{20}4^n - 2^n(n^2 + 1).
\end{aligned}
$$

We are left with the proof that K_{2^n} can be embedded into Q_n with congestion at most 2^{n-1}. Pick an arbitrary bijection between the vertices of the graph.

We need to specify a path between any two vertices u and v of Q_n. Think of the vertices of Q_n as binary strings of length n. Choose the path from u to v that, as i goes from 1 to n consists in turning the i-th bit of u into the i-th bit of v (if the bits differ, this corresponds to traversing an edge of the path). Consider a particular edge ab of dimension i. For a path from u to v to traverse ab, the first $i - 1$ bits of a and v have to be the same, as well as the last $n - i$ bits of b and u. Hence, there are at most $2^{i-1}2^{n-i} = 2^{n-1}$ paths that traverse edge ab, which is what we had to show. $\square$

If we are only interested in asymptotic lower bounds, we can work with $\lim_{n\to\infty} \mathrm{cr}(K_n)/Z(n) \geq 0.8594$, from Corollary 1.30, to get that $\lim_{n\to\infty} \mathrm{cr}(Q_n)/4^n > 1/19$. We note that the (asymptotic) truth of Hill's or Zarankiewicz's conjecture would yield a lower bound of $1/16\,n^4 - o(n^4)$, which does not match the upper bound of $5/32$ (see Theorem 2.19).

Using a result by Leighton and Rao [226] on uniform multicommidity flows, Kolman and Matoušek gave a bound on the congestion of embedding the complete graph into a host graph. To state the bound, we need the notion of (edge) expansion: the *(edge) expansion*, $\beta(G)$, of a graph G is the minimum $|E(V_1, V_2)|/\min(|V_1|, |V_2|)$ over all partitions $V_1 \cup V_2 = V(G)$.

Theorem 3.10 (Kolman and Matoušek [213]). *If G has edge expansion β, then $\phi(K_n \hookrightarrow G) = O(n\log(n)/\beta)$.*

Combining Lemma 3.8 with Theorem 3.10 yields the following lower bound on graphs with edge expansion β.

Corollary 3.11. *If G has edge expansion β, then $\mathrm{cr}(G) \geq (n\beta/\log n)^2 - \mathrm{ssqd}(G)$.*

For example, it is known that there are 3-regular graphs with positive edge expansion (a random 3-regular graph will do, but there are also explicit constructions [193]). Such a graph has crossing number $\Omega(n^2/\log^2 n)$, a lower bound which does not follow from the crossing lemma, since the graph is too sparse.

3.3 Measures of Planarity

Crossing numbers are not the only parameters which measure how close a graph is to planarity; in this section, we briefly review skewness, edge crossing number, and thickness. We do not study coarseness and splitting number, and leave genus for Section 3.5. Book thickness will be discussed in Section 8.1.1.

3.3.1 Skewness

The *skewness* of a graph, $\text{sk}(G)$, is the smallest number of edges that need to be removed from G to make it planar.

Lemma 3.12. *We have* $\text{sk}(G) \leq \text{cr}(G)$ *and there are graphs* G_k *with* $\text{sk}(G) = 1$ *and* $\text{cr}(G) = k$.

The proof is Exercise (3.5). The skewness of a graph can function as a (often rough) lower bound for its crossing number. The following lemma is implicitly used by Guy [169], the first explicit statement (for arbitrary surfaces) may be by Kainen [201]. It has been rediscovered several times.

Lemma 3.13 (Guy [169]; Kainen [201]). *If G is a graph of girth g on n vertices and m edges, then*

$$\text{sk}(G) \geq m - \frac{g}{g-2}(n-2).$$

Proof. As long as $m > g/(g-2)(n-2)$, G must have at least one crossing by Corollary A.5, so the result follows by induction on $m - g/(g-2)(n-2)$. $\square$

The results in the following corollary are attributed to Kotzig by Harary [171], but they do not appear in the often referenced [215].

Corollary 3.14 (Guy [169]). *The following equalities are true.*

(*i*) $\text{sk}(K_n) = \binom{n}{2} - 3(n-2)$, *for* $n \geq 5$.

(*ii*) $\text{sk}(K_{m,n}) = mn - 2(m+n-2)$, *for* $m, n \geq 2$.

(*iii*) $\text{sk}(Q_n) = 2^{n-1}n - 2(2^n - 2)$, *for* $n \geq 3$.

Proof. All the lower bounds follow from Lemma 3.13, so we only need to show the upper bounds. (*i*) A C_n with both its interior and exterior triangulated has $3(n-2)$ edges. (*ii*) Take a plane drawings of $G_1 = K_{2,n}$, and $G_2 = K_{m-2,2}$; then G_1 has two right-partition vertices on its outer face, and G_2 has two left-partition vertices on it outer face. We can identify the two pairs to get a planar graph with $2(m+n-2)$ edges, which occurs as a subgraph of $K_{m,n}$. (*iii*) We have to show that Q_n contains a planar subgraph with $2(2^n - 2)$ edges. This is true for $n = 3$, since Q_3 is planar itself. If it is true for n, then Q_n has a planar subgraph of $2(2^n - 2)$ edges. Take two copies of that subgraph and draw them together in the plane so that one of them is mirrored. The outer face of each graph has 4 vertices, so we can add 4 edges while keeping the drawing planar. Hence Q_{n+1} has a subgraph of $4(2^n - 2) + 4 = 2(2^{n+1} - 2)$ edges. $\square$

Skewness is more closely related to the Euler genus of a graph than its crossing number. One bound is obvious:

$$2\,\text{eg}(G) \leq \text{sk}(G),$$

since we can add the sk(G) edges whose removal makes G planar each using its own handle. The other direction is non-trivial, and is covered by the following result, which we will not prove.

Theorem 3.15 (Djidjev and Venkatesan [115])**.** *If G is a graph on m edges, with Euler genus eg $=$ eg(G), and maximum degree Δ, then*

$$\mathrm{sk}(G) = O((\Delta \, \mathrm{eg} \cdot m)^{1/2}).$$

3.3.2   Edge Crossing Number

A parameter whose definition sounds very similar to that of skewness is the *edge crossing number*, ecr(G), the smallest number of edges involved in crossings in any drawing of G. By definition, sk(G) $\leq$ ecr(G), and the numbers can be arbitrarily far apart, see Exercise (3.6). As the following theorem shows, the edge crossing number is more closely bound to the crossing number than skewness.

Theorem 3.16. *We have $\frac{\mathrm{ecr}(G)}{2} \leq \mathrm{cr}(G) \leq \binom{\mathrm{ecr}(G)}{2}$.*

Proof. In a cr-minimal drawing of G, there are at least ecr(G) edges involved in crossings; since a crossing belongs to exactly two edges, this implies that there are at least ecr(G)/2 crossings. To see the second inequality, take an ecr-minimal drawing of G. Temporarily remove the crossing-free edges from the drawing, and apply Lemma 1.3 to obtain a good drawing of the crossing edges. Note that edges are redrawn arbitrarily close to the existing (crossing) edges. So we can ensure that the redrawing does not affect the crossing-free edges, which we add back. At this point, we have a good drawing of G in which at most ecr(G) edges are involved in crossings, so the crossing number is at most $\binom{\mathrm{ecr}(G)}{2}$. $\square$

Both upper and lower bounds are asymptotically tight, e.g., take nK_5 and K_n (for the latter, recall Theorem 2.5, it implies that ecr(K_n) $= \binom{n}{2}-(2n-2)$). Mengersen [243] determined ecr($K_{x_1,\ldots,x_n}$) for all complete n-partite graphs.

3.3.3   Thickness

The *thickness*, $\theta(G)$, of a graph is the smallest number of planar graph on $V(G)$ so that G is the union of the planar graphs. Obviously, we can assume that the planar graphs are disjoint.

Lemma 3.17. *If G is a graph of girth g on n vertices and m edges, then $\theta(G) \geq m/\lfloor (n-2)g/(g-2) \rfloor$.*

Proof. By Corollary A.5 a planar graph has at most $g/(g-2)\,(n-2)$ edges. $\square$

Theorem 3.18. *The following equalities hold.*

(i) $\theta(K_n) = \lfloor (n+7)/6 \rfloor$ *for* $n \neq 9, 10$, *and* $\theta(K_9) = \theta(K_{10}) = 3$. (Beineke and Harary [53]; Alekseev and Gončakov [26])

(ii) $\theta(K_{m,n}) = \lceil \frac{mn}{2(m+n-2)} \rceil$, *unless both* m *and* n *are odd, and there an integer* k *so that* $n = \lfloor 2k(m-2)/(m-2k) \rfloor$. (Beineke, Harary, and Moon [54])

(iii) $\theta(Q_n) = \lceil (n+1)/4 \rceil$. (Kleinert [206])

Proof. The lower bounds in all cases follow from Lemma 3.17 and some arithmetic, except for $\theta(K_9) = \theta(K_{10}) \geq 3$ which requires a separate proof (sketched in Battle, Harary, and Kodama [50], also Tutte [339]). For the upper bound in (i) we refer to Beineke and Harary [53] and Alekseev and Gončakov [26], for (ii) to Beineke, Harary, and Moon [54], and for (iii) to Kleinert [206]. □

The values of $\theta(K_{m,n})$ are mostly, but not completely known, but require a more complicated construction [54]. Graphs with thickness at most two are known as *biplanar*, and graphs with thickness at most k as k-*planar*; this terminology conflicts with the use of k-planar for the local crossing number (see Chapter 7), so we will mostly avoid it, except in Chapter 10 on the k-planar crossing number.

There also is an upper bound on the thickness in terms of the number of edges m; the example of K_n shows that this dependency must be $O(m^{1/2})$, and indeed this can be achieved.

Theorem 3.19 (Dean, Hutchinson, and Scheinerman [107]). *If G is a graph on m edges, then*

$$\theta(G) \leq \left\lfloor \left(\frac{m}{3}\right)^{1/2} + \frac{1}{6} \right\rfloor.$$

We leave the proof to Exercise (3.13).

3.4 Chromatic Number and Albertson's Conjecture

At a first glance, chromatic and crossing number do not seem to have a strong relation; if G is a K_5 in which every edge has been replaced by n parallel paths of length 2, then $\chi(G) = 2$, since the graph is bipartite, but $\mathrm{cr}(G) \geq n^2$, so the crossing number of a graph is not bounded in its chromatic number. The reverse direction turns out to be more interesting.

Theorem 3.20 (Schaefer [24]). *If $\chi(G) \geq 9$, then*

$$\chi(G) \leq 1 + 5\,\mathrm{cr}^{1/4}(G),$$

and that bound is asymptotically tight for $G = K_n$.

We follow the proof given by Alberston, Cranston, and Fox [25].

Proof. The second claim is immediate, since $\chi(K_n) = n$, and $\mathrm{cr}(K_n) = \Omega(n^4)$. For the upper bound, let G' be a vertex-minimal subgraph of G with $\chi(G') = \chi(G)$. Then $|E(G')| \geq (\chi(G) - 1)n/2$: if any vertex of G' has degree less than $\chi(G') - 1$, we can remove it, $(\chi(G) - 1)$-color the remaining graph inductively, and add back the removed vertex, assigning it one of the $\chi(G) - 1$ colors not used by its neighbors. If $\chi(G) \geq 9$, then $|E(G')| \geq 4n$, and we can apply the crossing lemma, Theorem 1.8, to obtain

$$
\begin{aligned}
\mathrm{cr}(G) &\geq \mathrm{cr}(G') \\
&\geq \frac{1}{64} \frac{|E(G')|^3}{n^2} \\
&\geq \frac{1}{512}(\chi(G) - 1)^3 n \\
&\geq \frac{1}{512}(\chi(G) - 1)^4,
\end{aligned}
$$

which implies that $\chi(G) \leq 1 + 5\,\mathrm{cr}^{1/4}(G)$. $\qquad\square$

Using an improved crossing lemma, Theorem 7.5, the bound in Theorem 3.20 can be improved to $\chi(G) \leq 1 + 4\,\mathrm{cr}^{1/4}(G)$ for $\chi(G) \geq 13$. The bound on $\chi(G)$ can be lowered by increasing the constant term in the upper bound.

The proof of Theorem 3.20 showed that $\mathrm{cr}(G) \geq 1/512\,(\chi(G) - 1)^3 n \geq 1/512\,\chi(G)(\chi(G) - 1)^3$, assuming that $\chi(G) \geq 9$. If we compare the last term to $Z(k)$ for $k = \chi(G)$, we obtain $\mathrm{cr}(G) \geq 1/8\,Z(k) \geq 1/8\,\mathrm{cr}(K_k)$. Albertson conjectured that the $1/8$ coefficient can be dropped.

Conjecture 3.21 (Albertson [24]). *If $\chi(G) \geq k$, then $\mathrm{cr}(G) \geq \mathrm{cr}(K_k)$.*

The first interesting case of the conjecture occurs for $k = 5$, which is the (highly non-trivial) four-color theorem. We will show that the conjecture is also true for $k = 6$, a result due to Oporowski and Zhao [254].

Theorem 3.22 (Oporowski and Zhao [254]). *If $\mathrm{cr}(G) \leq 2$, then G is 5-colorable.*

Proof. Let G be a vertex-minimal counterexample to the claim; in particular, $n = |V(G)| \geq 6$, and the minimum degree of G is 5 (any vertex of degree less than 5 can be removed to get a 5-coloring of the smaller graph, at which point we can add back the removed vertex, giving it one of the 5 colors different from the colors of its at most four neighbors). Lemma 1.7 implies that $2 \geq \mathrm{cr}(G) \geq m - 3(n - 2)$, so $m \leq 3n - 4$. Hence, G must contain a vertex v of degree at most 5, and, by our observation on the minimum degree, exactly 5. Let the neighbors of v be $v_1, \ldots, v_5$. Suppose, for a contradiction, that there is $1 \leq i < j \leq 5$ so that there is no path in $G - v$ between v_i and v_j. Fix a 5-coloring of $G - v$. In that coloring, the neighbors of v must take 5 different colors. Since there is no path between v_i and v_j in $G - v$ we can take the component containing v_i and in that component swap the colors belonging to v_i and v_j. This leads to no conflict in the component, and does not change the color of v_j, since it does not belong to that component. We now have a 5-coloring of $G - v$ in which v_i and v_j have the same color, so we can extend this to a coloring of G, assigning v the color not used by its neighbors (the color v_i used to have).

We conclude that there is a path in $G - v$ between any two neighbors of v. In the drawing of G replace each such path with an edge; this gives us a drawing of a K_6 on $\{v, v_1, \ldots, v_5\}$. If two edges in this drawing cross, this crossing must be one of the original (at most) two crossings, or it must occur between two adjacent edges (since the corresponding paths must have intersected in a vertex of the same color, meaning that the two paths shared a common endpoint of that color). Hence, we have a drawing of K_6 witnessing $\mathrm{cr}_-(K_6) \leq 2$, which we know to be impossible, see Exercise (1.3). □

Alberston, Cranston, and Fox [25] were able to establish the truth of Albertston's conjecture for $k \leq 12$; their result was strengthened to cover cases up to $k \leq 16$ by Barát and Tóth [47], by pushing the approach we used to show a $1/8 \, \mathrm{cr}(K_k)$ lower bound.

3.5 Beyond the Plane

3.5.1 Genus

Genus and crossing number seem to be only weakly related at a first glance: take a K_5 and replace each edge with k paths of length two. The resulting graph has $10k + 5$ vertices and $20k$ edges; it is *toroidal*, meaning it can be embedded in the torus, so it has genus 1, but its crossing number in the plane is k^2, so arbitrarily large. The picture changes if we also take into account Δ, the maximum degree of the graph.

Theorem 3.23 (Djidjev and Vrt'o [117]). *If G is a graph on n vertices with maximum degree Δ and genus $\gamma = \gamma(G)$, then*

$$\mathrm{cr}(G) \leq c\Delta\gamma n,$$

for some constant $c > 0$.

The bound is asymptotically tight, as shown by Djidjev and Vrt'o [117]; it is easy to reexpress in terms of ssqd as well.

Corollary 3.24 (Djidjev and Vrt'o [117]). *If G is a graph of genus $\gamma = \gamma(G)$, then*

$$\mathrm{cr}(G) \leq c\gamma \, \mathrm{ssqd}(G),$$

for some constant $c > 0$.

Proof. Take G and replace each vertex v with a grid H_v of size $\deg(v) \times \deg(v)$, and attach edges to the boundary of that grid in the same order as in the embedding of G on a surface with genus γ; call the resulting graph G'. By Theorem 3.23, there is a drawing D' of G' with $\operatorname{cr}(D') \leq 4c\gamma\operatorname{ssqd}(G)$. From D' we obtain a drawing of G by selecting in each H_v a vertex v' and $\deg(v)$ vertex-disjoint paths in H_v from v' to the edges attached to H_v. We can then draw v as v' and the edges incident to v by extending the edges attached to H_v along the paths we chose. Then $\operatorname{cr}(G) \leq \operatorname{cr}(D') \leq 4c\gamma\operatorname{ssqd}(G)$. $\qquad\square$

Instead of Theorem 3.23 we prove a weaker bound—with c^γ replacing $c\gamma$—due to Pach and Tóth.

Theorem 3.25 (Pach and Tóth [267]). *If G is a graph of genus $\gamma = \gamma(G)$, then*

$$\operatorname{cr}(G) \leq c^\gamma \operatorname{ssqd}(G),$$

where $c > 0$ is a constant.

Lemma 3.26. *If G can be embedded in S_γ, where $\gamma \geq 1$, then G can be drawn in $S_{\gamma-1}$ with at most $c\operatorname{ssqd}(G)$ crossings, where $c > 0$ is a constant.*

Proof. If the embedding of G contains no non-contractible cycle, then G is planar, and the bound immediate, so let C be a shortest non-contractible cycle in the embedding of G in S_γ, and suppose C has length k.

Let G^* be the topological dual of G with edges crossing C removed. Let v_L^* and v_R^* be two vertices of G^* which correspond to faces on the left and right side of C in the embedding of G.

We claim that there are at least k edge-disjoint paths between v_L^* and v_R^* in G^*. If there were not, then by Menger's Theorem [114, Corollary 3.3.5(ii)], there is a set $S^* \subseteq E(G^*)$ of fewer than k edges so that S^* separates v_L^* from v_R^*; that is, every v_L^*-v_R^* path in G^* contains an edge in S^*. Let $S \subseteq E(G)$ be the edges of G crossed by edges in S^*. Then S must contain a non-separating cycle. If it did not, then removing S from the surface would not separate v_L^* from v_R^* in the dual. Since $|S| < k$, this contradicts the choice of C.

By Corollary A.17, G has at most $3(n + 2\gamma - 2)$ edges, so one of the k edge-disjoint paths between v_L^* and v_R^* crosses at most $3(n + 2\gamma - 2)/k$ edges of G. Let α be such a curve on the surface. Cut the surface close to one side of C, and fill in the two holes with disks. This leaves us with a drawing of G on $S_{\gamma-1}$ with at most Δk severed edges. We can route the severed edges along α, creating at most $3(n + 2\gamma - 2)\Delta$ crossings, and then reconnect them to their original ends.

Using the same construction as in Corollary 3.24, this implies a bound of $c\operatorname{ssqd}(G)$ for $\operatorname{cr}_{S_{\gamma-1}}(G)$. $\qquad\square$

Proof of Theorem 3.25. Suppose G is embeddable in a surface of genus γ. By Lemma 3.26, G has a drawing on $S_{\gamma-1}$ with at most $c\operatorname{ssqd}(G)$ crossings. Let G' be the "planarization" of this drawing (replace each crossing by a vertex of degree 4), so $\mathrm{ssqd}(G') = 16c\,\mathrm{ssqd}(G) + \mathrm{ssqd}(G) \leq 17c\,\mathrm{ssqd}(G)$, where we assume that $c \geq 1$. Since G' is embedded, we can apply Lemma 3.26 again. Each step adds a factor of $17c$, so we conclude that $\mathrm{cr}(G) \leq (17c)^\gamma \mathrm{ssqd}(G)$.

$\square$

3.5.2 Crossings on Surfaces

All of the parameters we have defined for the plane can be generalized to other surfaces. For example, we write cr_Σ for the smallest number of crossings in a drawing of G on surface Σ and sk_Σ for the smallest number of edges that need to be removed from G so it becomes embeddable in Σ.

Lemma 3.27 (Kainen [201]; Kainen and White [202]). *For a graph of girth g on $n \geq 3$ vertices and m edges, and a surface Σ with Euler genus $\mathrm{eg} = \mathrm{eg}(\Sigma)$, we have*

$$\mathrm{sk}_\Sigma(G) \geq m - \frac{g}{g-2}(n + \mathrm{eg}-2).$$

Proof. As long as $m > g/(g-2)(n+\mathrm{eg}-2)$, G must have at least one crossing by Corollary A.5, so the result follows by induction on $m - g/(g-2)(n + \mathrm{eg}-2)$. $\square$

Corollary 3.28 (Kainen [201]; Kainen and White [202]). *For a surface Σ with Euler genus $\mathrm{eg} = \mathrm{eg}(\Sigma)$, we have*

(i) $\mathrm{sk}_\Sigma(K_n) \geq \binom{n}{2} - 3(n + \mathrm{eg}-2)$ *if $n \geq 3$.*

(ii) $\mathrm{sk}_\Sigma(K_{m,n}) \geq mn - 2(m + n + \mathrm{eg}-2)$ *if $m + n \geq 3$.*

(iii) $\mathrm{sk}_\Sigma(Q_n) \geq 2^{n-1}n - 2(2^n + \mathrm{eg}-2)$ *if $n \geq 2$.*

The bounds follow immediately from Lemma 3.27; they are not tight, for example $\mathrm{sk}_{S_1}(K_5) = 0$, but the right-hand side in (i) evaluates to -5.

We have a crossing lemma for cr on a surface Σ.

Theorem 3.29 (Crossing Lemma. Shahrokhi, Székely, Sýkora, and Vrt'o [309]). *For a surface Σ with Euler genus $\mathrm{eg} = \mathrm{eg}(\Sigma)$, we have*

$$\mathrm{cr}_\Sigma(G) \geq \begin{cases} \frac{5}{432}\frac{m^3}{n^2} & \text{if } \mathrm{eg} \leq n^2/m, \text{ and} \\ \frac{5}{432}\frac{m^2}{\mathrm{eg}} & \text{if } n^2/m \leq \mathrm{eg} \leq m/36, \end{cases}$$

as long as $m \geq 6n$, where $m = |E(G)|$, $n = |V(G)|$.

Proof. By the Euler-Poincare formula (Corollary A.17) a graph embedded on Σ has at most $3(n+\mathrm{eg}-2)$ edges, so $\mathrm{cr}_\Sigma(G) \geq m-3(n+\mathrm{eg}-2) \geq m-3(n+\mathrm{eg})$. Fix a cr-minimal drawing D of G on Σ. The usual probabilistic proof (see Theorem 1.8) yields

$$\mathrm{cr}(D) \geq \frac{m}{p^2} - \frac{3n}{p^3} - \frac{3\,\mathrm{eg}}{p^4},$$

where p is the probability of keeping an edge. In the case that $eg \leq n^2/m$, we let $p = \lambda n/m$, which gives us

$$\mathrm{cr}(D) \geq \frac{m^3}{n^2}(\lambda^{-2} - 3\lambda^{-3}) - 3\lambda^{-4}\frac{m^4}{n^4}\,eg\,.$$

Using $eg \leq n^2/m$, we conclude that

$$\mathrm{cr}(D) \geq \frac{m^3}{n^2}(\lambda^{-2} - 3\lambda^{-3} - 3\lambda^{-4}).$$

For $\lambda = 6$ we get the first bound. To ensure that p is a probability, we need $m \geq 6n$.

If $n^2/m \leq eg \leq m/36$, we let $p = \lambda(eg\,/m)^{1/2}$. With this, we have

$$\mathrm{cr}(D) \geq \frac{m^2}{eg}(\lambda^{-2} - 3\lambda^{-3}\frac{n}{(m\,eg)^{1/2}} - 3\lambda^{-4}).$$

Using $n^2/m \leq eg$ we get

$$\mathrm{cr}(D) \geq \frac{m^2}{eg}(\lambda^{-2} - 3\lambda^{-3} - 3\lambda^{-4}).$$

Letting $\lambda = 6$ gives us the second bound. For p to be a probability, we need $\lambda(eg\,/m)^{1/2} \leq 1$ which is equivalent to $eg \leq m/\lambda^2 = m/36$. $\qquad\square$

The crossing lemma implies that $\mathrm{cr}_\Sigma(K_n) = \Omega(n^4/\,eg)$ for $eg = eg(\Sigma) \leq 1/36\binom{n}{2}$. There is a nearly matching upper bound.

Theorem 3.30 (Shahrokhi, Székely, Sýkora, and Vrt'o [309]). *If Σ is a surface of Euler genus $eg = eg(\Sigma) \geq 4$, then*

$$\mathrm{cr}_\Sigma(K_n) = O(n^4\frac{\log^2 eg}{eg}).$$

We do not include a proof; it is based on embedding K_n in a Cartesian product of a hypercube and a perfect shuffle graph. The proof is constructive in the sense that it creates a drawing realizing the bound in polynomial time. The same is true of the following result (but requires more work).

Corollary 3.31 (Shahrokhi, Székely, Sýkora, and Vrt'o [309]). *If Σ is a surface of Euler genus $eg = eg(\Sigma) \geq 4$, then*

$$\mathrm{cr}_\Sigma(G) = O(\frac{m^2\log^2 eg}{eg}).$$

Proof. Let $n = |V(G)|$. By Theorem 3.30 there is a drawing of K_n on Σ with $cn^4\frac{\log^2 eg}{eg}$ crossings, for some constant $c > 0$. Consider a random injection of $V(G)$ into $V(K_n)$. Such an injection induces a drawing of G as a subgraph of K_n. The probability that two independent edges in G cross in the induced drawing is at most

$$\frac{cn^4 \frac{\log^2 \text{eg}}{\text{eg}}}{3\binom{n}{4}} \leq c' \frac{\log^2 \text{eg}}{\text{eg}}.$$

Hence, the expected number of crossings is at most $c'\binom{m}{2}\frac{\log^2 \text{eg}}{\text{eg}}$, and there must be drawing with at most that many crossings, which proves the corollary. $\square$

3.5.3 Crossing Sequences

Corollary 3.31 implies that increasing the genus of the underlying surface decreases the crossing number of a graph at least proportionally. For bounded-degree graphs, the reverse is also true. For reasons we will see presently, we work with orientable surfaces only in this section.

Lemma 3.32. *If Σ is a surface of genus $\gamma = \gamma(\Sigma)$, then*

$$\text{cr}_\Sigma(G) \geq c\left(\frac{\text{cr}(G)}{\gamma} - \text{ssqd}(G)\right).$$

Proof. Fix a drawing of G on Σ with $k = \text{cr}_\Sigma(G)$ crossings. Replace each crossing by a degree-4 vertex, obtaining a new graph G' embedded in Σ. By Corollary 3.24 we know that there is a constant $c > 0$ so that $\text{cr}(G') \leq c\gamma\,\text{ssqd}(G') = c\gamma(\text{ssqd}(G)+16k)$, where $n = |V(G)|$. Since $\text{cr}(G) \leq \text{cr}(G')+k$, we conclude that $k \geq c'(\text{cr}(G)/\gamma - \text{ssqd}(G))$. $\square$

Lemma 3.32 may suggest that the crossing number decays linearly as the genus increases; to study this more closely, we consider the so-called *crossing sequence* $(c_0, c_1, c_2, \ldots, c_k)$, where $c_i := \text{cr}_{\Sigma_i}(G)$, where Σ_i has genus i, $0 \leq i \leq k$. The natural question then is, which sequences are crossing sequences.[1] Every crossing sequence is strictly decreasing (we can use a handle to remove at least one crossing), and it seems intuitive that it is *convex* if the decreases do not decrease, that is, $c_i - c_{i+1} \leq c_{i-1} - c_i$ for all $0 \leq i < k$; for example, in the case $k = 2$ convexity means that for a graph on the double torus, removing a handle removes at least as many crossings as removing the second handle.

Theorem 3.33 (Širáň [317]). *Every convex, strictly decreasing sequence is a crossing sequence.*

For the construction, we use a common tool, which often simplifies the argument: *(edge)-weighted graphs*, graphs equipped with a weight function on edges: $w : E(G) \to \mathbb{R}_{\geq 0}$. A crossing between two edges of weight w_1 and w_2 counts as $w_1 w_2$; this determines $\text{cr}(G, w)$. If $w : E(G) \to \mathbb{N}$ we call G an *integer weighted graph*.

---

[1] Here is the reason we do not work with Euler genus in this section: a graph embeddable in a surface of genus g is also embeddable in a surface of genus $g' > g$ (and similarly for non-orientable genus), but this type of monotonicity does not hold for Euler genus. There are projective-planar graphs which are not embeddable on any orientable surface.

Lemma 3.34 (Kainen [201]). *For every integer-weighted graph (G, w) there is an unweighted graph G' so that $\mathrm{cr}_\Sigma(G, w) = \mathrm{cr}_\Sigma(G')$ on any surface Σ. The size of G' is bounded by a polynomial in the maximum weight (represented in unary) and the size of G.*

Proof. We obtain G' from G by adding $w(e) - 1$ paths of length 2 between the endpoints of e for every edge $e \in E(G)$. If we have a drawing of G on Σ realizing $\mathrm{cr}_\Sigma(G, w)$, we can insert these paths close to the edge they duplicate showing that $\mathrm{cr}_\Sigma(G') \leq \mathrm{cr}_\Sigma(G, w)$. For the other direction fix a cr-minimal drawing of G' on Σ. Let e be an edge in G' and P one of the paths we inserted parallel to e. Then e and the two edges in P are involved in the same number of crossings (otherwise we could reroute one along the other reducing the number of crossings), so we can reroute P arbitrarily close to e. Repeating this for all P parallel to e shows that we can replace e and the parallel paths we added by a single edge of weight $w(e)$ without changing the crossing number of the drawing. Repeating this for all edges $e \in E(G)$ shows that $\mathrm{cr}_\Sigma(G, w) \leq \mathrm{cr}_\Sigma(G')$.   □

Proof of Theorem 3.33. Let $(c_0, c_1, c_2, \ldots, c_k)$ be a convex, strictly decreasing sequence. We can assume that $c_k = 0$, if not, we add 0 at the end of the sequence. Let $d_i = c_i - c_{i+1}$, for $i = 0, \ldots, k-1$. By assumption, the d_i are a positive, non-increasing sequence.

By Lemma 3.34, it is sufficient to construct a weighted graph realizing $(c_0, c_1, c_2, \ldots, c_k)$; we build G as follows: take k copies $G_1, \ldots, G_k$ of K_5; in each G_i pick a pair of independent edges, give weight d_i to one of them, weight 1 to the other, and weight $w := c_0 + 1$ to the remaining edges. Let G be the disjoint union of the G_i (if we want G to be connected, we can easily do so by adding edges).

Each G_i has a plane drawing with d_i crossings, so G has a drawing with $c_0 = \sum_{0 \leq i < k} d_i$ crossings in the plane and every other surface. This implies that in a cr-minimal drawing of G in any surface, none of the edges of weight $w > c_0$ are involved in crossings, which, in turn, implies that no two of the G_i intersect in a cr-minimal drawing in any surface; if they did, we can pick in each of the two G_i a cycle all of whose edges, except possibly the one involved in the crossing, have weight w. Since two disjoint cycles have to cross at least twice, an edge of weight w would have to be involved in a crossing, which we already ruled out.

We will show that the crossing number of G on a surface with j handles is c_j. The upper bound is immediate. We can embed $G_1, \ldots, G_j$ without crossings using the j handles; the remaining G_i can be drawn with $d_j + \cdots + d_k = c_j - c_k = c_j$ crossings.

To establish the lower bound, we will show that $H_\ell := \bigcup_{\ell \leq i < k} G_i$ has crossing number at least $\sum_{i=\ell+j}^{k-1} d_i$ on a surface with j handles, using induction on ℓ and j. Letting $\ell = 0$ completes the proof, since $\sum_{i=j}^{k-1} d_i = c_j - c_k = c_j$.

Note that for each G_i there are two options only: either, it is drawn in the plane, incurring d_i crossings, or it is embedded, in which case it contains a nonseparating cycle (by Lemma A.20). If $j = 0$, then there are no nonseparating curves in the surface, so each G_i incurs d_i crossings, and H_ℓ requires at least $d_\ell + \cdots + d_{k-1}$ crossings.

It also follows that if $j > 0$, then at least one of the G_i in H_ℓ must contain a nonseparating cycle, and is therefore embedded. Let α be a nonseparating curve in that G_h. Remove G_h, then no other G_i crosses α, since, as we saw, the G_i do not intersect, so we can cut the surface along α, and add two closed disks to close the boundary; this results in a surface of smaller genus.

Since the smallest number of crossings of $H_\ell - G_h$ on any surface is obtained for $h = \ell$, we can assume that $h = \ell$, so we have a drawing of $H_{\ell+1}$ on a surface of genus at most $j - 1$, which, by induction, requires at least $\sum_{i=\ell+1+j-1}^{k-1} d_i = \sum_{i=\ell+j}^{k-1} d_i$ crossings. Since the drawing of $H_\ell - G_h$ was not changed, this means that the drawing of H_ℓ had at least that many crossings, which is what we had to show. $\qquad\square$

Archdeacon, Bonnington, and Širáň [34] proved later, that convexity is not a necessary condition for being a crossing sequence. To explain their example, we introduce another tool: graphs with patches. A *patch* in a graph is a trail which can occur as a face boundary (every edge occurs at most twice, and every vertex which occurs more than once is a cutvertex). In a drawing of a patched graph on a surface, we require each patch to be a face (bound a disk).

Lemma 3.35. *Given a graph with a collection of disjoint patches, we can construct a simple graph which has the same crossing number as the patched graph on any surface Σ on which the patched graph can be drawn.*

Proof. Let G be the given graph, and $\mathcal{P}$ a collection of disjoint patches. For every patch in $\mathcal{P}$ of length ℓ we add a wheel graph W_ℓ with edges of weight $w := \binom{|E(G)|}{2} + 1$, and identify the ℓ vertices on the trail with the ℓ outer vertices of the wheel graph, in the order that the vertices occur on the trail; the edges of G have weight 1. Call the resulting, weighted graph G'.

If G has a patched drawing on Σ, each patch bounds a face, and we can insert a wheel for each patch, giving us a drawing of G' with the same number of crossings as G, which is at most $w - 1$.

On the other hand, if G' has a drawing in Σ with (weighted) crossing number at most $w - 1$, then no wheel edge is involved in a crossing; we can then reroute the outer edges of the wheel along its spokes, so the wheel is embedded in a plane disk, and we can then route the edges of the patch along the outer edges of the wheel, showing that G has a patched embedding with the same number of crossings as G'.

Finally, using Lemma 3.34, there is a simple graph G'' with the same crossing number as G'. $\qquad\square$

Theorem 3.36 (Archdeacon, Bonnington, and Širáň [34])**.** *For every $a \geq b \geq c \geq 0$ there is a graph G with crossing sequence $(a + b + c, b + c, 0)$.*

Choosing $a = b = c = 1$ yields the non-convex sequence $(3, 2, 0)$. Archdeacon, Bonnington, Širáň conjecture that *every* strictly decreasing sequence is a crossing sequence of some graph. DeVos, Mohar, and Šámal [110] showed that the conjecture is true for all sequences of length 3, that is, $(a, b, 0)$ with $a > b > 0$.

Proof. By Lemma 3.35 it is sufficient to create a patched graph with the given crossing sequence. Consider the patched graph in Figure 3.1. Clearly, a plane

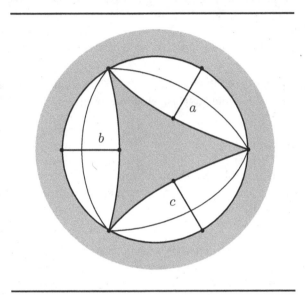

FIGURE 3.1: A patched graph with a non-convex crossing sequence. Patched areas are gray. Illustration is based on [34].

patched drawing requires $a + b + c$ crossings, and the graph has a patched embedding in the double torus. It remains to show that the crossing number of a patched drawing on the torus is $b + c$. The upper bound we can achieve by using the handle to remove the crossing with the edge of weight a. For the lower bound note that at least one of the edges weighted a, b, or c must lie on a non-separating curve. Cutting along that curves leaves us with a drawing in the plane with $a+b$, $b+c$, or $a+c$ crossings. Since those crossings were present in the toroidal drawing, the smallest of these, $b + c$, is a lower bound on the number of crossings in a patched drawing of the graph on the torus.   □

Theorem 3.36 answers another natural question. Suppose we have two disjoint graphs G and H, is it possible that $\mathrm{cr}_{S_k}(G) + \mathrm{cr}_{S_k}(H) < \mathrm{cr}_{S_{2k}}(G \cup H)$, that is, can we save crossings when combining two surface drawings? The answer is yes: let G be the graph from Theorem 3.36 for $a = b = c = 1$. Then $2\,\mathrm{cr}_{S_1}(G) = 2$, while $\mathrm{cr}_{S_2}(2G) = 3$, since we can embed one of the copies of G on S_2, and add a plane drawing with 3 crossings of the other copy.

Conjecture 3.37 (DeVos, Mohar, and Šámal [110]). If G and H are disjoint graphs, then in a cr-minimal drawing of $G \cup H$ on any surface, G and H do not intersect.

The conjecture is true for the sphere, since we can move the drawings of G and H apart if they intersect, reducing the number of crossings. It has only been shown for the projective plane [110] and the Klein bottle [51]. There is also a translation of the problem into an integer vector space problem which may be more amenable to computer analysis [81]. The conjecture is open for the torus.

Theorem 3.38 (DeVos, Mohar, and Šámal [110]). *If G and H are disjoint graphs, then in a cr-minimal drawing of $G \cup H$ in the projective plane, G and H do not intersect.*

The proof will apply Lins' theorem [305, Corollary 74.1b]: the largest number of edge-disjoint, one-sided cycles contained in an Eulerian graph embedded in the projective plane equals the smallest number of crossings between G and any one-sided curve in the projective plane.

Proof. Fix a cr-minimal drawing of G and H on the projective plane. Replace crossings within G and within H by dummy vertices, so each graph by itself is embedded in the projective plane, and double each edge, resulting in two Eulerian multigraphs, call them G and H for simplicity. The drawing of $G \cup H$ is still cr-minimal (if there is a better drawing, there would have been a better initial drawing of G and H). We have to show that there are no crossings. Let c_G and c_H be the largest number of edge-disjoint, one-sided cycles in G and H, respectively. We can assume that $c_G \leq c_H$. By Lins' theorem, the number of crossings between G and H is at least $c_g c_H$. On the other hand, a plane drawing of G requires at most $\binom{c_G}{2}$ crossings: again by Lins' theorem, there is a one-sided curve γ in the projective plane, which has c_G crossings with G. Cutting the surface along γ results in a drawing of G in the plane, with boundary, with c_G severed pairs of edges. We glue a disk into the hole, and reconnect the edge ends using the disk, resulting in at most $\binom{c_G}{2}$ crossings. Since we started with a cr-minimal drawing, we must have $\binom{c_G}{2} \geq c_G c_H$, which, since $c_G \leq c_H$ is only possible if $c_G = 0$, and the drawing of G is plane, and, therefore, does not cross H. $\square$

3.5.4   Minors

Recall that H is a *minor* of G if it can be obtained from G by a sequence of vertex- and edge-deletions, and edge-contractions. With respect to minors, the standard crossing number has one serious defect: it is not minor-monotone, that is, if G is minor of H, then we do not necessarily have $\mathrm{cr}(G) \leq \mathrm{cr}(H)$. For example, Figure 3.2 shows a graph which has K_6 has a minor, but can be drawn with two crossings.

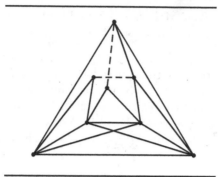

FIGURE 3.2: A K_6-minor with crossing number two; contracting the dashed edges yields the K_6.

Why would it be advantageous if the crossing number were minor-monotone? We do not have to look far to find some answers; $\mathrm{cr}(G) = 0$, that is, G being planar, is downward closed under minors, and the graph minor machinery then implies some general results, for example, that planarity can be characterized by a finite set of excluded minors (we know that already, they are K_5 and $K_{3,3}$), and that it can be tested in polynomial time (by verifying that none of the forbidden minors occurs). For $\mathrm{cr}(G) = 2$ or $\mathrm{cr}(G) \leq 2$, the example of K_6 above shows that these two properties are not closed under minors, so they cannot be characterized by forbidden minors.

We can deal with this, by forcibly closing the crossing number with respect to minors.

Definition 3.39. The *minor crossing number*, $\mathrm{mcr}(G)$, is the smallest crossing number of any H so that G is a minor of H.

Figure 3.2 shows that $\mathrm{mcr}(K_6) \leq 2$; in Exercise (3.10) we will see that equality holds.

Since the set of graphs with $\mathrm{mcr}(G) \leq k$ is, by definition, (downward) closed under minors, the graph minor machinery implies a finite set of forbidden minors, and a polynomial-time algorithm for recognizing these graphs (for each, fixed k). For $k = 1$, this class is known as the *single-crossing* graphs, and there is an explicit list of 41 forbidden minors for it, due to Robertson and Seymour [293].

Lemma 3.40 (Bokal, Fijavž, and Mohar [73])**.** *If G is a graph on $n \geq 3$ vertices and m edges, then*

$$\mathrm{mcr}(G) \geq \frac{g-2}{g}m - n + 2,$$

where g is the girth of G, unless G is a tree, in which case we let $g = 3$

Proof. Choose H so that $\operatorname{cr}(H) = \operatorname{mcr}(G)$ and G is a minor of H. Fix a drawing of H with $k := \operatorname{cr}(H)$ crossings. By adding k crosscaps to the plane, we can remove all crossings, so $\operatorname{cr}_{N_k}(H) = 0$, and, since embeddability on a surface is closed under minors, $\operatorname{cr}_{N_k}(G) = 0$. By Corollary A.17, $m \leq g/(g-2)(n+k-2)$, using $\operatorname{eg}(N_k) = k$. Solving for k gives us $k \geq \frac{g-2}{g}m - n + 2$. $\square$

As a corollary, we get that $\operatorname{mcr}(K_n) \geq 1/3 \binom{n}{2} - n + 2$. The currently best bounds are $1/2 \leq \lim_{n \to \infty} \operatorname{mcr}(K_n)/\binom{n}{2} \leq 1$ [73]. For a systematic study of how traditional methods, such as crossing lemma, bisection width, and the embedding method extend to the minor crossing number, see [72].

We turn from the minor crossing number to the effect of minors on the crossing number.

Theorem 3.41 (Garcia-Moreno and Salazar [145]; Bokal, Fijavž, and Mohar [73]). *If H is a minor of G, then $\operatorname{cr}_\Sigma(H) \leq \lfloor \Delta(H)/2 \rfloor^2 \operatorname{cr}_\Sigma(G)$ on any surface Σ.*

Garcia-Morena and Salazar proved Theorem 3.41 for $\Delta(H) \leq 4$, and Bokal, Fijavž, and Mohar [73] proved the more general version. Theorem 3.41 implies that cr and mcr are polynomially related for bounded-degree graphs, since it implies that $\operatorname{mcr}_\Sigma(G) \leq \operatorname{cr}_\Sigma(G) \leq \lfloor \Delta(G)/2 \rfloor^2 \operatorname{mcr}_\Sigma(G)$.

Lemma 3.42. *Given a tree with positive vertex weights, there is a vertex whose removal splits the tree into pieces each of which has weight at most half the total weight.*

Proof. Pick an arbitrary vertex of the tree, and root the tree at that vertex. If all children have weight at most half the total weight, we are done. Otherwise, the subtree rooted in one of the children has weight more than half the total weight, and we continue recursively in the subtree. Note that since the child has weight more than half, the subtree (after removal of the child) which the parent is contained in has weight at most half. Since our graph is a finite tree, the process must stop. $\square$

Proof of Theorem 3.41. Fix a cr-minimal drawing of G on Σ. Since H is a minor of G, it can be obtained from G by deleting vertices and edges, and contracting edges. We can rearrange these operations, so all deletions come first. Then, after the deletions, each vertex v of H is the result of contracting the edges of a tree $T(v) \subseteq G$, where the trees are vertex-disjoint; an edge $uv \in E(H)$ corresponds to an edge between $T(u)$ and $T(v)$ in the drawing of G. Let us look at a specific vertex v and its tree $T(v)$. To each vertex of $T(v)$ assign as a weight the number of edges of H it is incident to; then the total weight of $T(v)$ is $\deg_H(v)$. By Lemma 3.42, there is a vertex in $T(v)$, call it v as well, so that each of the trees in $T(v) - v$ has weight at most $\Delta(H)/2$. We can then remove the edges and vertices of $T(v)$ except v, and route curves from v to the ends of edges of H incident to $T(v)$ by following $T(v)$ closely. Each crossing that $T(v)$ is involved in, results in at most $\Delta(H)/2$ crossings with the newly drawn edges of $N(v)$. If we perform this redrawing for all $v \in V(H)$, each crossing in the drawing of G belongs to at most two trees, so causes at most $\lfloor \Delta(H)/2 \rfloor^2$ crossings in the drawing of H, establishing the upper bound.   □

We conclude this section with an interesting consequence of Theorem 3.41. The *facewidth*, also known as *representativity*, of an embedding of a graph in a surface is the smallest number of points in which a closed, noncontractible curve intersects the embeddding.

Theorem 3.43 (Garcia-Moreno and Salazar [145]). *If a graph G can be embedded on a torus with facewidth $r \geq 6$, then $\mathrm{cr}_{N_1}(G) \geq \lfloor 2r/3 \rfloor^2/36$.*

Proof. Fix an embedding of G on the torus with facewidth $r \geq 6$. By [100], G contains a $C_k \square C_k$ minor, with $k \geq \lfloor 2/3r \rfloor$. Theorem 3.41 then implies that $\mathrm{cr}_{N_1}(G) \geq \mathrm{cr}_{N_1}(C_k \square C_k)/4$. Since $C_3 \square C_3$ does not embed in the projective plane, N_1, by Euler-Poincaré, Corollary A.17, the standard counting argument implies that $\mathrm{cr}_{N_1}(C_k \square C_k) \geq k^2/9 \geq \lfloor 2/3r \rfloor^2/9$.   □

The projective plane, N_1 can be replaced with the Klein bottle, N_2, by using that $C_4 \square C_4$ does not embed on the Klein bottle [199]; the lower bound then becomes $\lfloor 2r/3 \rfloor^2/64$. For the crossing number of projective-planar graphs, the following result holds.

Theorem 3.44 (Gitler, Hliněný, Leaños, and Salazar [151]). *If a graph G can be embedded on a projective plane with facewidth $r \geq 6$, then $\mathrm{cr}(G) \geq r^2/36$.*

A more general lower bound for arbitrary surfaces was found by Hliněný and Chimani [186].

3.6   Notes

The bisection width and embedding methods and their relevance to crossing number research are studied in depth by Shahrokhi, Sýkora, Székely, and Vrt'o [313]. Liebers [228] survey on planarizing graphs surveys many planarity measures we did not discuss in this chapter. For a survey specifically on graph thickness, see [253]. For a closer look at the crossing number on surfaces, Shahrokhi, Sýkora, Székely, and Vrt'o [309] is a good starting point. The study of the relationship between (Euler) genus and crossing number has received intense attention recently. See [76, Section 2.6] and [89].

3.7 Exercises

(3.1) Show that $\operatorname{cr}(G) + n \geq \operatorname{bw}(G)^2/c$, for some constant c depending on Δ, the maximum degree of G. *Hint:* You can do so directly, as a consequence of the planar (vertex) separator theorem, or as a consequence of Theorem 3.1.

(3.2) Show that the grid graph $P_n \square P_n$ has bisection width $\Omega(n)$.

(3.3) Let $A_r := \overbrace{P_n \square \cdots \square P_n}^{r}$ be the r-dimensional grid (array) with sidelength n. Show that $\operatorname{cr}(A_r) = \Omega(n^{2(r-1)})$ for $r \geq 3$. *Hint:* Use bisection width or embedding method.

(3.4) Let φ be an embedding of H into G. The *vertex-congestion* of φ is the largest number of paths $\varphi(uv)$, $uv \in E(H)$ containing a specific vertex. Let D be a drawing of G. Show that if φ has edge-congestion η and vertex congestion ν, then there is a drawing D' of H for which $\operatorname{cr}(D') \leq \eta^2 \operatorname{cr}(D) + n/2 \, \nu^2$.

(3.5) Prove Lemma 3.12.

(3.6) Construct graphs G with $\operatorname{sk}(G) = 1$ and $\operatorname{ecr}(G) = n$.

(3.7) Determine $\operatorname{sk}(C_m \times C_n)$.

(3.8) Find (and establish) sk for K_n, $K_{m,n}$, and Q_n on a surface of genus g. Use this to determine the genus of these graphs.

(3.9) Show that there is a graph G with $\operatorname{mcr}(G) = 1$ and $\operatorname{cr}(G) > 1$.

(3.10) Show that $\operatorname{mcr}(K_6) = 2$.

(3.11) Show that $\operatorname{mcr}(K_7) = 3$.

(3.12) Show that $\operatorname{mcr}(Q_n) = \Omega(4^n/n^2)$. *Hint:* Use Corollary 3.9.

(3.13) Show that there is a constant c so that $\theta(G) \leq \lfloor (m/3)^{1/2} + c \rfloor$. *Hint:* You can get $c = 1/6$ if you use the exact value of $\theta(K_n)$.

(3.14) Show that K_7 and K_8 are single-crossing graphs, that is, they are minors of a graph with crossing number 1.

Chapter 4

Complexity and Algorithms

4.1 The Hardness of Crossing Number Problems

The crossing number problem is hard, in a precise sense, as computational complexity theory tells us. Determining whether $cr(G) \leq k$ for a graph G and an integer k is **NP**-complete, a result first shown by Garey and Johnson [146] via a reduction from the optimal linear arrangement problem.

Theorem 4.1 (Garey and Johnson [146]). *The crossing number problem is* **NP**-*complete.*

Instead of the original proof, we present a recent construction due to Cabello [78] which turns out to be more powerful and versatile.

Proof. To show that the crossing number problem lies in **NP**, we need to exhibit a polynomial-sized witness for $cr(G) \leq k$ which can be verified in polynomial time. This is easy: we orient each edge e of G and associate with it a list $L(e)$ of at most k distinct edges of G. This list is meant to encode the sequence of crossings along the oriented edge. We can easily check whether these lists witness $cr(G) \leq k$: if $e \in L(f)$, then we also need to have $f \in L(e)$, and the total lengths of the lists, $\sum_{e \in E(G)} |L(e)|$ must be $2k$. If this is the case, we can construct from G and L a graph G' by replacing each crossing by a vertex of degree 4 connected to other crossings or vertices, as determined by the lists. If G' is planar, then the embedding of G' can be turned into a drawing of G with at most k crossings. On the other hand, if $cr(G) \leq k$, we let $L(e)$ be the list of crossings along e; this will be a witness.

To prove hardness we use the **NP**-complete problem *multiway cut* [99]: given a graph G, three vertices $t_1, t_2, t_3 \in V(G)$ and an integer k, is there a set of edges $E' \subseteq E(G)$ of size at most k so that no two of the vertices t_1, t_2, t_3 lie in the same component of $G - E'$?

Start with a $K_5 - e$, and let T be the triangle which separates the two non-adjacent vertices. Subdivide each edge of the triangle, creating vertices t_1, t_2, t_3. We call this is the *frame* of the graph we are constructing, see Figure 4.1. Assign weight n^5 to all edges of the frame, where $n = |V(G)|$; these are the *heavy* edges. Then identify t_1, t_2, t_3 with their namesakes in G, assigning weight 1 to edges of G, the *light* edges. For the resulting graph H we have

$$kn^5 \leq \mathrm{cr}(H, w) \leq kn^5 + n^4,$$

where k is the smallest integer for which G has a (t_1, t_2, t_3)-multiway cut of size k.

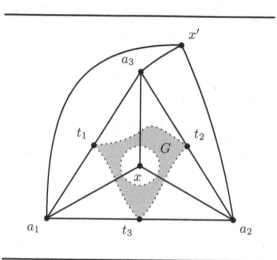

FIGURE 4.1: Framework for **NP**-hardness of crossing number problem. The two non-adjacent vertices are x and x', and triangle T consists of $a_1 a_2 a_3$. Graph G is drawn in gray.

Let $E' \subseteq E$ be a multiway cut of G with $|E'| \leq k$. Then $G - E'$ consists of three graphs G_i with $t_i \in G_i$, $1 \leq i \leq 3$. Draw G_i inside the 4-face incident to t_i with t_i in G_i and the frame identified. This may force crossing in G_i, pick a drawing with the smallest number of crossings. Then add the edges in E' so as to minimize additional crossings. The drawing of G (by itself) can have at most $\binom{|E(G)|}{2} \leq n^4$ crossings. Only the edges of E' are involved in additional crossings with edges of the frame, and since each such edge connects two of the G_i it crosses exactly one of the heavy edges accounting for kn^5 crossings. This shows that $\mathrm{cr}(H, w) \leq kn^5 + n^4$. Suppose that $\mathrm{cr}(H, w) < kn^5$, and fix a drawing of H realizing $\mathrm{cr}(H, w)$. Let $E' \subseteq E$ be the set of edges crossing heavy edges in the drawing. All edges in E' are light, since a crossing between two heavy edges would contribute at least $n^{10} > kn^5$ (weighted) crossings ($k \leq n^2$ since E' is a cut). Also, $|E'| < k$ since each crossing with a heavy edge has weight n^5 and we assumed the weighted crossing number of the drawing is less than kn^5. But then $G - E'$ is free of crossings with heavy edges, so G splits into three graphs G_i with $t_i \in G_i$ enclosed by the 4-face incident to t_i. But then G has a (t_1, t_2, t_3)-multiway cut of size less than k contradicting the definition of k. So $\mathrm{cr}(H, w) \geq kn^5$.

By Lemma 3.34 we can now find a simple graph H' so that G has a (t_1, t_2, t_3)-multiway cut of size k if and only if $kn^5 \leq \mathrm{cr}(H') \leq kn^5 + n^4$.

It follows that $\lfloor \mathrm{cr}(H')/n^5 \rfloor$ is the size of a smallest (t_1, t_2, t_3)-multiway cut of G showing that the crossing number problem is **NP**-hard. $\square$

We claimed that Cabello's proof is more powerful than the original **NP**-hardness proof. As a first piece of evidence we will see that the new construction shows that the crossing number remains hard to approximate within a constant factor. We say a function $f(G)$ *c-approximates* some parameter $\psi(G)$ if $f(G)/c \leq \psi(G) \leq cf(G)$ for all graphs G and a constant $c \geq 1$. An algorithm computing $f(G)$ is then called a *approximation algorithm* for $\psi(G)$.

Theorem 4.2 (Cabello [78]). *There is a $c > 1$ so that $\mathrm{cr}(G)$ cannot be approximated to within a factor of c by a polynomial-time algorithm, unless* **P = NP**.

Proof. The proof of Lemma 4.1 showed how to take G and three vertices t_1, t_2, t_3 and construct a simple graph H in polynomial time so that $\lfloor \mathrm{cr}(H)/n^5 \rfloor$ is the size of a smallest (t_1, t_2, t_3)-multiway cut of G. If $f(H)$ is a c-approximation of $\mathrm{cr}(H)$, then $\lfloor f(H)/n^5 \rfloor$ is a c-approximation of the size of the smallest (t_1, t_2, t_3)-multiway cut. Since it is known that there is a constant $c > 1$ so that the multiway-cut problem cannot be approximated to within a factor of c in polynomial time, unless **P = NP** [99], $\mathrm{cr}(H)$ cannot be approximated to within a factor of c, unless **P = NP**. $\square$

The construction in Theorem 4.1 can be modified to show that the crossing number problem remains **NP**-complete for cubic graphs with a given rotation system, see Exercise (4.1).

Corollary 4.3 (Hliněný [188]; Pelsmajer, Schaefer, and Štefankovič [279]). *The crossing number problem is* **NP**-*complete for cubic graphs with a given rotation system.*

Theorem 4.1 also implies that nearly all known crossing number variants are **NP**-hard, see Exercise (4.3) for a more formal statement and which crossing number variants it applies to. Other applications include the **NP**-hardness of 1-planarity, see Theorem 7.14.

One hardness result that does not seem to follow easily from Theorem 4.1 is that the crossing number problem remains **NP**-hard for *near-planar* graphs, that is, graphs of skewness one (removing one edge makes them planar). This result was shown as a consequence of another hardness result of interest in its own right. For a list of vertices A, an *A-anchored* drawing of a graph G is a drawing of G in which the vertices of A are incident to the outer region of the drawing, and occur along the boundary in the same order as in A. The list A may contain vertices not belonging to G. We write $\mathrm{cr}(G, A)$ for the *A-anchored crossing number* of G, that is the smallest number of crossings in any A-anchored drawing of G. We say G is *A-planar* if its A-anchored crossing number is zero.

Theorem 4.4 (Cabello and Mohar [80]). *The anchored crossing number problem is* **NP**-*complete even if the graph is the union of two A-planar, vertex-disjoint and connected graphs.*

We will not prove the theorem, but we include some interesting consequences.

Corollary 4.5 (Cabello and Mohar [80]). *The crossing number problem is* **NP**-*complete for near-planar graphs.*

Proof. By Theorem 4.4 we can reduce from the anchored crossing number problem, where we can assume that G_1 and G_2 are connected, A-planar graphs with $G = G_1 \cup G_2$ and $V(G_1) \cap V(G_2) = \emptyset$. If there is a vertex $a \in A$ which has degree more than one, we move a out of A and replace it with $\binom{|V(G)|}{4}$ new, and consecutive, vertices along A; connect each of these vertices to a by a path of length 2. This modification of G does not change the A-anchored crossing number of the graph. Hence, we can assume that all vertices in A have degree one. Since both graphs are anchored in A, we can now pick two consecutive vertices $a_i \in V(G_i) \cap A$. Let v_i be the (unique) neighbor of a_i in G_i.

Let $A = (a_0, \ldots, a_k)$. Between a_i and a_{i+1}, for $0 \le i < k$, as well as a_k and a_0 we add $|V(G)|^4$ paths of length 2. Finally, we add the edge $v_1 v_2$. The resulting graph is G'.

Any A-anchored drawing of G can be turned into a drawing of G' without adding any crossings; this includes edge $v_1 v_2$, which can be routed close to $v_1 a_1 a_2 v_2$ (if necessary, we can move v_1 and v_2 close enough to a so that they are crossing-free, since v_1 and v_2 have degree 2, that is not a problem).

For the other direction, let D be a cr-minimal drawing of G'. Since $\mathrm{cr}(G, A) < |V(G)|^4$, there must be a path between a_i and a_{i+1}, for every $0 \le i < k$, and a path between a_k and a_0, all of which are free of crossings. Let C be the union of these paths, so C is a crossing-free drawing of a cycle. Since G_1 and G_2 are connected by edge $v_1 v_2$, they must lie on the same side of that cycle. Moreover, the vertices of A occur along C in the right order, so D contains an A-anchored drawing of G, showing that $\mathrm{cr}(G, A) \le \mathrm{cr}(G')$. $\quad\square$

Richter and Salazar [284] asked whether "computing the planar crossing number of projective planar graphs [is] **NP**complete?" The construction from Corollary 4.5 contains the answer.

Corollary 4.6. *The crossing number problem is* **NP**-*complete for projective-planar graphs.*

Proof. We claim that the graph G' in the proof of Corollary 4.5 is projective planar. This is easy to see though: $G - \{v_1 v_2\}$ is planar, we can simply draw G_1 and G_2 on both sides of the anchoring cycles. We can then connect v_1 to v_2 using a single crosscap to pass by the edges between a_1 and a_2 which separate v_1 from v_2. $\quad\square$

We mention that most other measures of planarity we saw are also **NP**-complete, including skewness [231] and thickness [236]. The complexity of the edge crossing number appears to be open.

4.2 Drawing Graphs with Rotation

4.2.1 How to Draw $K_{2,n}$

Given two cyclic permutations π and π', we can define a function $d(\pi, \pi')$ as the smallest number of swaps of adjacent elements to turn π into π'. A *swap* changes the order of i and $\pi(i)$ in the permutation. For example, if $\pi = (42531)$, in the usual cycle notation, we could swap 5 and 3 to get $\pi' = (42351)$ in a single swap; similarly, $d((42531), (12354)) = 2$. The function d satisfies the metric axioms (check!), so we can use it as a distance measure for cyclic permutations.

While our goal is to study the crossing number of the complete bipartite graph $K_{2,n}$ with rotation system, we start by looking at 2-vertex multigraphs with rotation system, since this leads to a cleaner treatment.

Theorem 4.7. *If π and ρ are the rotations of the vertices in a 2-vertex multigraph, then the crossing number of the graph with the specified rotations equals $d(\pi, \rho^{-1})$.*

Proof. It is easy to see that the crossing number is at most $d(\pi, \rho^{-1})$: draw the vertex with rotation π together with initial segments of the edges incident to the vertex. Let $\sigma_1, \sigma_2, \ldots, \sigma_k$ be a sequence of $k = d(\pi, \rho^{-1})$ swaps turning π into ρ^{-1}. Extend the edges and for every σ_i, $i = 1 \ldots, k$, swap the edges in σ_i. Since for each swap, the edges being swapped are consecutive in the rotation, this can be done in the drawing. Eventually, the ends are in order ρ^{-1}, as seen from the first vertex, at which point we can connect them to the second vertex with rotation ρ.

The other direction we show by induction on the crossing number. If the crossing number is zero, then the permutations at the two vertices must be inverses of each other, since no two edges can cross. So we can assume that the crossing number is greater than zero; fix a drawing D realizing the crossing number. If there are two edges that cross more than once, then, by Corollary A.14, they contain two arcs ending in crossings which are otherwise free of crossings (with the two edges). We can swap the two arcs, reducing the number of crossings overall. Since we assumed that D was cr-minimal, this cannot happen, so any two edges cross at most once. It follows that a crossing of two edges creates two empty bigons (Lemma A.13), one at each vertex. Pick two crossing edges e and f with a bigon whose ends have minimal distance at the first vertex (counting the number of ends between them). Then

e and f must be consecutive, since any edge starting within the bigon has to cross either e or f, once, leading to a smaller bigon at the first vertex. Swapping the arcs of e and f incident to the first vertex, with rotation π, corresponds to exchanging two consecutive edges in the rotation, yielding a new rotation π', and it reduces the number of crossings of the drawing by one. We can then apply induction to show that the crossing number is at least $1 + d(\pi', \rho^{-1}) = d(\pi, \rho^{-1})$. $\square$

The next lemma shows how to move from a 2-vertex multigraph to the complete bipartite graph $K_{2,n}$. To simplify the presentation, we assume that the two edges incident to the right-partition vertices in $K_{2,n}$ have the same name, so the permutations on the two left-partition vertices are on the same set.

Lemma 4.8. *For any drawing of a 2-vertex multigraph with given rotation system in which every pair of edges crosses at most once, there is a good drawing of $K_{2,n}$ with the same rotation at the two left-partition vertices, and the same number of crossings.*

Proof. Fix a drawing D of the 2-vertex graph, with $d(\pi, \rho^{-1})$ crossings. As we argued in the proof of Theorem 4.7, any two edges cross at most once. Draw a curve γ close to one of the edges so that every edge crosses γ at most once. Redraw the graph in a rectangle, with the bottom and top sides representing the two vertices u and v, and the left and right sides identified as γ. We draw the ends of edges separately along the bottom and top, let $p_w(e)$ be the x-coordinate of edge e at vertex $w \in \{u, v\}$. We perturb the ends so that all x-coordinates are distinct. Finally, we add a virtual line ℓ in the middle between u and v. See Figure 4.2.

Whether two edges cross only depends on their endpoints and whether they cross γ. We will show how to draw $K_{2,n}$ without changing this information, which will allow us to conclude that the drawing realizes $d(\pi, \rho^{-1})$ crossings.

Classify edges by whether they turn left or right after leaving u using γ as the guide. Edges crossing γ are easily classified, since they cross γ from either left to right or right to left, determining a direction for the edge. For edges not crossing γ, we compare $p_u(e)$ to $p_v(e)$. If $p_u(e) < p_v(e)$, then we say e turns right going from u to v, otherwise it turns left.

For edges turning right, we start e at $p_u(e)$ on the bottom, and connect it to $p_v(e)$ at the top by moving up vertically from u beyond ℓ, and then horizontally to the right to $p_v(e)$, unless we run into a previously drawn edge turning right, at which point we start following its shape until we reach $p_v(e)$, or the other edge ends and we keep moving right (this may happen several times). If we hit the right margin, we wrap around above v and continue on the left. Once we have reached x-coordinate $p_v(e)$, we make the endpoint of the curve a vertex, and connect it to the line-segment representing v at $p_v(e)$ by a horizontal straight-line segment. All crossings are between an edge incident to u and an edge incident to v, so the crossings are independent. We now add

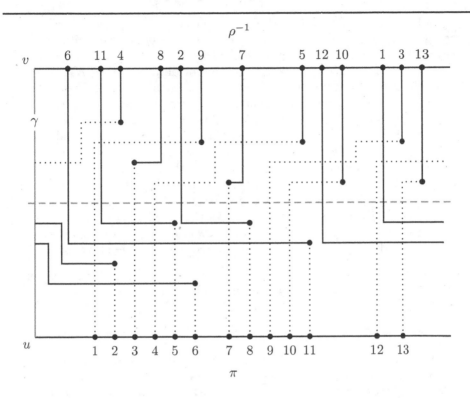

FIGURE 4.2: How to draw $K_{2,n}$. Identify the left- and right-hand sides of the drawing.

drawings of edges turning left, by mirroring the drawing along ℓ (exchanging the role of u and v), and drawing these edges as edges turning right starting at v using the same technique. Within each half, crossings can only occur between a vertical edge incident to the vertex on that side, and a horizontal segment, which is incident to the vertex on the other side, so all crossings are independent. □

As a consequence, we can reduce the calculation of the crossing number of a $K_{2,n}$ with rotation to a purely combinatorial problem.

Corollary 4.9. *If π and ρ are the rotations of the two left-partition vertices of a $K_{2,n}$, then the crossing number of $K_{2,n}$ with the specified rotation equals $d(\pi, \rho^{-1})$. If n is odd, then $\mathrm{cr}(D) \equiv d(\pi, \rho^{-1}) \bmod 2$ for any good drawing D of $K_{2,n}$ with the given rotation.*

Proof. Let u, v and π, ρ be the vertices and rotations of the 2-vertex graph obtained by suppressing the degree-2 vertices of $K_{2,n}$. Then $d(\pi, \rho^{-1})$ is a lower bound on the crossing number of the 2-vertex and, thereby, the $K_{2,n}$ with the given rotations. On the other hand, $K_{2,n}$ can be realized with $d(\pi, \rho^{-1})$ crossings by Lemma 4.8 and Theorem 4.7.

To show that $\mathrm{cr}(D)$ mod 2 is the same for all drawings of $K_{2,n}$ with the given rotation, we use a similar argument as in the proof of Theorem 1.12; first we suppress all the degree-2 vertices to get a drawing of a 2-vertex graph with rotation system, and with the same number of crossings. Compare two ways of drawing an edge e, as γ_e and as γ'_e. Since we are not allowed to change rotation, we can assume that γ_e and γ'_e agree close to their endpoints. In particular, they form a closed curve γ which avoids the two vertices. We can then 2-color $\mathbb{R}^2 - \gamma$. The two vertices either have the same or opposite colors, but in either case, there is an even number of edges between them (since n is odd), so the crossing number of the drawing with e as γ_e and with e as γ'_e differs by en even number. This shows that the crossing number of the drawing modulo 2 does not change as we redraw one edge at a time.   □

The question then becomes whether we can calculate $d(\pi, \rho)$ efficiently. The answer is yes, as was first proved by Jerrum [197] and rediscovered by Pelsmajer, Schaefer, and Štefankovič [279].

Theorem 4.10 (Jerrum [197]). *We can calculate $d(\pi, \rho)$ in polynomial time.*

We will prove this result later, as a consequence of Theorem 11.23. The proof, somewhat surprisingly, uses the crossing number characterization of $d(\pi, \rho)$ and requires linear programming. We are not aware of any simple, combinatorial solution to the problem (using dynamic programming, for example). Combining Theorem 4.10 with Corollary 4.9 then allows us to conclude that we can calculate $\mathrm{cr}(K_{2,n})$ for a given rotation system.

Corollary 4.11. *The crossing number of a $K_{2,n}$ with rotation system can be computed in polynomial time.*

In spite of this result, analyzing particular examples often gets complicated; we present one special case which we will use later on.

Theorem 4.12 (Pelsmajer, Schaefer, and Štefankovič [276]). *Consider a 2-vertex graph with rotation $(n(n-1)\ldots 1)$ and $(n_1(n_1-1)\ldots 1(n_1+n_2)(n_1+n_2-1)\ldots(n_1+1)\ldots n(n-1)\ldots(n-n_k+1))$, where $n = \sum_{i=1}^{k} n_k$. Assume that $n \leq 2\max\{n_i : 1 \leq i \leq k\}$. The crossing number of this graph is $\sum_{i=1}^{k} \binom{n_i}{2}$.*

More intuitively, the graph can be described as taking a planar 2-vertex graph with n edges, splitting it into k block with n_i edges in the i-th block, and flipping the rotation of the i-th block at one vertex. Figure 4.3 illustrates this graph.

Proof. We saw in the proof of Theorem 4.7 that no two edges cross more than once. Pick a curve connecting the two vertices, starting between edge 1 and n at the first vertex, and ending between n_1 and $n - n_k + 1$ at the other vertex.

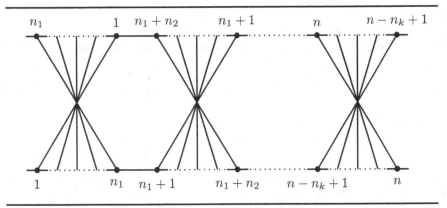

FIGURE 4.3: Flipping blocks of size n_i.

Every edge either crosses this curve, or it does not. So there are essentially two ways of drawing an edge. Suppose in the i-th block $0 \leq \ell_i \leq n_i$ edges cross this curve, while the remaining $n_i - \ell_i$ edges do not. The total number of crossings in this drawing is

$$\sum_{i=1}^{k} \binom{\ell_i}{2} + \binom{n_i - \ell_i}{2} + \sum_{i \neq j} \ell_i(n_j - \ell_j).$$

We can simplify this to

$$= \sum_{i=1}^{k} \binom{n_i}{2} + \sum_{i=1}^{k} \ell_i(\ell_i - n_i) + \sum_{i \neq j} \ell_i(n_j - \ell_j)$$

$$= \sum_{i=1}^{k} \binom{n_i}{2} + \sum_{i=1}^{k} \ell_i(-w_i + \sum_{j \neq i} w_j) + \sum_{i=1}^{k} \ell_i^2 - \sum_{i \neq j} \ell_i \ell_j$$

$$\geq \sum_{i=1}^{k} \binom{n_i}{2},$$

using $n_i \leq \sum_{j \neq i} n_j$ (obviously true for any but the largest n_i, and true for the largest n_i by assumption) and $0 \leq (\sum_{i=1}^{k} \ell_i)^2 = \sum_{i=1}^{k} \ell_i^2 + \sum_{i \neq j} \ell_i \ell_j$ for the estimate in the last line. Since the graph can easily be realized with $\sum_{i=1}^{k} \binom{n_i}{2}$ crossings (no edge crosses the special curve, so $\ell_i = 0$ for $1 \leq i \leq k$), this is the crossing number of the graph. □

If in Theorem 4.12 instead of starting with a planar 2-vertex graph, we start with a non-planar drawing, we still get a lower bound.

Corollary 4.13. *Consider a 2-vertex graph with rotation $(12 \ldots n)$ and ρ,* *where ρ is a cyclic permutation of $\{1, \ldots, n\}$ extending all of $(1 \ldots n_1)$, $((n_1 + 1) \ldots n_2)$, $((n_2+1) \ldots n_3)$, $\ldots$, $((n-n_k+1) \ldots n)$, where $n = \sum_{i=1}^{k} n_k$. Assume that $n \leq 2 \max\{n_i : 1 \leq i \leq k\}$. The crossing number of this graph is at least $\sum_{i=1}^{k} \binom{n_i}{2}$.*

Proof. In the proof of Theorem 4.12 we show that the number of crossings is

$$\sum_{i=1}^{k} \binom{\ell_i}{2} + \binom{n_i - \ell_i}{2} + \sum_{i \neq j} \ell_i(n_j - \ell_j).$$

If in the initial drawing the i-th and j-th block cross, the number of crossings with each other changes from $\ell_i(n_j - \ell_j) + \ell_j(n_i - \ell_i)$ to $n_i n_j + \ell_i(\ell_j - n_j) + \ell_j(\ell_i - n_i)$ which is at least as large, so

$$\sum_{i=1}^{k} \binom{\ell_i}{2} + \binom{n_i - \ell_i}{2} + \sum_{i \neq j} \ell_i(n_j - \ell_j)$$

remains a lower bound, and we can apply the rest of the analysis.   □

The next step would be to analyze $K_{3,n}$ and 3-vertex graphs with rotation; as far as we know, there are no results in this direction.

4.2.2   The Cyclic-Order Graphs

Theorem 4.7 suggests studying the *cyclic-order* graph CO_n whose vertex set are the cyclic permutations of n elements, with an edge between two permutations if they can be obtained from each other by swapping two adjacent elements in the permutation. Figure 4.4 shows CO_4.

Since the names of elements are arbitrary, we can turn any permutation into any other permutation, so CO_n is vertex-transitive. We collect some more facts about this graph.

Theorem 4.14 (Woodall [352]). *The following are true.*

(i) *$d(\pi, \pi^{-1}) = X(n) = \lfloor n/2 \rfloor \lfloor (n-1)/2 \rfloor$, for any cyclic permutation π of n elements.*

(ii) *CO_n has diameter $X(n)$, and two elements have maximal distance if and only if their permutations are inverses of each other.*

(iii) *CO_n is bipartite for odd n.*

Proof. (i) By renaming the elements of the permutation, we can assume that $\pi = (123 \ldots n)$, so we are calculating $d((12 \ldots n), (n(n-1) \ldots 1))$. Let us analyze the corresponding crossing number problem for a 2-vertex graph with edges $1, \ldots, n$ and rotations π and π^{-1}. Fix a cr-minimal drawing of the graph. As we argued in Theorem 4.7, every pair of edges crosses at most once, so there

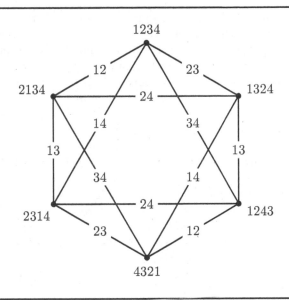

FIGURE 4.4: The cyclic-order graph CO_4; edges are labeled with transpositions. Based on Woodall [352, Figure 1].

are only two choices for each edge: connecting directly, or wrapping around the other vertex. Suppose ℓ of the n edges wrap around, the others do not. Then the drawing has

$$\binom{\ell}{2} + \binom{n - \ell}{2}$$

crossings, since exactly the edges belonging to the same group cross. This expression is minimal for $\ell = \lfloor n/2 \rfloor$ and, in that case, equals $\frac{1}{2}(\lfloor n/2 \rfloor(\lfloor n/2 \rfloor - 1) + \lceil n/2 \rceil(\lceil n/2 \rceil - 1)) = \frac{1}{2}((\lfloor n/2 \rfloor + \lceil n/2 \rceil)^2 = \lfloor n/2 \rfloor - \lceil n/2 \rceil) = \binom{n}{2} - \lfloor n/2 \rfloor \lceil n/2 \rceil = X(n)$. This analysis gives us the lower bound, but it also shows us how to achieve that lower bound: route half the edges directly, and half the edges around the other vertex.

(ii) Surprisingly, there does not seem to be an elegant, short proof of this fact; we have to refer the reader to the four-page proof given in the appendix of [352].

(iii) We use a parity argument in the style of Theorem 1.12. We first show that for n odd, the parity of the number of crossings in any drawing of a 2-vertex graph with n edges and a fixed rotation system is the same. Again, it is sufficient to see what happens if we redraw a single edge, say e along a new curve e'. Let us assume that the ends of e and e' are consecutive at both vertices; together e and e' form a closed curve γ, which may not be simple. Let γ' be the result of removing any self-crossings of γ. Then γ' is a simple closed curve, with both vertices on its boundary. Now all edges either start and end on the same side of γ' or on opposite sides of γ' (this depends on how often γ' winds around the vertices). If all edges start on the same side, then every edge crosses γ' an even number of times, but then the same is true for γ. That implies that the parity of crossing of every edge with e is the same as e', so we can redraw e as e' without changing the parity of crossing. If all edges end on opposite sides of γ', then every edge crosses γ' oddly, so the number of times an edge crosses e plus the number of times it crosses e' is odd. So the parity of the number of crossings changes for each edge as we go from e to e'. Since there are an even number of edges other than e in the drawing, this means the parity of crossing does not change (this is the only place we use that n is odd).

So if CO_n contains an odd cycle, suppose it passes through ρ and π. Then the two paths of the cycle connecting ρ to π correspond to drawing a 2-vertex graph with rotations ρ and π, one with an even, the other with an odd number of crossings. This contradicts that we established a parity theorem for 2-vertex graphs with an odd number of edges.   □

Parts (i) and (ii) of the theorem, together with Lemma 4.8, give us the largest number of crossings in a $K_{2,n}$.

Corollary 4.15. *The crossing number of a $K_{2,n}$ with rotation system is at most $X(n) = \lfloor n/2 \rfloor \lfloor (n-1)/2 \rfloor$ and this bound is achieved by fixing the rotation of the two degree n vertices to be the same.*

Woodall [352] used cyclic-order graphs to find lower bounds for $K_{m,n}$ using a computer program.

Theorem 4.16 (Kleitman [207]). *We have $\mathrm{cr}(K_{5,5}) = 16$ and $\mathrm{cr}(K_{5,7}) = 36$.*

Proof. We saw in the proof of Theorem 1.14 that $\mathrm{cr}(K_{5,n}) < Z(5,n)$ implies that there is a cr-minimal drawing of $K_{5,n}$ in which there are two vertices v_1, v_2 in the right partition for which $\mathrm{cr}(v_1, v_2) = 2$ and $\mathrm{cr}(v_1, v) = \mathrm{cr}(v_2, v) = 1$ for the remaining $v \neq v_1, v_2$ in the right partition, and $\mathrm{cr}(v, v') \leq 2$ for all v, v' in the right partition. (Recall that $\mathrm{cr}(u, v)$ denotes the number of crossings between edges incident to u and edges incident to v.) Let $n = 5$, and call the remaining vertices of the right partition v_3, v_4, and v_5. We can assume that the rotation at v_1 is $\pi_1 = (54321)$. Then the rotations π_i at v_i, for $i = 3, 4, 5$ must satisfy $d(\pi_1, \pi_i^{-1}) = 1$ and must therefore belong to the set $R = \{(21345), (13245), (12435), (12354), (23415)\}$. Moreover, the rotations are distinct, since two vertices with the same rotation span a $K_{2,n}$ with at least $Z(3,5) = 4$ crossings, which contradicts $\mathrm{cr}(v, v') \leq 2$. Now π_2^{-1} has distance 2 from π_1, so π_2^{-1} must be, without loss of generality, (32145) or (21435). Neither of these has distance one from three distinct elements in the set R, so there is no drawing of $K_{5,n}$ with less than $Z(5,n)$ crossings.

For $n = 7$ a similar, but lengthier, case analysis could be done by hand. It is simpler, and faster, to use the computer to test all 7-element multisubsets $\{\pi_1, \ldots, \pi_7\}$ of the 7! possible rotations, and verify that $\sum_{i<j} d(\pi_i, \pi_j^{-1}) \geq 36$.

By Corollary 4.9, the value $\sum_{i<j} d(\pi_i, \pi_j^{-1})$ is a lower bound on $\mathrm{cr}(K_{5,7})$, so the result follows. Woodall [352] describes a more refined algorithm, which he uses to establish the lower bounds for $K_{7,7}$ and $K_{7,9}$. $\qquad\square$

4.3 Good Drawings of the Complete Graph

4.3.1 Testing Goodness

We saw earlier that to build cr-minimal drawings of K_n recursively, we have to include good drawings; if we want to do so algorithmically, we may wonder how easy it is to check whether a given rotation system of a K_n can be realized by a good drawing. Kynčl, surprisingly, showed that the problem can be solved in polynomial time.

Theorem 4.17 (Kynčl [222]). *Given a rotation system for a complete graph, we can check in polynomial time whether there is a good drawing realizing the rotation system.*

The core idea behind the proof is that the way edges intersect a maximal star-subgraph of the complete graph is fully determined by the rotation system, and we can exploit this information. We will use the same idea repeatedly, most notably in the next section to prove Gioan's theorem, Theorem 4.20.

If we orient edges of a graph (arbitrarily), we can distinguish whether an edge crosses another edge from left to right, or from right to left, see Figure 4.5, we call this the *direction* of a crossing (and we will study this idea in more detail in Section 11.4).

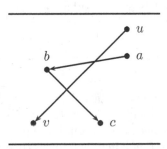

FIGURE 4.5: Direction of crossing. Edge uv is crossed by ab from left to right, and by bc from right to left.

We say that two drawings of an edge are *equivalent with respect to a vertex* w if the order and direction of crossings with $S(w)$, the maximal star with center w, is the same along both drawings of the edge.

Lemma 4.18. *Suppose we have an embedded star $S(w)$ together with two drawings of an edge between two outer vertices of the star so that the two drawings of the edge are equivalent with respect to w, and $S(w)$ together with either drawing of the edge is good. Then the two drawings of the edge form an empty bigon.*

If we puncture the surface at the outer vertices of the star, then this lemma states that two simple curves with the same endpoints, and which are *homotopic rel boundary*, that is, homotopic without changing the boundary, form a contractible (sometimes called singular) bigon. We prove this directly for our purposes without taking recourse to homotopy theory.

Proof. Let $e = uv$ be the edge, and e_1 and e_2 the two drawings of e. Both e_1 and e_2 are split into arcs by their consecutive intersections with the *star edges*, the edges of $S(w)$. By assumption, the j-th arcs of e_1 and e_2 connect the same sides of the same star edges. Suppose there is a j so that the two j-th arcs of e_1 and e_2 cross each other, and let j be minimal with this property. Let c be the first crossing of the j-th arcs along the j-th arc of e_1. This gives us two subarcs $\gamma_i \subseteq e_i$ between u and c. If these two arcs cross, then, by choice of c, there must be a crossing between γ_2 and an i-th arc of e_1, for $i < j$, or between γ_1 and an i-th arc of e_2, for $i < j$. Since the two i-th arcs do not cross each other (since $i < j$ and j is minimal), the region they bound (together with the star edges they touch) does not contain a vertex, since the arcs do not cross star edges, and attach on the same sides of star edges. Since parts of γ_1 or γ_2 cross an i-th arc, there must be a bigon in the region bounded by the two i-th arcs, and, since it is contained in the region bounded by the i-th arcs, it cannot contain a vertex. If, on the other hand, γ_1 and γ_2 do not cross, they bound a bigon with endpoints u and c, and this bigon does not contain a vertex since both arcs cross exactly the same set of star edges. In both cases we have found bigons of subarcs of the e_i which do not contain vertices. The full drawings of e_1 and e_2 may cut through these bigons, but we can take a smallest bigon contained in the bigon we found. That bigon is a bigon of e_1 and e_2, and since it is contained in a vertex-free bigon, it must also be empty. □

Given a complete graph with rotation system, and a vertex w of the graph, we define $c(e)$, for e not incident to w, to be the smallest number of self-crossings of any arc (not necessarily simple), equivalent to e with respect to w, and $c(e, f)$, for e and f not incident to w, to be the smallest number of crossings between any two drawings e' and f' which are equivalent to e and f with respect to w, and which have the same rotation as e and f at their endpoints.

Lemma 4.19 (Kynčl [222]). *Suppose we are given a complete graph and a vertex w of the graph. Let $c(e, f)$ be as defined above.*

(i) We can test whether $c(e) = 0$ in polynomial time.

(*ii*) *We can test whether* $c(e, f) = 0$ *and* $c(e, f) = 1$ *in polynomial time.*

(*iii*) *If* e' *and* f' *are equivalent drawings of* e *and* f, *respectively, with the same rotation at their end-vertices, and the number of crossings between* e' *and* f' *is larger than* $c(e, f)$, *then* e' *and* f' *form an empty bigon.*

(*iv*) *If* $0 \leq c(e, f) \leq 1$ *for every pair of edges* e *and* f *not incident to* w, *then there is a good drawing of the complete graph in which* e *and* f *cross* $c(e, f)$-*times, and the rotation remains unchanged.*

Again, these results follow from stronger, well-known results in homotopy theory, but we prove them directly here.

Proof. For (*i*) note that $c(e) = 0$ means that e can be realized as a simple arc; since the rotation system determines the order and direction of crossing with all edges incident to w, we can replace each crossing with a vertex (with induced rotation) and test whether $S(w) \cup \{e\}$ with rotation system is planar, something that can be done in linear time (see Theorem A.21).

For (*ii*), $c(e, f) = 0$ means there are simple arcs equivalent to e and f which do not cross. We already know the order and direction of crossings of e and f with respect to all edges incident to w. If e and f do not cross, we also know the order in which they cross along edges incident to w. So, as in (*i*), we can test realizability by replacing each crossing with a vertex, and testing realizability of $S(w) \cup \{e\} \cup \{f\}$ with the induced rotation system. For $c(e, f) = 1$ there is exactly one crossing between e and f. Both edges are split into at most n arcs by the $n - 1$ edges incident to w, so the crossing can occur in one of n^2 ways. We can test each of them using the same method as for $c(e, f) = 0$; arcs after the crossing have their order exchanged along edges incident to w.

For (*iii*), suppose that e' and f' do not form an empty bigon. Let e'' and f'' be equivalent to e and f so that e'' and f'' cross $c(e, f)$ times. Since e' and e'' are equivalent, by Lemma 4.18, they form an empty bigon with arcs $\gamma_{e'}$ and $\gamma_{e''}$. The arc $\gamma_{e'}$ cannot cross f', since then e' and f' would form an empty bigon, so any crossings of f' with the empty bigon either cross both arcs, or only $\gamma_{e''}$; hence we can reroute $\gamma_{e'}$ close to $\gamma_{e''}$ without decreasing the number of crossings between e' and f'. We can do so until e' and e'' no longer cross. We use the same method to redraw f' so that it looks like f'', and, again without decreasing the number of crossings. In a final step, we redraw e' as e'' and f' as f'' without decreasing the number of crossings, but this contradicts the fact that e'' and f'' crossed $c(e, f)$ times, which was assumed to be smaller than the number of times e' and f' crossed.

For (*iv*), fix a drawing with the given rotation system, and minimizing the total number of crossings. If two edges e and f cross more than $c(e, f)$ times, they form an empty bigon by part (*iii*). Pick a smallest bigon contained in this empty bigon. That bigon is still empty, and all edges cross the bigon *transversally*, that is, after crossing into the bigon, they cross the other arc, before crossing the same arc again. We can then swap the two arcs, reducing the number of crossings overall, which is a contradiction. Hence, every pair of edges crosses $c(e, f)$ times. Since we assumed that number to be at most 1, the drawing is good.   □

Proof of Theorem 4.17. We are given a rotation system of a complete graph. Pick an arbitrary vertex w, and fix a drawing of $S(w)$ as determined by the rotation system. By Lemma 4.19, we can test whether all other edges can be drawn without self-crossings, and that every pair of them crosses at most once (in the given rotation system). If all tests succeed, then, by part (iv), there is a realization in which every pair of edges crosses at most once, so there is a good drawing. Otherwise, there is no good drawing.   □

For general graphs, testing goodness of a graph with fixed rotation system is **NP**-complete, see Exercise (4.6). By a more careful analysis of obstructions to goodness, Kynčl was able to show that if a given rotation does not describe a good drawing of a complete graph, then there already is a K_5-subgraph whose induced rotation cannot be realized in a good drawing [223]. For weak isomorphism, it is enough to look at K_6-subgraphs.

4.3.2   Gioan's Theorem

We saw in Theorem 2.4 that there are $2^{\Theta(n^4)}$ non-isomorphic, good drawings of a K_n. For the lower bound, we used slide moves (Reidemeister moves of type 3), see Figure 2.2. Gioan's theorem shows that not only do slide moves not change the weak isomorphism type of a drawing of K_n, but also they generate it.

Theorem 4.20 (Gioan's Theorem). *Two weakly isomorphic good drawings of a complete graph on the sphere can be made isomorphic by slide moves.*

The result was first announced in 2005, but proofs only appeared more recently [38, 150]. Our new proof is based on ideas by Kynčl [222]. A *detour* for an edge e in a good drawing of a graph is a curve δ_e which has both its endpoints on e and does not intersect e otherwise. If the interior of the bigon formed by δ_e and the subarc $\gamma_e \subseteq e$ connecting the endpoints of δ_e on e does not contain any vertices of the graph, we call the detour *homotopic*.

Lemma 4.21. *For a good drawing of a complete graph, together with a homotopic detour δ_e for subarc γ_e on edge e, let $E_{\geq 2}$ contain γ_e and all edges which cross δ_e at least twice. Then there is a sequence of slides and homeomorphisms which only redraw curves in $E_{\geq 2}$ and so that after the redrawing γ_e follows δ_e closely, without crossing it, and the drawing of the graph is still good.*

Proof. Since $E_{\geq 2}$ contains γ_e, it is not empty; pick an edge f and a subarc $\alpha_f \subseteq f$ which forms a smallest bigon (in terms of enclosed region) with a subarc $\beta_e \subseteq \delta_e$; see Figure 4.6. Since the given drawing is good except for δ_e, α_f intersects every edge at most once, except β_e, which it intersects twice (at

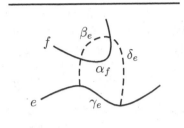

FIGURE 4.6: Detouring e along δ_2.

its endpoints). Therefore, if an edge has two consecutive intersections with $\alpha_f \cup \beta_e$, these intersections must belong to different arcs (two consecutive intersections with β_e would form a smaller bigon). If the bigon $\alpha_f \cup \beta_e$ does not contain any crossings in its interior, we can redraw α_f along β_e. If either of the two intersections between β_e and α_f is a vertex, that vertex can be kept in the redrawing. If the bigon does contain a crossing in its interior, we claim there must be such a crossing which can be moved out of the bigon by sliding α_f across it: pick an edge g_1 involved in a crossing within the bigon, and pick a crossing c_1 on this edge closest to α_f (distance being the number of crossings along g_1). Then there is no crossing between c_1 and α_f on the edge. The other edge g_2 involved in c_1 must also cross α_f, but that arc may contain crossings. Pick a crossing c_2 along that arc of g_2 closest to α_f. Since the arc between c_1 and α_f is free of crossings, the other edge, g_3, involved in c_2 must cross α_f between g_1 and g_2. This process must end eventually with a crossing where both arcs connecting it to α_f are crossing-free. We can then slide α_f over that crossing. Repeating this process as necessary, we can redraw α_f so that the bigon formed by α_f and β_e contains no crossings, a case we have already dealt with. □

Proof of Theorem 4.20. Pick an arbitrary vertex v and apply homeomorphisms so that v and the edges E_v incident to it are drawn identically in the two good drawings, call them D_1 and D_2; in particular the vertex locations in both drawings are the same. Since the drawings are weakly isomorphic, they have the same rotation system, by Lemma 2.1. Orient all edges (arbitrarily), so we can distinguish left-to-right from right-to-left crossings. We claim that the order and direction in which any edge intersects the edges of E_v is determined by the rotation system. Since edges in E_v do not cross each other, we can consider $e \notin E_v$ and $f, g \in E_v$. The five endpoints of these edges form a K_5 whose isomorphism type is determined by the rotation system (as we saw in the proof of Lemma 2.1). This determines the order in which e crosses f and g and the direction in which it crosses them.

Starting with D_1 we will redraw edges one at a time turning them into their D_2 version. We must do so only using slides and homeomorphisms while maintaining the goodness of the drawing. Let E_2 be the set of edges already drawn as in D_2. Initially, $E_v \subseteq E_2$. Suppose there is an edge $e \notin E_2$, and let $e_i = D_i(e)$, $i \in \{1, 2\}$, be the two drawings of edge e. We show that we can turn e_1 into e_2 by a sequence of slide moves. By the argument we made earlier, e_1 and e_2 cross edges in E_v in the same order and in the same direction. By Lemma 4.18, e_1 and e_2 form a bigon not containing a vertex, meaning that there is a subarc δ of e_2 which forms a homotopic detour with e_1. We can then apply Lemma 4.21 to redraw e_1 so it follows δ closely, only using slides and homeomorphisms. Only parts of edges which cross δ at least twice, and e_1 get changed in the redrawing. Since δ is part of e_2, and edges crossing δ twice cannot be drawn as in D_2 yet—so they cannot belong to E_2—we are not changing drawings of edges in E_2. Repeat this process until e_1 is drawn close to e_2 (when both together form a bigon). At this point, we can remove e_1 and add e to E_2. Once $E_2 = E(G)$, we are done. □

For an application of Gioan's theorem, see [21].

4.4   Computing the Crossing Number

In principle, solving the crossing number problem algorithmically is not difficult; **NP**-membership always guarantees an exponential-time algorithm. We saw in Theorem 4.1 that if $\mathrm{cr}(G) \leq k$, then there is a polynomial-size witness of that fact (an encoding of a planarization of a drawing of G with at most k crossings), which can be verified in polynomial time. Given a graph G and k, we can then test all potential polynomial-size for whether they witness $\mathrm{cr}(G) \leq k$, taking exponential time. This naïve idea does not perform very well in practice, but with (significant) additional work it can be turned into a form which lends itself to an efficient integer linear programming formulation[1].

What are our options if we want to avoid exponential-time algorithms? We can try to make the problem easier, by restricting graphs or drawings, though, as we saw, in many cases the problem remains hard. We can also give up on optimal solutions and settle on approximations, remembering that constant-factor approximations are still hard.

One of the most elegant computational results on the crossing number is that it is *fixed-parameter tractable*, that is, for every fixed k there is a polynomial-time algorithm which tests whether $\mathrm{cr}(G) \leq k$, and the exponent of the polynomial does not depend on k.

Theorem 4.22 (Grohe [160]). *Testing* $\mathrm{cr}(G) \leq k$ *can be performed in quadratic time for every fixed* $k \in \mathbb{N}$.

---

[1]Implemented as part of the Open Graph Drawing Framework, and at `http://crossings.uos.de/`.

The result was later improved to linear time [205]. Both proofs rely heavily on graph minor theory and Courcelle's theorem, a powerful tool from logic. While this gives depth to the proofs, it unfortunately makes them unsuitable for implementation.

We next turn to approximation algorithms for $cr(G)$. We saw in Theorem 3.2 that every graph has a convex, rectilinear drawing with $O(\log(n)(cr(G) + ssqd(G)))$ crossings, and such a drawing can be found in polynomial time. We can therefore approximate $cr(G) + ssqd(G)$ to within a factor of $\log(n)$ in polynomial time. If $cr(G)$ is small compared to $ssqd(G)$, this does not work well as an approximation of cr. If the degree of G is bounded, and the graph is sufficiently dense, then $cr(G)$ dominates $ssqd(G)$ and we do have an approximation algorithm for $cr(G)$, as first observed by Leighton and Rao [226], with a weaker bound.

Theorem 4.23 (Leighton and Rao [226]). *For bounded-degree graphs with n vertices and $m \geq 4n$ edges, the crossing number can be approximated to within a factor of $O(\log n)$ in polynomial time.*

Proof. Using Theorem 3.2 gives an $O(\log n)$-approximation of $cr(G)+ssqd(G)$. Since G has bounded degree, $ssqd(G) \leq cn$, for some constant c. On the other hand, $cr(G) \geq 1/64\, m^3/n^2 \geq m/64 \geq c/64\, ssqd(G)$, so we have an $O(\log n)$-approximation of $cr(G)$. $\square$

Without the density condition $m \geq 4n$ we only get an $O(n \log n)$-approximation algorithm for $cr(G)$.

Chimani and Hliněný [91] established a connection between cr and the size of a *planarizing set*, a set of edges whose removal leaves the graph planar.

Theorem 4.24 (Chimani and Hliněný [91]). *If $G - E'$ is planar, for a multigraph G and a set E' of k edges of G, then we can construct in polynomial time a drawing D of G in the plane with at most*

$$cr(D) \leq \Delta k(cr(G) + k)$$

crossings.

The theorem would give us a good approximation to $cr(G)$ for bounded-degree graphs, if we could find a small planarizing set E'. Unfortunately, we already know that computing the skewness of a graph is **NP**-complete. How else can we find a small E'? What about an embedding of G on a surface?

Theorem 4.25 (Gitler, Hliněný, Leaños, and Salazar [151]). *We can approximate the crossing number of projective-planar graphs to within a factor of $36\Delta^2$ in polynomial time, where Δ is the maximum degree of G.*

In particular, we have a constant-factor approximation for the crossing number of bounded-degree, projective-planar graphs (recall that without the degree bound the problem is **NP**-complete by Corollary 4.6).

Proof. If G is planar, $\mathrm{cr}(G) = 0$, and we are done (since we can easily test planarity). So we can assume that $\mathrm{cr}(G) > 0$. Find an embedding of G in the projective plane in polynomial time [247, 294]. Compute the facewidth r of that embedding, again in polynomial time [79]. Let γ be the closed simple curve that intersects the embedding of G in r points (vertices). Perturbing γ slightly, we can assume that it does not pass through any vertices, this increases the number of intersections with G by at most a factor of Δ, the maximum degree of D. Let E' be the set of at most Δr edges crossing γ. Since γ is non-contractible, removing these edges results in a planar graph. Using Theorem 4.24, we can find a drawing of G in the plane with at most $\Delta^2 r(\mathrm{cr}(G) + r)$ crossings. If $r < 6$, then, because $\mathrm{cr}(G) \geq 1$, this gives us a $(\Delta r)^2$-approximation of $\mathrm{cr}(G)$. On the other hand, if $r \geq 6$, we know, by Theorem 3.44, that $\mathrm{cr}(G) \geq r^2/36$, and our drawing is within a factor of $(6\Delta)^2$ of that value.   □

These results were the beginning of a series of papers that extended the approximation results to graphs embedded on an arbitrary surface, see [95, 186]. Combining this approach with an approximation algorithm for the Euler genus of a graph, Chekuri and Sidiropolous [89] have announced an approximation algorithm for the crossing number of bounded-degree graphs. Along a slightly different route, Chuzhoy earlier found a randomized polynomial-time approximation algorithm for bounded degree graphs, which achieves an approximation ratio of $O(n^{9/10} \log^c n)$ for bounded-degree graphs, which beats the $O(n \log n)$ bound significantly. Proofs of both results are beyond the scope of this book.

4.5   Notes

Many of the early papers on $\mathrm{cr}(K_n)$ contain proofs based on backtracking, Eggleton's thesis [120] implements such an approach; more recently, Ábrego et al. [1] enumerated (and generated) all good drawings of small complete graphs (up to K_8).

For an excellent overview of algorithmic techniques for computing the crossing number, including heuristics and exact algorithms, see the chapter on "Crossings and Planarization" by Buchheim, Chimani, Gutwenger, Jünger, and Mutzel [76] in the *Handbook of Graph Drawing and Visualization* [331].

4.6 Exercises

(4.1) Show Corollary 4.3. *Hint:* First show **NP**-completeness for graphs with a given rotation system, and then use hexagonal grids to obtain cubic graphs.

(4.2) Show that the crossing number problem remains **NP**-complete for graphs partially embedded in the plane. (A graph is *partially embedded* in the plane, if some of its edges are already drawn in the plane without crossings; the given drawing cannot be changed, but it can be involved in crossings with other edges of the graph). *Hint:* The multiway cut problem remains **NP**-complete for planar graphs and an unbounded number of terminals.

(4.3) Show that if γ is any graph parameter for which $\text{iocr}(G) \leq \gamma(G) \leq \overline{\text{cr}}(G)$, then approximating γ to within a constant factor is **NP**-hard. This implies that nearly all well-known crossing number variants are **NP**-hard to approximate. *Hint:* This problem requires some basic tools for dealing with the independent odd crossing number, $\text{iocr}(G)$, discussed in Chapter 11, and the rectilinear crossing number, $\overline{\text{cr}}(G)$, from Chapter 6.

(4.4) Show that calculating cr_Σ is **NP**-complete for every surface Σ. *Hint:* An embedding of a K_5 on a surface must contain a non-separating cycle, that is, cutting the surface along that cycle reduces the genus or Euler genus.

(4.5) Show that the crossing number of a single-vertex multigraph on m edges with rotation system can be computed in time $O(m \log m)$.

(4.6) Show that testing whether a graph with rotation system has a good drawing realizing the rotation system is **NP**-complete. *Hint:* Kratochvíl [219] showed that testing whether the weak isomorphism type of a matching can be realized by a good drawing is **NP**-complete (such graphs are called 1-*string graphs*, see Section 9.3). Use that result.

(4.7) Show that there is a rotation system for K_5 which cannot be realized by a good drawing, but so that all K_4-subgraphs with their induced rotation systems do have good drawings.

(4.8) Show that there is a weak isomorphism type for K_6 which cannot be realized by a good drawing, but so that all K_5-subgraphs with their induced weak isomorphism type do have good drawings.

Part II

Crossing Number Variants

Chapter 5

The Rectilinear Crossing Number: Rectilinear and Pseudolinear Drawings

5.1   A First Look

Recall that in a *rectilinear* (also called *geometric* or *straight-line*) drawing of a graph, edges are realized by straight-line segments (without bends). The *rectilinear crossing number*, $\overline{cr}(G)$, of a graph G is the smallest number of crossings in any rectilinear drawing of G. If $\overline{cr}(G) = 0$, then G is planar, and, as it turns out, the reverse is also true, a result known by many names.

Theorem 5.1 (Fary's (Wagner's, Stein's) Theorem [132, 320, 347]). *Every planar graph has a rectilinear embedding.*

We use a simple geometric fact: every polygon of length at most 5 is *star-shaped*; that is, it contains a point in its interior which can *see* all points in the polygon, in the sense that it can be connected by a straight-line segment—lying within the polygon—to any other point inside the polygon. (Triangulate the polygon; there are at most three triangles, so there must be a vertex belonging to all triangles; that vertex, and points in a wedge close to it will do.)

Proof. We prove by induction on the number of vertices that every plane graph has a rectilinear embedding with the same rotation scheme. The result is clear for graphs with $n \le 3$ vertices, so we can assume that there are at least 4 vertices. By Euler's formula, Corollary A.5, there is a vertex v of degree at most 5. Inductively find a rectilinear embedding of $G - v$. We need to argue that we can add v back into the face from which it was removed (the face still exists in the rectilinear embedding, since we maintained the rotation system). But this is possible by the geometric fact we outlined before the proof. $\square$

The proof establishes more than we claimed: the rectilinear embedding we construct is isomorphic to the original embedding. A drawing which is isomorphic to a rectilinear drawing is called *stretchable*. So Fary's theorem states that every plane embedding is stretchable. If the graph is 3-connected, we can even guarantee that the rectilinear embedding is *strictly convex*, that is, all inner faces are strictly convex polygons (all inner angles strictly less than π).[1] This is Tutte's celebrated spring theorem.

Theorem 5.2 (Tutte [338]). *Every plane, 3-connected graph is isomorphic to a strictly convex, rectilinear embedding.*

Tutte's result implies Fary's theorem (Exercise (5.1)), but its proof is more intricate; we recommend [333]. Does Fary's Theorem remain true in the presence of crossings? The answer is a qualified yes.

Theorem 5.3 (Bienstock and Dean [65]). *We have* $\overline{cr}(G) = cr(G)$ *if* $cr(G) \leq 3$.

The strategy for the proof is simple: add crossing-free edges to G to ensure that G without the crossing edges is a plane, 3-connected graph. Take a strictly convex embedding of this graph, and reinsert the crossing edges. This approach works for $cr(G) = 1$ ($cr(G) = 0$ is Fary's theorem), but even for $cr(G) = 2$ more work is required as demonstrated by the graph in Figure 5.1.

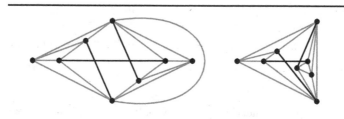

FIGURE 5.1: Graph with crossing number 2; in a drawing realizing $\overline{cr}(G) = 2$, the face containing the crossings cannot be convex.

The case $cr(G) = 1$ follows from Theorem 7.15, as we will see in Exercise (7.13). For the remaining cases, we refer the reader to the original paper [65] (which side-steps the case $cr(G) = 3$ with the words that the proof is "similar, but longer").

For $cr(G) = 4$ it is no longer true that $\overline{cr}(G) = 4$; it can be arbitrarily large.

Theorem 5.4 (Bienstock and Dean [65]). *For every k there is a graph G for which* $cr(G) = 4$, *but* $\overline{cr}(G) \geq k$.

We follow the simplified proof of Theorem 5.4 due to Hernández-Vélez, Leaños, and Salazar [184]. The proof is based on an obstacle to stretchability first identified by Thomassen [332]. The W-configuration shown in Figure 5.2 is not isomorphic to a rectilinear drawing.

---

[1] We will use strict convexity only for embeddings, to avoid confusion with the notion of a convex drawing, which is a global property.

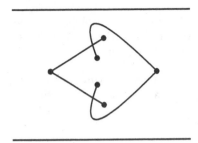

FIGURE 5.2: The W-configuration.

The W-configuration is not isomorphic to a rectilinear drawing in the plane, so whenever it occurs in a drawing, we know that drawing cannot be rectilinear.

Proof. Consider a C_8 with a diagonal edge connecting two vertices at distance 4 in the cycle. This graph has a unique embedding in the plane once the outer face is fixed (by Theorem A.9). Add to the graph all diagonal edges at distance 3, the *short diagonals*, while not incident to the endpoints of the original diagonal, see the left side of Figure 5.3. Consider a drawing of this graph

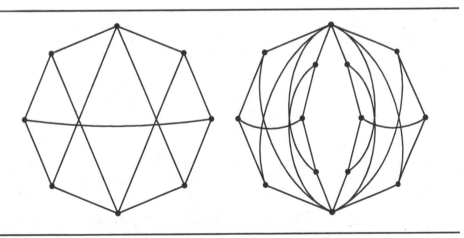

FIGURE 5.3: Separating $\overline{\mathrm{cr}}$ from cr. *Left:* C_8 with diagonals. *Right:* Combining two copies by identifying degree 4 vertices.

in which the C_8 and the original diagonal are crossing-free. If the original diagonal is not in the outer face, then the four short diagonals form a W-configuration with the C_8, so the graph has no rectilinear drawing in which only the short diagonals are involved in crossings. Taking two of these graphs and identifying their vertices of degree 4 pairwise yields a graph in which at least one of the two copies of the original graph has no such drawing (see the right side of Figure 5.3), that is, the original diagonal lies inside the C_8. By replacing all edges except for edges of the short diagonals by k parallel paths of length 2 each, we obtain the result. □

The maximum degree of the separating graph in the proof of Theorem 5.4 grows with $\overline{\mathrm{cr}}$, and this is unavoidable as the following result shows.

Theorem 5.5 (Bienstock and Dean [64]). *For every graph G,*

$$\overline{\mathrm{cr}}(G) = O(\Delta\,\mathrm{cr}(G)^2),$$

where Δ is the maximum degree of G.

We will not prove this result, but we can use the bisection width method to give us another way of bounding $\overline{\mathrm{cr}}(G)$ in terms of $\mathrm{cr}(G)$ and the maximum degree of G.

Theorem 5.6 (Shahrokhi, Sýkora, Székely, and Vrt'o [308]). *If G is a graph on n vertices, and $m \geq 4n$ edges, then*

$$\overline{\mathrm{cr}}(G) = O(\Delta\,\mathrm{cr}(G)\log n),$$

where Δ is the maximum degree of G.

Proof. We know from Theorem 3.2 that G has a convex rectilinear drawing with at most $c\log(n)(\mathrm{cr}(G) + \mathrm{ssqd}(G))$ for some constant c. Since $m \geq 4n$, the crossing lemma, Theorem 1.8, implies that $\mathrm{cr}(G) \geq \frac{1}{64}m^3/n^2 \geq m/4 \geq n$. Then $\mathrm{ssqd}(G) = \sum_{v \in V(G)} \deg^2(v) \leq 2\Delta m \leq 8\Delta\,\mathrm{cr}(G)$. Altogether,

$$c\log(n)(\mathrm{cr}(G) + \mathrm{ssqd}(G)) \leq c'\log(n)\Delta\,\mathrm{cr}(G).$$

□

Theorem 5.6 allows us to turn any approximation algorithm for cr into an approximation algorithm for $\overline{\mathrm{cr}}$. For example, using Theorem 4.23 gives us the following.

Corollary 5.7. *For bounded-degree graphs with n vertices and $m \geq 4n$ edges, the rectilinear crossing number can be approximated to within a factor of $O(\log^2 n)$ in polynomial time.*

Recently, a much better, additive approximation bound was shown by Fox, Pach, and Suk [138]: a drawing of G with at most $\overline{\mathrm{cr}}(G) + o(n^4)$ crossings can be found in polynomial time.

5.2 Pseudolines and the Pseudolinear Crossing Number

5.2.1 Pseudolines and Wiring Diagrams

Rectilinear drawings are hard to handle; from a drawing perspective, because they often require high precision—see, for example, the drawing of K_9 in Figure 6.1 of Chapter 6; from a proof perspective, because they are geometric, not combinatorial. Pseudoline arrangements are a very successful tool for analyzing rectilinear drawings combinatorially. There are various ways of defining arrangements of pseudolines, all of them essentially equivalent. We choose a very permissive definition, compare it to a stricter definition—wiring diagrams—and show that the two definitions are the same (for our purposes).

Definition 5.8. A *pseudoline* is a curve which is isomorphic to a straight line in the plane. A *pseudoline arrangement* is a family of pseudolines, every two of which cross exactly once, and, otherwise, do not intersect.

Figure 5.4 shows an example of a pseudoline arrangement. A pseudoline

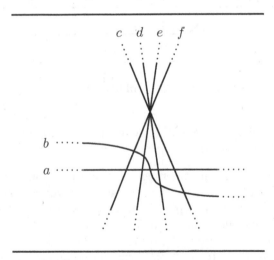

FIGURE 5.4: A pseudoline arrangement.

behaves like a straight line in many respects. It has no self-crossings, and it separates the plane into two halves, one on either side of it. Arrangements of pseudolines capture the fact that any two straight lines which are not parallel cross exactly once.

Definition 5.9. A *wiring diagram* is a family of horizontal, straight lines except that groups of neighboring lines are allowed to cross in a single crossing, forming a ∗-like shape. Every pair of lines crosses exactly once.

See Figure 5.5 for a sample wiring diagram. Both pseudoline arrangements and wiring diagrams may contain points in which more than two lines cross. If at most two lines cross in every point, the arrangement or diagram is called *simple* or *uniform*.

Lemma 5.10. *Every pseudoline arrangement is isomorphic to a wiring diagram.*

The reverse is obvious, since a wiring diagram is a pseudoline arrangement. The wiring diagram in Figure 5.5 is isomorphic to the pseudoline arrangement in Figure 5.4.

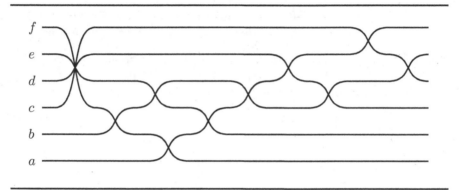

FIGURE 5.5: Wiring diagram isomorphic to pseudoline arrangement in Figure 5.4.

The proof combines a sweep of the arrangement with an argument we used in the proof of Lemma 4.21.

Proof. Suppose we are given an arrangement of n pseudolines. Fix a disk large enough to contain all crossings of the arrangement. Outside the disk, there are no crossings, but every pair of pseudolines crosses exactly once, so if we read the order of crossings along the boundary, it has the form ww for some word w over the alphabet consisting of pseudoline names. Erase everything outside the disk and move the intersections along the boundary of the disk, so that half of them are on the far left, and the other half are on the far right at the same y-coordinates as the intersections on the left. Connect each intersection on the boundary with a horizontal line to infinity (on its side of the disk). This is the first step in turning the pseudoline arrangement into a wiring diagram. We next need to straighten out the inside of the disk.

We sweep the pseudoline arrangement: let ℓ be a vertical line far enough to the left so that to the left of ℓ all the pseudolines are parallel horizontal lines. We claim that there is a crossing c so that all arcs connecting c to ℓ are free of crossings. Pick the first crossing c_1 along the top pseudoline crossing ℓ. The bottom arc connecting c_1 with ℓ may have crossings, so pick the first crossing on that arc from ℓ to c_1, and call it c_2. As we continue with the top/bottom arc connecting c_i to ℓ, we get a sequence of crossings which eventually has to stop, since the area enclosed by the top and bottom arcs incident to c_i gets strictly smaller, and there are only finitely many crossings. Once we stop, we have a crossing c for which both the bottom and top arc connecting it to ℓ are free of crossings. All other arcs connecting c to ℓ must be free of crossings as well, since these arcs cross through c, so they cannot cross each other within the area bounded by ℓ, and the top and bottom arcs connecting ℓ to c. We can now shorten the arcs to move c close to ℓ, past all other crossings between it and ℓ, make the crossing look like an $*$, and move ℓ beyond it. Continuing the sweep, we eventually end up with a wiring diagram, isomorphic to the original arrangement. □

The order of crossings (from left to right) in a wiring diagram is not uniquely determined, but we can use it to describe a wiring diagram, and therefore a pseudoline arrangement, up to isomorphism.

We note that Lemma 5.10 implies that every pseudoline arrangement is isomorphic to an x-monotone drawing, since wiring diagrams are x-monotone.

5.2.2 Pseudolinear and Good Drawings

We say a drawing of a graph is *pseudolinear* (sometimes *extendable*) if there is a pseudoline arrangement so that each edge is a subarc of a pseudoline. Edges in pseudolinear drawings are sometimes called *pseudosegments*. Every rectilinear drawing is pseudolinear, and we will see later that the reverse is not true.

In a pseudolinear drawing, triangles of vertices are convex in the sense that for any two vertices lying inside the triangle (including the boundary), the edge connecting them also lies inside the triangle. A *bad* K_4 is a drawing of a K_4 for which this is not the case, see Figure 5.6. Two edges cross, and one of the vertices lies in the interior of the other three vertices (making the edge crossing the triangle boundary non-convex).

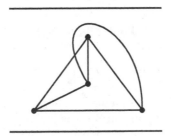

FIGURE 5.6: A bad K_4.

As it turns out this is the only obstacle to pseudolinearity in a good drawing, a result which we will not prove. For monotone drawings such a characterization was proved by Balko, Fulek, and Kynčl [42].

Theorem 5.11 (Aichholzer et al. [21]). *A good drawing of K_n is pseudolinear if and only if it does not contain a bad K_4.*

5.2.3   Structure in Pseudolinear Drawings

Recall that an empty triangle is a triangle which does not contain a vertex in its interior. Katchalski and Meir [204] showed that there can be as few as $O(n^2)$ empty triangles in a rectilinear drawing of K_n.

Theorem 5.12 (Katchalski and Meir [204]). *There are rectilinear drawings of K_n with at most $O(n^2)$ empty triangles.*

We present a later construction using so-called Horton sets. Some work has gone into optimizing the constant factor. The currently best bound is $1.62n^2$ [46].

Proof. Define the k-th *Horton set* recursively as follows: $H_1 := \{(1,1)\}$, $H_2 := \{(1,1),(2,2)\}$, and, for $k \geq 3$,

$$H_k := \{(2x-1, y), (2x, y+d_k) : (x,y) \in H_{k-1}\},$$

where d_k is sufficiently large (3^{k-1} is the traditional choice). By construction, $|H_k| = 2^{k-1}$. Choose $k = \lceil \log n \rceil + 1$. Then a complete graph on the points in H_k contains a K_n, so it is sufficient to show that $h(k)$, the number of empty triangles in H_k is of order $O(n^2)$.

We can split H_k into the points at even, H_k^+, and odd coordinates, H_k^-. These two sets are congruent to each other and scaled translates of H_{k-1}. By choosing sufficiently large d_k, we can ensure that any line through two points in H_k^- has all of H_k^+ below it (then any line through two points in H_k^+ has all of H_k^- above it). Let $h^+(k)$ be the number of segments in H_k that have no points in H_k below them (a point is *below a segment*, if an upwards vertical ray starting at the point crosses the segment), and $h^-(k)$ be the number of segments in H_k that have no points in H_k above them. Then $h^+(k) = 0$ for $k = 1$, $h^+(k) = 1$ for $k = 2$, and

$$h^+(k) \leq h^+(k-1) + 2^{k-1} - 1,$$

since there are $2^{k-1} - 1$ segments connecting points with adjacent x-coordinates. So $h^+(k) \leq 2^k$, and, using the same argument, $h^-(k) \leq 2^k$.

Any empty triangle in H^k lies either in one of H_k^+ or H_k^-, or has one point in one of the sets, and two in the other. The latter case splits into two cases, depending on whether the two points belong to H_k^+ or H_k^-. Suppose the two points belong to H_k^+; then there can be no points below the segment spanned by the two points in H_k^+, so there are at most $|H_k^-|h^+(k-1) = 2^{2k-3}$ such triangles. Overall, we have

$$h(k) \le 2h(k-1) + |H_k^-|\,h^+(k-1) + |H_k^+|\,h^-(k-1) = 2h(k-1) + 2^{2k-2},$$

which implies that $h(k) \le 2^{2k}$. And this number is $O(n^2)$. $\qquad\square$

Horton sets have exponential area, and this cannot be improved to polynomial area [48]; as far as we know, there is no bound on empty triangles in polynomial grid drawings of the complete graph.

The $O(n^2)$ upper bound is nearly met by a lower bound due to Dehnhardt [109] originally proved for rectilinear drawings; a careful reading shows that his proof works for pseudolinear drawings as well.[2]

Theorem 5.13 (Dehnhardt [109])**.** *Any pseudolinear drawing of a K_n, $n \ge 3$, contains at least $n^2 - 5n + 7$ empty triangles.*

For a pseudoline drawing of a triangle abc, let $A(abc)$ be the unbounded region of the plane bounded by edge bc (in the letter A, it is the region below the bar if a is at the top), see Figure 5.7.

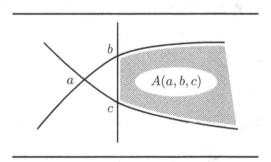

FIGURE 5.7: The region $A(abc)$ relative to triangle abc.

Lemma 5.14 (Dehnhardt [109])**.** *If abc is an empty triangle in a pseudolinear drawing of K_n, and $A(abc)$ contains n' vertices of the K_n, then a is part of $2n' + 1$ empty triangles.*

Proof. We prove this result by induction on n'. If $n' = 0$, a is part of abc and we are done. If $n' > 0$, then there must be a point d in $A(abc)$ which forms an empty triangle with bc (choose $d \in A(abc)$ so that the pseudolinear triangle bcd is minimal with respect to containment). Then abd and adc are empty triangles, and $A(abd)$ and $A(adc)$ partition the $n'-1$ vertices in $A(abc)$ different from d. If $A(abd)$ contains n_1' and $A(adc)$ contains n_2' vertices, then $n_1' + n_2' + 1 = n'$, and a belongs to $(2n_1'+1) + (2n_2'+1) + 1 = 2(n'-1) + 3 = 2n' + 1$ empty triangles, where the $+1$ counts abc. $\qquad\square$

---

[2]There is a better-known, but weaker, bound due to Bárány and Füredi [45], which also extends to pseudolinear drawings as shown in [18].

Proof of Theorem 5.13. We inductively prove the slightly stronger statement that any pseudolinear drawing of K_n, $n \geq 3$, contains at least

$$n^2 - 5n + 4 + k,$$

where k is the size of the convex hull C of the drawing. Since $k \geq 3$ if $n \geq 3$, this implies the claim.

In the base case $n = 3$, and we have one empty triangle. So assume $n > 3$, and let $v_1, \ldots, v_k$ be the edges of the convex hull C. Then edge $v_2 v_k$ is contained inside the convex hull, so every edge on the convex hull must lie on one of the sides of the pseudoline through $v_2 v_k$. Hence the convex hull after removing v_1 must contain the vertices $v_2, \ldots, v_k$, and any other vertices on the convex hull must lie inside the triangle $v_1 v_2 v_k$. Let those vertices be $v_{k+1}, \ldots, v_{k'}$ so that the convex hull after removal of v_1 is $v_2, \ldots, v_k, v_{k+1}, \ldots, v_{k'}$. To simplify notation, let $v_{k'+1} = v_2$, see Figure 5.8.

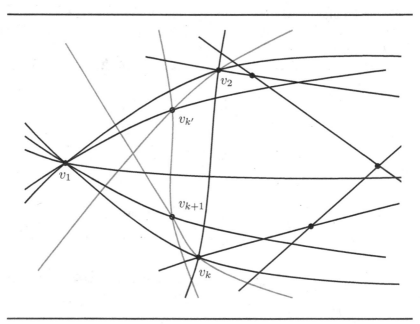

FIGURE 5.8: Two convex hulls: with and without v_1.

Triangles $v_1 v_i v_{i+1}$ are empty for $k + 1 \leq i \leq k'$ (since edges $v_i v_{i+1}$ are on the convex hull after removal of v_1), so $A(v_1 v_i v_{i+1})$ partition the $n - k' - 3$ vertices not belonging to one of these triangles (any two of these sets are disjoint, since pseudolines through $v_1 v_i$ and $v_1 v_j$ cross once only, namely in v_1). Applying Lemma 5.14 to each of these triangles gives us $2(n - k' - 3) + k' + 1 = 2n - k' - 5$ empty triangles involving v_1. By induction, the drawing of K_{n-1} obtained after removing v_1 contains $(n-1)^2 - 5(n-1) + 4 + (k-1) + k' =$

$n^2 - 7n + 9 + k + k'$. Together with the $2n - k' - 5$ triangles involving v_1, this gives us $n^2 - 5n + 5 + k$ empty triangles, which is what we needed. $\qquad \square$

5.2.4   The Pseudolinear Crossing Number

As we relaxed rectilinear drawings to pseudolinear drawings, we can similarly relax the definition of $\overline{cr}$.

Definition 5.15. The *pseudolinear crossing number*, $\widetilde{cr}(D)$ of a pseudoline drawing of a graph is the number of crossings occurring between pairs of pseudosegments. The *pseudolinear crossing number*, $\widetilde{cr}(G)$, of a graph G is the smallest $\widetilde{cr}(D)$ of any pseudolinear drawing D of G.

Since pseudolinear drawings are intermediate between good and rectilinear drawings, we have $cr(G) \leq \widetilde{cr}(G) \leq \overline{cr}(G)$. The separation of cr from $\overline{cr}$ in Theorem 5.4 only relied on the fact that W-configurations have no rectilinear drawings. Since they also do not have pseudolinear drawings, we immediately get the following result.

Theorem 5.16 (Hernández-Vélez, Leaños, and Salazar [184]). *For every k there is a graph G for which $cr(G) = 4$, but $\widetilde{cr}(G) \geq k$.*

This opens up the possibility that $\widetilde{cr}(G) = \overline{cr}(G)$, but this is not the case. The difference $\overline{cr}(G) - \widetilde{cr}(G)$ is unbounded [184], but it is possible that $\overline{cr}$ can be bounded in $\widetilde{cr}$. Although we know that $\overline{cr}(G)$ and $\widetilde{cr}(G)$ differ, many results which are true for rectilinear drawings, already hold for pseudolinear ones. The following theorem is one example.

Theorem 5.17 (Balogh, Leaños, Pan, Richter, and Salazar [43]). *The outer face in a $\widetilde{cr}$-minimal drawing of K_n is a triangle.*

We will show this for rectilinear drawings in Theorem 6.6. It is not difficult to see that Theorem 4.1 implies that $\widetilde{cr}$, as well as $\overline{cr}$ are **NP**-hard. In the case of $\widetilde{cr}$ it can also be shown that the problem lies in **NP**, since pseudolinear drawings can be encoded efficiently.

Theorem 5.18 (Hernández-Vélez, Leaños, and Salazar [184]). *Testing whether $\widetilde{cr}(G) \leq k$ is **NP**-complete.*

For the rectilinear crossing number, it is not at all obvious that $\overline{cr}(G) \leq k$ can be tested in **NP**. A closer look at that question reveals an unexpected connection with real algebraic geometry as we will now see.

5.2.5   Stretchability and Complexity

We opened this chapter with Fáry's theorem, stating that every plane graph is stretchable, that is, isomorphic to a rectilinear drawing. What about pseudolinear drawings, or pseudoline arrangements? In the previous section we mentioned that $\overline{\mathrm{cr}}(G) - \widetilde{\mathrm{cr}}(G)$ can be unbounded [184], so not every pseudolinear drawing is isomorphic to a rectilinear drawing, and the same is true for pseudoline arrangements, a fact first mentioned by Levi [227]; also see Ringel [289].

Theorem 5.19 (Levi [227]). *There are non-stretchable pseudoline arrangements.*

The proof is based on the Pappus configuration shown on the left in Figure 5.9. There are nine triples of points along nine lines. Pappus' theorem states that if eight of these triples are collinear, then all nine of them are. We can use this to construct a pseudoline arrangement that looks just like the Pappus configuration with one exception: the middle line passes just above the crossing of the two diagonal lines (right half of Figure 5.9). By Pappus' theorem, this pseudoline arrangement is not stretchable.

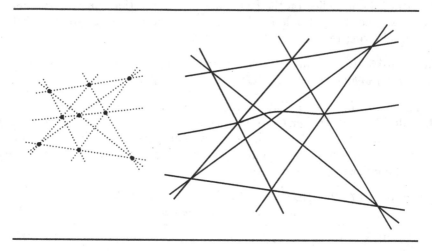

FIGURE 5.9: *Left:* Pappus configuration. *Right:* Levi's non-stretchable pseudoline arrangement based on Pappus configuration.

By judiciously replacing each triple crossing by three crossings, we can even construct a simple pseudoline arrangement which is not stretchable (Ringel [289] explains this modification). Levi's non-stretchable arrangement consists of nine pseudolines, and this is optimal; all pseudoline arrangements with fewer pseudolines are stretchable [154].

The following theorem (which we will not prove, since it requires some projective geometry) gives us a first hint of what makes stretchability hard. If a simple pseudoline arrangement is stretchable, we can slightly perturb the crossings in a realization, so all coordinates are rational, without changing the isomorphism type of the drawing. We can then scale the result to ensure all coordinates of crossing points are integers. Let us call a straight-line drawing of a pseudoline arrangement in which all crossings have integer coordinates a *grid drawing.*

Theorem 5.20 (Goodman, Pollack, and Sturmfels [156]). *For every n there is a stretchable simple arrangement of n pseudolines which requires a grid of size $2^{2^{cn}}$ in a straight-line grid drawing, for some constant $c > 0$.*

Fortunately, it can be shown that it does not get worse than that. A double-exponential size grid is always sufficient to embed a stretchable simple pseudoline arrangement. This is a consequence of a result on roots of multivariate polynomials due to Vorobjov [345].

Theorem 5.21 (Vorobjov [345]; Grigoriev and Vorobjov [158, Lemma 9]). *Suppose we have a family of m polynomials $f_1, \ldots, f_m : \mathbb{R}^n \to \mathbb{R}$, all of degree at most d and with integer coefficients bounded by 2^M. If there is an $x \in \mathbb{R}^n$ so that $f_1(x) = \cdots = f_m(x) = 0$, then there is such an x within distance $2^{Md^{cn}}$ of the origin, where $c > 0$ is a constant.*

To capture the geometry of a vertex set, we work with the *chirotope* function, $\chi(a, b, c)$, for points $a = (a_x, a_y)$, $b = (b_x, b_y)$, and $c = (c_x, c_y)$ in the plane, which is defined as

$$\chi(a, b, c) = \operatorname{sgn} \det \begin{pmatrix} 1 & a_x & a_y \\ 1 & b_x & b_y \\ 1 & c_x & c_y \end{pmatrix}.$$

By definition, $\chi(a, b, c) \in \{-1, 0, 1\}$ and some basic analytic geometry shows that $\chi(a, b, c)$ is 1 if $\angle(a, b, c) < \pi$, 0 if a, b, and c are collinear, and -1 if $\angle(a, b, c) > \pi$. The determinant of the matrix is a quadratic polynomial in the coordinates of the three points, namely $p(a, b, c) = a_x b_y - a_x c_y - a_y b_x + a_y c_x + b_x c_y - b_y c_x$; so $p(a, b, c)$ is positive if $\angle(a, b, c) < \pi$, 0 if a, b, and c are collinear, and negative if $\angle(a, b, c) > \pi$.

Theorem 5.22 (Goodman, Pollack, and Sturmfels [156]). *Every stretchable simple arrangement of n pseudolines has a straight-line grid drawing on a grid of size $2^{2^{cn}}$, for some constant $c > 0$.*

Proof. Let $(x_i, y_i)_{i=1}^{\ell}$ encode the coordinates of the $\ell := \binom{n}{2}$ crossings. The arrangement is specified completely, up to isomorphism, by whether (x_k, y_k) lies on the left or right side of the line through (x_i, y_i), (x_j, y_j), for all tripes $1 \le i < j < k \le \ell$. Using the determinant in the chirotope function, this can be encoded as a condition $p_{i,j,k}(x_i, y_i, x_j, y_j, x_k, y_k) > 0$, where $p_{i,j,k}$ is a quadratic polynomial. Since $p_{i,j,k}(x_i, y_i, x_j, y_j, x_k, y_k) > 0$ is equivalent to there being a $z_{i,j,k}$ and a $z'_{i,j,k}$ so that $p_{i,j,k}(x_i, y_i, x_j, y_j, x_k, y_k) - z^2_{i,j,k} = 0$ and $z_{i,j,k} z'_{i,j,k} - 1 = 0$, we can construct a family of quadratic polynomials f_ℓ in $x = (x_1, \ldots, x_\ell)$, $y = (y_1, \ldots, y_\ell)$ and $z = (z_{i,j,k})$, $z' = (z'_{i,j,k})$ so that the pseudoline arrangement is stretchable if and only if there are x, y, z, and z' for which $f_\ell(x, y, z) = 0$. By Theorem 5.21 this is equivalent to there being such a point with distance $2^{2^{cn}}$ from the origin for some constant $c > 0$ (since the degrees are at most 2, and the integer coefficients we need are small). In particular, each x_i and y_i has absolute value at most $2^{2^{cn}}$, and that remains true if we perturb the points slightly (while keeping the drawing isomorphic) to make them rational.   □

Theorem 5.22 implies that we can test stretchability of simple pseudoline arrangements, or, as it is called, *simple stretchability*, algorithmically. We only need to try all possible coordinate assignments to vertices of $V(G)$ with 2^{cn} many bits, a procedure that can be performed in exponential time. Can we do better? And how hard is it to test stretchability if the pseudoline arrangement is not simple?

Computational complexity gives us the answer. Testing stretchability, simple or not, is ∃ℝ-complete. In other words, it is as hard as deciding truth in the existential theory of the real numbers, the set ETR of existentially quantified sentences true over the field of real numbers (see Appendix B for definitions). This is a consequence of Mnëv's universality theorem [245], as shown by Shor [314], together with the result that simple stretchability and stretchability have the same complexity [302].

Theorem 5.23 (Mnëv [245]; Shor [314]; Schaefer and Štefankovič [302]). *Testing (simple) stretchability is ∃ℝ-complete.*

In particular, these problems are decidable, since the existential theory of the reals is decidable, as first shown by Tarski using quantifier elimination. It is now known that problems in ∃ℝ can be decided in polynomial space [83], a (probably proper) subclass of exponential time. Also, ∃ℝ-completeness implies **NP**-hardness, so testing (simple) stretchability is **NP**-hard.

5.2.6   The Complexity of the Rectilinear Crossing Number

With Theorem 5.23 as the starting point, we can show that the rectilinear crossing number problem is ∃ℝ-complete; the reduction was found by Bienstock [63].

Theorem 5.24 (Bienstock [63]). *Testing whether $\overline{cr}(G) \leq k$ is ∃ℝ-complete.*

Proof. We first show that given a graph G and an integer k, we can phrase $\overline{cr}(G) \leq k$ as a statement in the existential theory of the reals. We use the existential quantifiers to obtain coordinates for all vertices in G; we also quantify existentially over a new variable $c_{e,f}$ for every pair of independent edges e and f in G. We can test whether edges e and f cross as follows: the endpoints of e must lie on opposite sides of the line through f and the endpoints of f must lie on opposite sides of the line through e. Using the chirotope function as in the proof of Theorem 5.22, we can turn this into a polynomial inequality on the coordinates of the endpoints of e and f. We can then add a condition that ensures that if e and f cross, then $c_{e,f} = 1$, and $c_{e,f} = 0$ otherwise. Finally, we require that the sum of all $c_{e,f}$ is at most k. Then $\overline{\mathrm{cr}}(G) \leq k$ if and only if the quantified formula is true.

To see hardness, we reduce from the simple stretchability problem. Let $\mathcal{A}$ be a simple pseudoline arrangement. Draw a disk that contains all crossings of $\mathcal{A}$ and intersects each pseudoline twice along its boundary. As usual, we begin by constructing a frame for the reduction, which keeps the arrangement in place. To that end, create two copies of each pseudosegment inside the disk, one just to the left, the other just to the right of the pseudosegment. We let these copies be chords in a cycle along the boundary of the disk. Let F be the planarization of the graph consisting of the boundary cycle and the two chords for each pseudosegment (but not the pseudosegment itself); let L denote the union of subdivided chords in C. To F add the pseudosegments as edges, with each endpoint of the pseudosegment lying between its two copies. Let H be the resulting graph, and let the boundary cycle be C. Create a mirror copy H' of H and connect corresponding vertices along C and C' by an edge. Let M denote the matching between H and H'. Finally, replace all edges in C, C' and M with $10\binom{n}{2} + 1$ paths of length 2, where $n = |\mathcal{A}|$; let $G_{\mathcal{A}}$ be the resulting graph. See Figure 5.10 for an illustration of the construction.

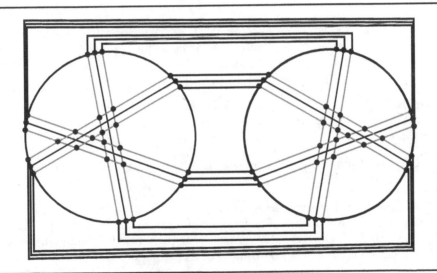

FIGURE 5.10: The graph $G_{\mathcal{A}}$ for a pseudoline arrangement $\mathcal{A}$. The heavy black edges belong to $C \cup C' \cup M$ and correspond to $10\binom{n}{2} + 1$ paths of length 2 each (to fit into the picture, some of them are drawn with bends); non-heavy black edges are the edges corresponding to pseudosegments from $\mathcal{A}$, and gray edges are the framework L, keeping the edges corresponding to pseudosegments in place.

The drawing of $G_{\mathcal{A}}$ suggested by its construction (and shown in Figure 5.10) has ten crossings for each pair of pseudolines, five in H and five in H', so $G_{\mathcal{A}}$ has a drawing with at most $10\binom{n}{2}$ crossings. If $\mathcal{A}$ is stretchable, then we can make the drawing of $G_{\mathcal{A}}$ rectilinear, so $\overline{\mathrm{cr}}(G_{\mathcal{A}}) \le 10\binom{n}{2}$ in this case.

This proves half of the claim that $\mathcal{A}$ is stretchable if and only if $\overline{\mathrm{cr}}(G_{\mathcal{A}}) \le 10\binom{n}{2}$. For the other half, fix a drawing D of $G_{\mathcal{A}}$ with $\overline{\mathrm{cr}}(D) \le 10\binom{n}{2}$. Since each edge of C, C' and M was replaced by $10\binom{n}{2} + 1$ paths, there must be at least one such path for each edge which is free of crossings. In particular, all of H lies on one side of $\hat{C}$, the subdivided C, and all of H' lies on the same side of $\hat{C}'$, a subdivided C'. Let e and f be edges in H corresponding to pseudosegments of $\mathcal{A}$. The order of the endpoints along C forces that e must cross f, as well as the two copies of f we added to H. Hence, each pair of pseudosegments forces at least 5 crossings in D. Since we have $2\binom{n}{2}$ such pairs, D must have at least $10\binom{n}{2}$ crossings. Since, we assumed that $\overline{\mathrm{cr}}(D) \le 10\binom{n}{2}$, and the particular crossings we accounted for were forced, $\hat{C} \cup L$ and $\hat{C}' \cup L'$ are plane graphs, and no pseudosegment crosses the two copies of itself. Since not both $\hat{C}$ and $\hat{C}'$ can contain the outer region, we can assume that $\hat{C}$ does not. Then the straight-line drawing of pseudosegments in H can be extended to a straight-line arrangement isomorphic with $\mathcal{A}$, which is what we had to show. $\qquad\square$

Bienstock's reduction in Theorem 5.24 is very geometric in the sense that a $\overline{\mathrm{cr}}$-minimal drawing of $G_{\mathcal{A}}$ contains a straight-line arrangement isomorphic to $\mathcal{A}$ if $\mathcal{A}$ is stretchable. This observation allows us to carry over the precision bounds for pseudoline arrangements to $\overline{\mathrm{cr}}$.

Corollary 5.25 (Bienstock [63]). *For every n there is an n-vertex graph G so that any $\overline{\mathrm{cr}}$-minimal drawing of G requires a grid of size $2^{2^{cn}}$, for some constant $c > 0$.*

Here we require that in a grid drawing of G all vertices (not crossings) lie on points of the grid.

Proof. The reduction in Theorem 5.24 shows that any $\overline{\mathrm{cr}}$-minimal drawing of $G_{\mathcal{A}}$ contains a drawing of a straight-line arrangement isomorphic to $\mathcal{A}$, assuming that $\mathcal{A}$ is stretchable. Suppose that there is a $\overline{\mathrm{cr}}$-minimal drawing of $G_{\mathcal{A}}$ on a grid of size g. The coordinates of a crossing between two edges can be expressed as a ratio of two numbers at most g. If we multiply the coordinates of all crossings by the denominator of all these ratios, we obtain coordinates for $G_{\mathcal{A}}$ in an integer grid of size $g^{\binom{n}{2}+1}$ in which all crossings, and all vertices, lie on the integer grid. By Theorem 5.20 we know that there are straight-line arrangements which require grids of size $2^{2^{cn}}$, for some constant $c > 0$ if all crossings have to lie on grid points. So we must have $g^{\binom{n}{2}+1} \ge 2^{2^{cn}}$, which implies that $g \ge 2^{2^{c'n}}$ for some constant $c' > 0$. $\qquad\square$

As we did in Theorem 5.22, the root bound in Theorem 5.21 can be used to show that a grid of size $2^{2^{cn}}$ is also always sufficient to draw a $\overline{cr}$-minimal drawing of any graph G, see Exercise (5.12).

On a more positive note, we already mentioned that Fox, Pach, and Suk [138] announced that $\overline{cr}$ can be approximated in polynomial time to within an additive constant of order $o(n^4)$ for an n-vertex graph.

5.3   Rectilinear Drawings

5.3.1   Enumeration of Rectilinear Drawings

In Theorem 2.4 we saw that K_n has $2^{\Theta(n^4)}$ non-isomorphic good drawings; the number of non-isomorphic rectilinear drawings is much smaller.

Theorem 5.26 (Goodman and Pollack [155]). *There are at most $(4n)^{2n}$ non-isomorphic rectilinear drawings of K_n.*

The proof uses Warren's theorem [348]: given m polynomials $f_1, \ldots, f_m : \mathbb{R}^n \to \mathbb{R}$ of degree at most d, a *sign pattern* is a vector

$$(\mathrm{sgn}(f_1(x)), \ldots, \mathrm{sgn}(f_m(x)))$$

for some $x \in \mathbb{R}^n$, where sgn is the *signum (sign)* function ($+1$ for positive arguments, 0 for 0 and -1 for negative arguments). By Warren's theorem, the number of sign patterns not containing a 0 entry is at most $(4emd)^n$.

Proof. The isomorphism type of a rectilinear drawing of K_n is determined by chirotope function χ applied to its vertices. We have $\chi(u, v, w) \in \{-1, 1\}$ for every triple u, v, w of vertices (it does not assign 0, since that would amount to three collinear points, which is not allowed in a rectilinear drawing of K_n). Consider a set of points $a_1, \ldots, a_n$ as locations for the vertices $v_1, \ldots, v_n$ of K_n. Recall that

$$\chi(a, b, c) = \mathrm{sgn}\det \begin{pmatrix} 1 & a_x & a_y \\ 1 & b_x & b_y \\ 1 & c_x & c_y \end{pmatrix},$$

so the chirotope condition for a triple of points can be expressed as a condition on the sign of a quadratic polynomial in the coordinates of the points. The isomorphism type of a rectilinear drawing can then be specified as a sign pattern of $\binom{n}{3}$ quadratic polynomials, not containing 0. By Warren's theorem, the number of sign patterns, and thus realizable chirotopes, is at most $(8e\binom{n}{3}/n)^n \leq (4n)^{2n}$, where e is Euler's constant. $\square$

5.3.2 Structure in Rectilinear Drawings

What unavoidable substructures occur in rectilinear drawings of graphs? In this section we do not restrict ourselves to $\overline{\mathrm{cr}}$-minimal drawings, so we are straying into the field of discrete geometry.

We saw earlier that any rectilinear drawing of a K_n with $n \geq 5$ must contain a convex quadrilateral, a convex drawing of a K_4. A natural generalization is the following: is it true that for any n there is an n' so that every rectilinear drawing of $K_{n'}$—with vertices in general position—contains a convex drawing of a K_n? Let $\mathrm{ES}(n)$ be the smallest such n' if it exists. The function is named after Erdős and Szekeres who first showed that $\mathrm{ES}(n)$ is finite, thereby creating a foothold for Ramsey theory in discrete geometry.

Theorem 5.27 (Erdős and Szekeres [127]). *For $n \geq 1$, we have*

$$\mathrm{ES}(n) \leq \binom{2n-4}{n-2} + 1.$$

There are many proofs of this theorem, and even the original paper contains two: one based on Ramsey's theorem, and one based on a geometric idea; here we present the second proof. We need a basic notion from discrete geometry. We say that a group of points $p_1, \ldots, p_k$, ordered from left to right, is a *cap* (*cup*) if the points are in convex position and lie above (below) $p_1 p_k$.

Lemma 5.28. *Let $f(k, \ell)$ be the smallest n so that any set of n points in general position contains a k-cup or an ℓ-cap. Then*

$$f(k, \ell) = \binom{k+\ell-4}{k-2} + 1.$$

The lemma implies Theorem 5.27, since $\mathrm{ES}(n) \leq f(n, n)$.

Proof. We claim that

$$f(k, \ell) = f(k-1, \ell) + f(k, \ell-1) - 1.$$

Together with the easily verified $f(k, 3) = k$ and $f(3, \ell) = \ell$ this implies the lemma by induction. We separate the proof of the claim into proving it as an upper bound and a lower bound.

The geometric idea at the core of the upper bound is the following: suppose we have a k-cup whose last (rightmost) vertex is the first (leftmost) vertex of an ℓ-cap. Then this configuration of points contains either a $k+1$-cup or an $\ell+1$-cap. If the angle formed by the edges at the shared point is greater than π, the k-cup can be extended by the second vertex of the ℓ-cap, and if it is less than π, the ℓ-cap can be extended by the penultimate vertex of the k-cup.

Let's see how to use this. Suppose we have $f(k-1, \ell) + f(k, \ell-1) - 1$ points in general position. Consider the leftmost $f(k-1, \ell)$ points. If they contain an ℓ-cap we are done, so they must contain a $(k-1)$-cup. Remove the last vertex of that $(k-1)$-cap, and add one of the remaining $f(k, \ell-1)-1$ vertices. Again, there is either an ℓ-cap, in which case we are done, or another $(k-1)$-cup. Again we delete the last vertex of that $(k-1)$-cup and keep going. If we never encounter an ℓ-cap, we instead found $f(k, \ell-1)$ vertices each of which is the last vertex of a $(k-1)$-cup. If these vertices contain a k-cup we are done, so we can assume that they contain an $(\ell-1)$-cap. But then this $(\ell-1)$-cap together with the $(k-1)$-cup that precedes it contain a k-cup or an ℓ-cap.

To prove the lower bound, we recursively construct sets of $f(k, \ell)-1$ points without either a k-cup or an ℓ-cap. For the base cases $k=3$ and $\ell=3$ we choose a $(k-1)$-cap and an $(\ell-1)$-cap. For the recursive step let X be a set of $f(k-1, \ell)-1$ points without a $(k-1)$-cup or an ℓ-cap, and let Y be a set of $f(k, \ell-1)-1$ points without a k-cup or an $(\ell-1)$-cap. Using scaling we can flatten both sets so all lines between two points are close to horizontal. Place X and Y so that all of X lies to the left and below Y and so that any line through a point for X and a point of Y is steeper than any line within those two sets. Then any k-cup can contain at most one point from Y, but then X would have to contain a $(k-1)$-cup, which it does not. Similarly, any ℓ-cap can contain at most one point of X, which would force an $(\ell-1)$-cap in Y which does not exist. We have constructed a set of $f(k-1, \ell)-1+f(k, \ell-1)-1$ points without k-cup or ℓ-cup, so $f(k, \ell) \geq f(k-1, \ell)+f(k, \ell-1)-1$, which is what we had to show. $\square$

Twenty-five years later, Erdős and Szekeres [128] supplied a lower bound.

Theorem 5.29 (Erdős and Szekeres [128]). *For $n \geq 1$, we have*

$$\mathrm{ES}(n) \geq 2^{n-2}+1.$$

Our proof is based on the presentation in [252].

Proof. By Lemma 5.28 there are sets X_i of $\binom{n-2}{i}$ points which contain neither an $i+2$-cup nor an $n-i$-cap. Scale each set so it becomes very flat (all lines between two points of the set are nearly parallel) and very short, and place them, from left to right, along a semi-arc so that no line through points in two of the sets passes through another set. Let $X = \cup_{i=0}^{n-2} X_i$, so $|X| = 2^{n-2}$. See Figure 5.11. Suppose there is a set Y in convex position. Let X_i and X_j be the first and last set containing a point of Y. If $i = j$, then $|Y| \leq \max(i+2, n-i)-1 \leq n-1$, so we can assume that $i < j$. Then $X_i \cap Y$ must be a cup, so it has size at most $i+1$, and $X_j \cap Y$ must be a cap, so it has size at most $n-j-1$. Every X_k with $i < k < j$ can contain at most one point in Y, so we have $|Y| \leq i+1+n-j-1+(j-i-1) = n-1$. $\square$

Erdős and Szekeres conjectured the lower bound of $2^{n-2}+1$ to be the correct value, but for decades the upper bound of Theorem 5.27 was only improved slightly. In an amazing breakthrough, Suk recently managed to nearly establish the conjecture. The proof is beyond the scope of this book.

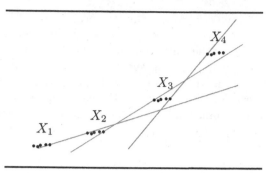

FIGURE 5.11: Placing the X_i along a semi-arc so they do not interfere with each other.

Theorem 5.30 (Suk [325]). *For $n \geq 1$, we have*

$$ES(n) \leq 2^{n+6n^{2/3}\log n}.$$

We can sharpen the Erdős-Szekeres problem by requiring the convex point sets to be empty, that is, not containing any other points of the set. We discussed empty triangles earlier. It turns out that there is no happy ending for this question, there are arbitrarily large pointsets without empty heptagons [195]. This is sharp, since any sufficiently large pointset in general position will contain an empty hexagon [148].

We turn to substructures defined by crossing patterns of edges. Two basic patterns are k pairwise crossing edges and k pairwise disjoint edges. Since a convex drawing of a K_{2k} contains k pairwise crossing edges—its diagonals—we know that any rectilinear drawing of a K_n with $n = ES(2k)$ vertices contains k pairwise crossing edges. This bound can be improved significantly.

Theorem 5.31 (Aronov, Erdős, Goddard, Kleitman, Klugerman, Pach, and Schulman [37]). *Every rectilinear drawing of K_n contains at least $\sqrt{n}/4$ pairwise crossing edges.*

For the proof we need a discrete version of the ham-sandwich theorem; this is a special case of Borsuk's original result, but can be proved directly as well; the following proof is due to Willard [349].

Lemma 5.32. *Given a set of $n = n_1 + n_2 + n_3 + n_4$ points in general position, there are two lines ℓ_1 and ℓ_2 so that the i-th quadrant of $\mathbb{R}^2 - \ell_1 - \ell_2$ contains n_i points, $i = 1, \ldots, 4$.*

The first quadrant is the quadrant containing (∞, ∞), and quadrants are labeled counterclockwise, and similarly for sextants.

Proof. Start with a horizontal line below the point set, and move it to a position above the pointset; at some intermediate stage the number of points below the line must be $n_3 + n_4$; stop there and let that line be ℓ_1. Slide a point p along ℓ_1, from $x = -\infty$ to $x = \infty$ and record $f(p)$, the smallest angle of a line ℓ with ℓ_1 which splits the pointset above ℓ_1 into n_2 points on the left and n_1 points on the right, and $g(p)$, the smallest angle of a line ℓ with ℓ_1 at p which splits the pointset below ℓ_1 into n_3 points on the left, and n_4 points on the right. Then $f(p)$ ranges from 0 to π, while $g(p)$ ranges from π to 0. Since both functions are continuous, there must be a p for which $f(p) = g(p)$ and the two lines coincide; let this be ℓ_2. By construction, the i-th quadrant contains n_i points, for $i = 1, \ldots, 4$. $\qquad\square$

We say that a set of points X *avoids* another set Y if no line through two points of X intersects the convex hull of Y; X and Y are *mutually avoiding* if they avoid each other.

Lemma 5.33. *If two pointsets of the same size n are mutually avoiding, then a complete bipartite graph $K_{n,n}$ on the two vertex sets contains n pairwise crossing edges.*

Proof. Let X and Y be the two pointsets of size $n \geq 2$, the case $n = 1$ being trivial. If we choose any two points from X and any two points of Y, their convex hull contains all four points (a point in the interior, together with its partner from the same set would separate the other two vertices). So no vertex lies in the convex hull of three other vertices, not all from the same set.

There is a line ℓ separating X from Y (since their convex hulls are disjoint) and the convex hull of $X \cup Y$ crosses ℓ at least twice. So there must be two X-Y-edges $x_1 y_1$ and $x_2 y_2$ which are on the convex hull and cross ℓ. All four vertices are distinct, since both X and Y contain at least two vertices, so the two edges are disjoint. See Figure 5.12.

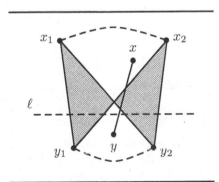

FIGURE 5.12: The convex hull of $X \cup Y$ with edges $x_1 y_1$ and $x_2 y_2$ on boundary crossing ℓ.

By the argument we made earlier, $x_1 y_2$ and $x_2 y_1$ must cross each other. We claim that they must also cross all other non-adjacent X-Y edges: let xy be such an edge; then x cannot lie in $x_1 y_1 y_2$ or $x_2 y_1 y_2$ and y cannot lie in

$y_1x_1x_2$ and $y_2x_1x_2$, so x and y lie on opposite sides of both x_1y_2 and x_2y_1 so xy has to cross both. Removing $\{x_1, x_2, y_1, y_2\}$ we can continue to pick edges recursively.   □

To complete the proof of Theorem 5.31 we need one more result, the *other* Erdős-Szekeres theorem which states that any sequence of $nm + 1$ real numbers contains either an increasing subsequence of n elements, or a decreasing subsequence of m elements [127].

Proof of Theorem 5.31. Fix a rectilinear drawing of K_n, so the vertices of K_n are in general position. To simplify notation, let us assume that n is a multiple of 12. By Lemma 5.33 it is sufficient to find two mutually avoiding sets of size $\sqrt{n/12}$. Use Lemma 5.32 to find two lines ℓ_1 and ℓ_2 with $n_1 = n_4 = 5n/12$ and $n_2 = n_3 = n/12$. After an affine transformation, we can assume that ℓ_1 and ℓ_2 are orthogonal. Slide a line parallel to ℓ_2 starting at ∞ towards ℓ_2 until one of the two sextants to the right of it contains $n/12$ points. Without loss of generality, we can assume this is the lower right sextant. Then the upper right sextant has fewer than $n/12$ points, so the upper middle sextant must contain more than $n/3$ points. See Figure 5.13. Let the points of the upper

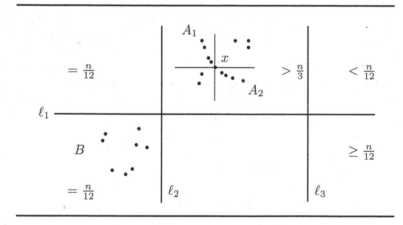

FIGURE 5.13: Partition of vertices into sextants.

middle sextant be $p_1, \ldots, p_k$, ordered from left to right. By the other Erdős-Szekeres theorem, there is a subsequence $p_1', \ldots, p_{k'}'$ of $k' = \lceil \sqrt{n/3} \rceil$ points whose y-coordinates are either increasing or decreasing. Let us assume they are decreasing (if they are increasing, we work with the lower right sextant). Note that any line through two points of A does not intersect the lower right sextant, so A avoids the set of points in that sextant. Let x be a point so that half the points, A_1, of A lie to the left and above x, and the other half, A_2, to the right and below x, so each half contains $\lceil \sqrt{n/12} \rceil$ points. Take the $n/12$ points in the lower left sextant, $q_1, \ldots, q_{n/12}$, ordered by shortest to longest distance from x. Again using the Erdős-Szekeres theorem we can find a subsequence B of length $\lceil \sqrt{n/12} \rceil$ points whose angles with x are either increasing or decreasing. If the angles are increasing, then no line through two points of B can intersect the convex hull of A_2, if they are decreasing, no such line intersects the convex hull of A_1. In either case B avoids a set of size at least $\lceil \sqrt{n/12} \rceil$ which also avoids it, so they are mutually avoiding. $\square$

While the bound on mutually avoiding pointsets is tight, Aronov et al. [37] conjecture that the number of pairwise crossings edges may be much larger, possibly linear. There are currently no non-trivial upper bounds.

For general graph, Valtr [341] shows that sufficiently many edges also force k pairwise crossing edges.

Theorem 5.34 (Valtr [341]). *If a rectilinear drawing of a graph does not contain any k pairwise crossing edges, then $m = O(n \log n)$, where $m = |E(G)|$, $n = |V(G)|$, and $k > 3$ is fixed.*

Valtr also shows that the result remains true for x-monotone drawings, a drawing style generalizing rectilinear drawings which we will study in Chapter 8.

We turn to pairwise disjoint edges.

Theorem 5.35 (Pach and Törőcsik [262]). *Every rectilinear drawing of a graph with more than $(k-1)^4 n$ edges contains k pairwise disjoint edges.*

The case $k = 2$ is extended in Corollary 12.15. For $k = 3$, Alon and Erdős [27] showed that $6n - 5$ edges are sufficient. The dependence on k can be made quadratic: Tóth [334] established an upper bound of $2^9 k^2 n$; it is true, like Theorem 5.35 for good, x-monotone drawings as well. (Felsner [136, Theorem 1.11] has improved the bound to $2^8 k^2 n$.)

For this result we need Dilworth's theorem, a generalization of the (other) Erdős-Szekeres theorem; recall that for a partial order $\prec$, a *chain* is a sequence of elements with $x_1 \prec x_2 \prec \cdots \prec x_k$. An *antichain* is a set of elements which are pairwise incomparable by $\prec$. Dilworth's theorem now states that if a partial order does not contain a chain of length k, then the underlying set of elements is the union of $k-1$ antichains [114].

Proof. For $k = 2$ this is the geometric thrackle conjecture, which we will prove in Theorem 12.23; so we assume that $k \geq 3$. Fix a rectilinear drawing of a graph; we can assume that no two vertices have the same x- or y-coordinates.

We need three relations on edges in a drawing: we say uv *precedes* $u'v'$, if u is to the left of u' and v to the left of v'; we say uv is *above* $u'v'$ if every vertical line which intersects both uv and $u'v'$ intersects uv above $u'v'$; finally, uv is *subsumed* by $u'v'$ if the extreme left and right endpoints among the four vertices are u' and v'.

With these three relations, we define four partial orders:

$$e \prec_1 f \quad \text{iff} \quad e \text{ precedes } f \text{ and } e \text{ is above } f$$
$$e \prec_2 f \quad \text{iff} \quad e \text{ precedes } f \text{ and } f \text{ is above } e$$
$$e \prec_3 f \quad \text{iff} \quad e \text{ is subsumed by } f \text{ and } e \text{ is above } f$$
$$e \prec_4 f \quad \text{iff} \quad e \text{ is subsumed by } f \text{ and } f \text{ is above } e$$

All four relations are transitive, so each defines a partial order. Moreover, for every pair of disjoint edges, at least one of the four orders compares them. If any of the orders contained a chain of k elements: $e_1 \prec_i e_2 \prec_i \cdots \prec_i e_k$, we have found k pairwise disjoint edges (since each relation is transitive). So none of the orders contains a chain of k elements, hence, by Dilworth's theorem, the set of edges E can be partitioned into $k-1$ sets $E_1^i, \ldots, E_{k-1}^i$ so that no two edges in the same set are comparable with respect to $\prec_i$, $1 \leq i \leq 4$. Call two edges equivalent, if they belong to the same partition set for each i. Then E is partitioned into at most $(k-1)^4$ equivalence classes, and no two equivalent edges are comparable. Since the result is true for $k = 2$, each equivalence class contains at most n edges, which gives us the required upper bound. $\quad\square$

Using topological sort and dynamic programming we can find a longest chain for each of the partial orders; this gives us an algorithm running in time $O(m^2)$ which finds $(m/n)^{1/4}$ pairwise disjoint edges.

5.4   Notes

The rectilinear crossing number is one of the oldest crossing number notions, first introduced in a paper by Harary and Hill [172]. As some of the results in this section show, it can often be easier to work with the pseudolinear crossing number instead. This is a relatively recent development, however, so not all results have been checked for if and how they transfer to the pseudolinear setting.

5.5   Exercises

(5.1) Show that Tutte's spring theorem, Theorem 5.2, implies Fáry's theorem, Theorem 5.1.

(5.2) The *Möbius ladder* M_n is a C_{2n} in which every two opposite vertices are connected by an edge. That is, if the vertices of C_n are $v_0, \ldots, v_{2n-1}$, we add all edges of the form $v_i v_j$ where $i - j \equiv 0 \bmod n$ and $i \neq j$. Determine the rectilinear crossing number of M_n.

(5.3) Show that every rotation system of a $K_{2,n}$ can be realized by a straight-line drawing.

(5.4) Show that the smallest number of crossing-free edges in a rectilinear drawing of a K_n is 5 for $n \geq 8$. *Hint:* If the outer face is a triangle, then there are at least six crossing-free edges.

(5.5) Show that an arrangement of n pseudolines splits the plane into $\binom{n}{2} + n + 1$ regions. *Note:* Equivalently, it splits the projective plane into $\binom{n}{2} + 1$ regions.

(5.6) Given a pseudoline arrangement and two points p and q which are not connected by a pseudoline of the arrangement, we can add a pseudoline to the arrangement which passes through p and q.

(5.7) In a pseudoline arrangement for which not all pseudolines pass through the same point, each pseudoline is incident to at least three triangular regions, where we consider regions in the projective plane (rather than the plane).

(5.8) Show that the number of weakly non-isomorphic rectilinear drawings of K_n is $2^{\Theta(n \log n)}$.

(5.9) Show that there are $\Omega(n^2)$ empty triangles in a $\sqrt{n} \times \sqrt{n}$ grid.

(5.10) Fill in the details of the proof of Theorem 5.18.

(5.11) Show that the existential theory of the reals is **NP**-hard. *Hint:* reduce from satisfiability; encode binary variables using real variables.

(5.12) Show that there is a $c > 0$ so that every n-vertex graph G has a $\overline{\text{cr}}$-minimal drawing on a grid of size at most $2^{2^{cn}}$.

(5.13) Sam Loyd created a puzzle called "When Drummers Meet" which asked how often n lines in the plane can cross [233]. To make it easier to give each drummer its route, let us require that the start and endpoint of each drummer has to be a grid-point on an $N \times N$-grid. Show that we can find routes for n drummers on a grid with $N = O(n^{1.5})$, assuming no more than two drummers can cross at any point. If we require the crossing points to be grid-points as well, how large does the grid have to be?

Chapter 6

The Rectilinear Crossing Number: Values and Bounds

6.1 A Look at the Complete Graph

6.1.1 Sylvester's Four Point Problem

Since $\overline{\mathrm{cr}}(K_5) = 1$, any five points in the plane always contain a convex quadrilateral, a result first proved by August Ferdinand Möbius [246, §255] and rediscovered by Esther Klein [168]. This fact points to a connection between the rectilinear crossing number, and Sylvester's Four Point Problem, first posed in 1864, which asks for the probability $q(R)$ that four randomly chosen points in an open region R of area 1 form a convex quadrilateral. It turns out that $q_* := \inf q(R)$, the smallest this probability can be, is the same as $\overline{\nu}^* := \lim_{n \to \infty} \overline{\mathrm{cr}}(K_n)/\binom{n}{4}$, the *rectilinear crossing constant*.[1]

Theorem 6.1 (Scheinerman and Wilf [304]). *We have $\overline{\nu}^* = q_*$.*

Proof. If we define $a_n := \overline{\mathrm{cr}}(K_n)/\binom{n}{4}$, then $(a_n)_{n \geq 4}$ is a monotonically increasing sequence, which is bounded from above, just as in the case of cr; we conclude that the limit of the sequence exists, call it $\overline{\nu}^*$.

Let R be a region of area 1. For any n points in this region, the complete geometric graph on these n points has at least $\overline{\mathrm{cr}}(K_n)$ crossings. Define a random variable X as the number of convex quadrilaterals formed by n randomly chosen points in R. Then $E(X) = \binom{n}{4}q(R)$. Since $E(X) \geq \overline{\mathrm{cr}}(K_n)$, this implies $q_* \geq \overline{\nu}^*$.

For the other direction, fix a $\overline{\mathrm{cr}}$-minimal drawing of K_n, and create a region R which consists of small disks of area $1/n$ centered at each vertex. By increasing the size of the drawing (but not the disks), we can assume that moving vertices within their disks does not change the drawing (up to isomorphism). Pick four points at random from R. If they belong to four different disks, then the probability that they form a convex quadrilateral is $\overline{\mathrm{cr}}(K_n)/\binom{n}{4}$. The probability that they do belong to four distinct disks is $(1-1/n)(1-2/n)(1-3/n)$, so $q_* \leq \overline{\mathrm{cr}}(K_n)/\binom{n}{4}(1-1/n)(1-2/n)(1-3/n) \to \overline{\nu}^*$ as $n \to \infty$. □

---

[1] The symbol $\overline{\nu}$ has often been used to refer to the rectilinear crossing number, based on the use of ν for the standard crossing number.

Bounding the value of $\overline{\nu}^*$ has driven much of the research on the rectilinear crossing number. The current best bounds are $0.379972 \leq 277/729 \leq \overline{\nu}^* \leq 0.380473$ (lower bound: [7], upper bound: [250]).

6.1.2   Upper Bounds

Several of the cr-minimal drawings of complete graphs we have produced were rectilinear, for example, the standard drawing of K_6. Below is a rectilinear drawing of K_9 with 36 crossings which shows that $\overline{cr}(K_9) = 36$, since we already know that $cr(K_9) \geq 36$.

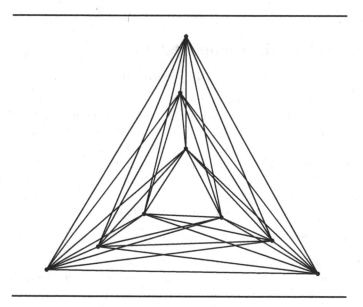

FIGURE 6.1: A $\overline{cr}$-minimal drawing of K_9 with 36 crossings (based on Guy [168].

Remark 6.2. Does $cr(K_9) = \overline{cr}(K_9)$ imply that cr and $\overline{cr}$ are equal for complete graphs on at most 9 vertices? It turns out that is not the case. A closer look at the drawing shows that all vertices have responsibility at most 17, so every induced drawing of K_8 has at least 19 crossings, which is the rectilinear crossing number of K_8 as we will prove in Theorem 6.12.

The drawing of K_9 is very suggestive: it consists of nested triangles, or, one can think of it as a single triangle on a line which has been rotated by $2\pi/3$ and $4\pi/3$. Can we get all $\overline{cr}$-minimal drawings this way? We analyze the underlying construction, first suggested by Singer [316].

Theorem 6.3 (Singer [316]). *For $n \geq 1$ we have*

$$\overline{cr}(K_{3n}) \leq 3\,\overline{cr}(K_n) + 3n\binom{n}{3} + 3\binom{n}{2}^2.$$

Proof. Let D be a $\overline{cr}$-minimal drawing of K_n. Scale the drawing in one direction so it becomes arbitrarily close to a line; take three copies of the scaled drawing and place them along three rays starting at the origin and forming angles or $2\pi/3$. Fill in the remaining edges. We count crossings by counting non-planar K_4s, each of which contributes a single crossing to the final drawing. There are $3\,\overline{cr}(K_n)$ non-planar K_4s with all vertices belonging to the same K_n, and $3\binom{n}{2}^2$ non-planar K_4s with two vertices in each of two of the K_n. This leaves the case that the K_4 has exactly three vertices in one of the K_n. Whether the resulting K_4 is planar or not then only depends on which of the other two K_n the fourth vertex belongs to, so there are $3n\binom{n}{3}$ non-planar K_4s like this, explaining the upper bound. □

For $n = 3$ this yields a bound of $9 + 27 = 36$ which is as in Figure 6.1. We can even obtain a $\overline{cr}$-minimal drawing of K_{10} from the K_9 drawing by adding the tenth vertex close to the middle of an edge of the innermost triangle, adding 26 crossings for a drawing of K_{10} with 62 crossings which is $\overline{cr}$-minimal (shown much later [75]). As Singer comments: "This drawing provides a clear warning about the asymmetric nature of minimal rectilinear drawings".

Asymptotically, Theorem 6.3 implies that $\overline{\nu}^* \leq 5/13 \leq 0.384616$ (Exercise (6.3)), so we already know that Singer's construction is not optimal. For K_{12} it yields a drawing with 156 crossings; the minimum is 153 as we will see shortly. The construction has been refined and modified many times over the years, different starting templates, and different replacement ideas have been tried (see [11] for a survey), but the unexpected outcome of all this has been that we do not yet have an analogue of Hill's conjecture for $\overline{cr}$. None of the upper bound methods suggested in the past have been optimal. This is stunning, since, compared to cr we have much more data on $\overline{cr}$. Table 6.1 shows how much we know about small complete graphs.

The reason for this success is a tool from discrete geometry, k-edges (also known as j-sets) which we will see in Section 6.1.3. It yields all the lower bounds in the table; for $n \leq 27$ see [7] (though some bounds were previously known), and for $n = 30$ [86]. While the upper bounds also benefitted significantly from the insights from discrete geometry, they also needed extensive computational work, and we may have reached a limit, see [17] for (descriptions of) rectilinear drawings realizing the upper bounds in the table.

One of the best available upper bound constructions is based on the idea of doubling each vertex; with certain assumptions on the initial drawing, this yields a highly competitive construction; for the result we borrow a notion from discrete geometry. A *halving line* ℓ for a set of points P, typically in general position, such as the vertices of a K_n, splits P in half in the sense that the number of points of P on the left and on the right differs by at most

n	$\overline{\mathrm{cr}}(K_n)$	$\mathrm{cr}(K_n)$		n	$\overline{\mathrm{cr}}(K_n)$		n	$\overline{\mathrm{cr}}(K_n)$
5	1	1		14	324		23	3077
6	3	3		15	447		24	3699
7	9	9		16	603		25	4430
8	19	18		17	798		26	5250
9	36	36		18	1029		27	6180
10	62	60		19	1318		28	?
11	102	100		20	1657		29	?
12	153	150		21	2055		30	9726
13	229	?		22	2528			

TABLE 6.1: The known values of $\overline{\mathrm{cr}}(K_n)$ and, for comparison, $\mathrm{cr}(K_n)$.

one (we do not count points on the line). If the halving line passes through two points of P and $|P|$ is even, then the number of points on each side of the line is the same. This is the case we are interested in here. We say P has a *halving-line matching* if we can assign to every $p \in P$ a halving line ℓ_p of P which passes through two points of P, including p, and $\ell_q \neq \ell_q$ for $p \neq q \in P$. Let us write p' for the second point of P on ℓ_p.

Lemma 6.4 (Ábrego and Fernández-Merchant [9]). *If n is even, and K_n has a rectilinear drawing D with a halving-line matching, then K_{2n} has a rectilinear drawing D' with a halving-line matching and so that*

$$\mathrm{cr}(D') = 16\,\mathrm{cr}(D) + \frac{1}{2}(n-1)n(2n-5).$$

Proof. Let $\varepsilon > 0$. For every vertex p in D pick two vertices p_1 and p_2 on ℓ_p at distance $\varepsilon > 0$ from p, so that p lies in the middle between p_1 and p_2, and the vertices occur in order $p_1 p p_2 p'$ on ℓ_p. We can choose ε small enough so that any line which intersects both $p_1 p_2$ and $p'_1 p'_2$ does not contain any vertices other than possibly p, p_1, p_2 and p', p'_1, p'_2 (the smaller ε, the closer such a line is to the halving-line ℓ_p in a compact region containing the drawing). Fix $\varepsilon > 0$ small enough so that it meets this constraint for all vertices p in D.

Let D' be the rectilinear drawing of K_{2n} on the resulting points. We need to show that D' has a halving-line matching, but this is easy. Consider the drawing of the K_4 graph on p_1, p_2, p'_1, p'_2. Then p_2 lies in the interior of the triangle $p_1 p'_1 p'_2$, so $p_2 p'_2$ is a halving-line, as is $p_1 p_2$. Choosing these two halving-lines for every p creates a halving-line matching.

We count crossings in D' by counting the number of non-planar K_4s in D'; we distinguish different types based on how many different points the vertices of the K_4 are associated with (we think of p_1 and p_2 as being associated with p), see Figure 6.2.

If the vertices of the K_4 belong to four different original points, then the K_4 is non-planar if and only if the K_4 on the original four points cause a crossing, which happens $\mathrm{cr}(D)$ many times; since there are 16 ways to choose

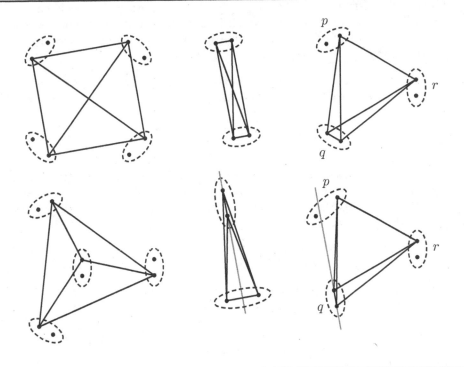

FIGURE 6.2: Types of crossings after doubling vertices in a K_n. *Left* Four endpoints. *Middle* Two endpoints; without halving line on top, and with halving line at bottom. *Right* Three endpoints; without halving line on top, and with halving line at bottom.

associated endpoints, this accounts for $16\,\mathrm{cr}(D)$ crossings in D'. If the vertices of the K_4 belong to two different original points, then the K_4 is non-planar if and only if the two points do not lie on a halving-line belonging to the matching. Since there are n such lines, these K_4 account for $\binom{n}{2} - n$ crossings. Finally, we have the case in which the endpoints of the K_4 belong to 3 different original points p, q and r, where two of the endpoints are associated with q. If either pq or qr is a halving line in the original matching, then two of the K_4 cause a crossing, accounting for $2n(n-2)$ crossings. Finally, if neither pq nor qr is a halving line, then the K_4 causes a crossing if p and r lie on the same side of ℓ_q, and there are four such K_4. Since ℓ_q is a halving line, there are $4n(2\binom{(n-2)/2}{2})$ such K_4. The total number of crossings then is $16\,\mathrm{cr}(D) +$ $\binom{n}{2} - n + 2n(n-2) + 4n(2\binom{(n-2)/2}{2})) = 16\,\mathrm{cr}(D) + \frac{1}{2}(n-1)n(2n-5)$. $\qquad\square$

The usual $\overline{\mathrm{cr}}$-minimal drawing of K_6 satisfies the conditions of Lemma 6.4. Since $\overline{\mathrm{cr}}(K_6) = 3$, we can immediately conclude that $\overline{\mathrm{cr}}(K_{12}) \leq 48+105 = 153$

which is the correct value. Repeating the procedure, we get a rectilinear drawing of K_{24} with 3702 crossings, which is only 3 above the minimum possible.

Let us see what impact the lemma has asymptotically, for the proof see Exercise (6.5).

Theorem 6.5 (Ábrego and Fernández-Merchant [9]). *Suppose there is a rectilinear drawing D of K_p with a halving-line matching. Then*

$$\overline{\mathrm{cr}}(K_n) \leq \frac{24\,\mathrm{cr}(D) + 3p^3 - 7p^2 + \frac{30}{7}p}{p^4}\binom{n}{4} + \Theta(n^3).$$

Theorem 6.5 used with $p = 6$ shows that $\overline{\nu}^* \leq 8/21 \leq 0.3809524$. Using an appropriate drawing of K_{30}, the bound can be improved to $\overline{\nu}^* \leq 29969/78750 \leq 0.380559$, which is pretty close to the current best bound of $\overline{\nu}^* \leq 0.380488$ [6].

The following theorem shows that it is no accident that the $\overline{\mathrm{cr}}$-minimal drawings we have seen in this section all have triangular outer faces.

Theorem 6.6 (Aichholzer, García, Orden, and Ramos [19]). *The outer face of a $\overline{\mathrm{cr}}$-minimal drawing of K_n is a triangle.*

It is tempting to conjecture that the second convex hull (the convex hull after removing the outer triangle) is a triangle as well, but this remains an open question.

For the proof we need some more observations on halving lines. Given a point p in a pointset P, there always is a halving line ℓ_p through p: start with an arbitrary line ℓ through p, and compute the difference between the number of points to the left and right of ℓ. Rotating ℓ changes that value by ± 2 when passing through a vertex. After a half-turn, a rotation of π radians, the difference has changed sign so at some point the difference must have been -1, 0, or 1, and we are done. If there are n vertices (including p), there are at least $\lceil n/2 \rceil - 1$ vertices on either side.

Proof. Fix a $\overline{\mathrm{cr}}$-minimal drawing of K_n and suppose there are two non-consecutive vertices p and q on its outer face. Pick halving lines ℓ_p and ℓ_q through p and q. Note that we can assume that ℓ_p and ℓ_q cross in the interior of the convex hull of the vertices of K_n: if not, they split the convex hull into three pieces, the middle one of which contains at most one point (since both lines are halving lines). We can then rotate the halving lines so they cross in the middle part and remain halving lines.

Note that moving p (and q) away from the convex hull along ℓ_p can only decrease the number of crossings: the crossing number of the drawing changes every time p crosses a line through two other points in the drawing, say a and b. Note that a and b must lie on the same side of ℓ_p (otherwise a line through them crosses ℓ_p inside the convex hull). Now moving p beyond the line through a and b is equivalent to moving a through pb (assuming the order of vertices is *pab* when they are collinear), but then we can calculate the change easily: a is incident to at least $\lceil n/2 \rceil - 1$ vertices beyond pb and at most $\lceil n/2 \rceil - 2$ vertices on its side (since a and b do not count), so the number of crossing changes by at least $\lceil n/2 \rceil - 2 - (\lceil n/2 \rceil - 1) = -1$, so it decreases. Now moving both p and q to a line parallel to pq and beyond the convex hull removes the vertex between them from the convex hull, while decreasing the crossing number of the drawing. Hence, the outer face is a triangle. $\qquad\square$

6.1.3   Lower Bounds and k-Edges

The amazing progress on lower bounds for $\overline{cr}(K_n)$ came through the use of k-edges. A *k-edge* in a rectilinear drawing D of K_n is an edge with exactly k points on one side (so $n - 2 - k$ points on the other side, we can always assume that no three points are collinear). Then $E_k(D)$ is defined as the number of k-edges in D. The following theorem establishes the connection between the rectilinear crossing number of a drawing, and the values E_k.

Theorem 6.7 (Ábrego and Fernández-Merchant [8]; Lovasz, Vesztergombi, Wagner, and Welzl [232]). *If D is a rectilinear drawing of K_n, then*

$$\overline{cr}(D) = 3\binom{n}{4} - \sum_{k=0}^{\lceil n/2 \rceil - 1} k(n - 2 - k)E_k(D).$$

Figure 6.3 shows a (non-minimal) drawing D with $(E_0, E_1, E_2) = (1, 9, 3)$. In this example $6 = \overline{cr}(D) = 45 - 3 \cdot 9 - 4 \cdot 3$.

Proof. Up to isomorphism (of the sphere) there are two good drawings of K_4: type $\triangle$ in which all regions are triangular, and type $\boxtimes$ in which there is a quadrangular region (in a rectilinear drawing the outer face). Type $\triangle$ is planar, type $\boxtimes$ contributes a single crossing to the drawing. Let $x_\triangle$ and $x_\boxtimes$ be the number of K_4s of the respective type. Then $\overline{cr}(D) = x_\boxtimes$ and $x_\triangle + x_\boxtimes = \binom{n}{4}$, the total number of K_4s.

We say an edge ab separates vertices c and d if c and d lie on opposite sides of the line through a and b. Then each K_4 of type $\triangle$ contributes 3 separations, while a K_4 of type $\boxtimes$ contributes 2 separations. On the other hand, a k-edge contributes $k(n - k - 2)$ separations, so

$$3x_\triangle + 2x_\boxtimes = \sum_{k=0}^{\lceil n/2 \rceil - 1} k(n - 2 - k)E_k(D).$$

Using $x_\triangle + x_\boxtimes = \binom{n}{4}$ and $\overline{cr}(D) = x_\boxtimes$, this implies

$$\overline{cr}(D) = 3\binom{n}{4} - \sum_{k=0}^{\lceil n/2 \rceil - 1} k(n - 2 - k)E_k(D),$$

which is what we had to show. $\qquad\square$

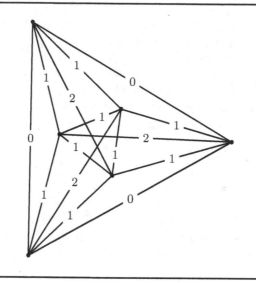

FIGURE 6.3: Non-minimal drawing of K_6 with triangular outer hull; k-edges are labeled by k.

Counting $E_k(D)$ directly is difficult; it is easier to work with $E_{\leq k}(D) := \sum_{i=0}^{k} E_k(D)$. The following lemma gives an easy bound on that number. An edge which has at most k edges on one side of it is known as a $(\leq k)$-*edge*.

Lemma 6.8 (Lovasz, Vesztergombi, Wagner, and Welzl [232]). *If D is a $\overline{\mathrm{cr}}$-minimal drawing of K_n and $k \leq \lceil n/2 \rceil - 1$, then*

$$E_{\leq k}(D) \geq 3\binom{k+2}{2}.$$

The result is tight for $k \leq \lfloor n/3 \rfloor - 1$: place $\lfloor n/3 \rfloor$ points along three rays separated by 120-degree angles; perturb them slightly, so that no line passing through two points close to the same ray separates two vertices on another ray.

There is a direct proof of the lemma using circular sequences; instead we give a somewhat simpler proof following an idea of Aichholzer, García, Orden, and Ramos [19]. Theorem 6.6 stated that the outer face of a $\overline{\mathrm{cr}}$-minimal drawing of K_n is a triangle. The redrawing technique in that proof can be used to establish a slightly stronger result.

Corollary 6.9 (Aichholzer, García, Orden, and Ramos [19]). *If D is a drawing of K_n in which the convex hull has more than three points, then D can be redrawn, without increasing $\overline{\mathrm{cr}}$, and so that $E_k(D)$ does not increase for any $k \leq \lfloor n/2 \rfloor - 1$.*

Proof. Let p, q, and ℓ_p, ℓ_q be as in the proof of Theorem 6.6. As in that proof, $\overline{cr}(D)$ and $E_k(D)$ do not change until we move p (or q) beyond a line through two points a and b. As before, both a and b must lie in the same side of ℓ_p. Since ℓ_p is a halving line, ab has at most $n/2 - 2$ points above it (it cannot have points above it on both sides of p for p to be on the convex hull). So moving p along ℓ_p may change ab from being a k-edge with $k \leq n/2 - 2$ to being a $k+1$-edge. So $E_k(D)$ decreases, but $E_{k+1}(D)$ remains the same. Since $k+1 \leq n/2 - 1$ is equivalent to $k \leq \lfloor n/2 \rfloor - 1$, this establishes the result. $\square$

Proof of Lemma 6.8. By Corollary 6.9, we can assume that the convex hull of the $\overline{cr}$-minimal drawing of K_n is a triangle. Then, $E_0(D) = 3$, proving the case $k = 0$. Let the three vertices be a, b, and c. Take a line through ab and rotate it around a towards c. This gives us a k-edge for every $k \leq n - 2$. Since we can repeat this for any of the six directed edges bounding the drawing, we get six k-edges for every $k \leq n - 2$ (each of them incident to a vertex on the outer face). This establishes the second base case, $E_{\leq 1} \geq 3 + 6 = 9$. Suppose $k \geq 2$. Remove the three vertices a, b, and c on the outer face. Any k-edge of the resulting drawing D' was a $k + 1$- or a $k + 2$-edge in D. Hence,

$$E_k(D) \geq E_{k-2}(D') + 6k \geq 3\binom{k}{2} + 6k = 3\binom{k+2}{2},$$

using induction. $\square$

Lemma 6.8 implies that Hill's conjecture holds for rectilinear drawings.

Theorem 6.10 (Ábrego and Fernández-Merchant [8]; Lovasz, Vesztergombi, Wagner, and Welzl [232]). *For $n \geq 1$ we have*

$$\overline{cr}(K_n) \geq Z(n).$$

Before we prove the theorem, we rewrite Theorem 6.7 in terms of $E_{\leq k}$.

Corollary 6.11 (Ábrego and Fernández-Merchant [8]; Lovasz, Vesztergombi, Wagner, and Welzl [232]). *If D is a rectilinear drawing of K_n, then*

$$\overline{cr}(D) = \sum_{k=0}^{\lceil n/2 \rceil - 2} (n - 2k - 3)E_{\leq k}(D) - \frac{1}{2}\binom{n}{2}\left(\left\lfloor \frac{n}{2} \right\rfloor - 1\right).$$

Proof. By Theorem 6.7 we know that

$$\overline{cr}(D) = 3\binom{n}{4} - \sum_{k=0}^{\lceil n/2 \rceil - 1} k(n - 2 - k)E_k(D).$$

We analyze the second term. Substituting $E_k(D) = E_{\leq k}(D) - E_{\leq k-1}(D)$, with $E_{\leq -1}(D) = 0$, we can rewrite

$$\sum_{k=0}^{\lceil n/2 \rceil - 1} k(n-2-k)E_k(D) = \sum_{k=0}^{\lceil n/2 \rceil - 1} k(n-2-k)E_{\leq k}(D) - E_{\leq k-1}(D)$$

$$= \left(\left\lceil \frac{n}{2} \right\rceil - 1 \right) \left(\left\lceil \frac{n}{2} \right\rceil - 1 \right) E_{\leq \lceil n/2 \rceil - 1}$$

$$+ \sum_{k=0}^{\lceil n/2 \rceil - 2} (k(n-2-k) - (k+1)(n-2-(k+1))) E_k(D)$$

$$= \left(\left\lceil \frac{n}{2} \right\rceil - 1 \right) \left(\left\lceil \frac{n}{2} \right\rceil - 1 \right) \binom{n}{2} - \sum_{k=0}^{\lceil n/2 \rceil - 2} (n-2k-3)E_k(D)$$

Also,

$$3\binom{n}{4} - \left(\left\lceil \frac{n}{2} \right\rceil - 1 \right) \left(\left\lceil \frac{n}{2} \right\rceil - 1 \right) \binom{n}{2} = -\frac{1}{2}\binom{n}{2}\left(\left\lceil \frac{n}{2} \right\rceil - 1 \right),$$

as can be verified by comparing the polynomials on both sides for n even and n odd. Using this in the original equation, yields the corollary. □

Proof of Theorem 6.10. Fix a $\overline{\mathrm{cr}}$-minimal drawing D of K_n. By Corollary 6.11, we have

$$\overline{\mathrm{cr}}(D) \geq \sum_{k=0}^{\lceil n/2 \rceil - 2} (n-2k-3)E_{\leq k}(D) - \frac{1}{2}\binom{n}{2}\left(\left\lfloor \frac{n}{2} \right\rfloor - 1 \right)$$

$$\geq \sum_{k=0}^{\lceil n/2 \rceil - 2} (n-2k-3)3\binom{k+2}{2} - \frac{1}{2}\binom{n}{2}\left(\left\lfloor \frac{n}{2} \right\rfloor - 1 \right).$$

If we calculate $\sum_{k=0}^{\lceil n/2 \rceil - 2}(n - 2k - 3)3\binom{k+2}{2} - \frac{1}{2}\binom{n}{2}\left(\left\lfloor \frac{n}{2} \right\rfloor - 1\right)$ for even n, we get $\frac{1}{64}n(n-2)(n^2 - 6n + 8) = Z(n)$; for odd n we get $\frac{1}{64}(n-1)(n((n-7)n + 7) - 9) + \frac{1}{4}\binom{n}{2} = \frac{1}{64}(n-1)(n-1)(n-3)(n-3) = Z(n)$. □

Theorem 6.10 invites two readings: as a lower bound on the rectilinear crossing number of K_n, in which case one may ask whether the analysis can be strengthened to give stronger lower bounds, a question to which we know the answer to be a resounding yes at this point, discussed below, but one can also see it as a lower bound technique for the crossing number under some drawing restrictions, in which case one will want to investigate whether we can achieve similar bounds for other drawing types than rectilinear drawings. It was known from the start that pseudolinear would be sufficient for many of the bounds, but we will later discuss the notion of bishellability, which is weaker than several other drawing notions.

For a reasonably recent review of progress on $\overline{\mathrm{cr}}(K_n)$, see [11]. The main line of attack consisted in improving the lower bound on k-edges in Lemma 6.8; since the bound is optimal for $k \leq \lfloor n/3 \rfloor - 1$ this had to be done for larger k, which is naturally harder, since we are analyzing edges deeper in the graph. Theorem 6.10 implies that the rectilinear crossing constant $\overline{\nu}^*$ is at least $3/8 = 0.375$. The current best lower bound, combining results from various papers, is $\overline{\nu}^* \geq 277/729 > 0.37997$, see [11].

With the k-edge machinery in place we can now give relatively short and combinatorial proofs of lower bounds, below is one of the simplest examples, and an important one, since it shows that $\overline{\mathrm{cr}}$ and cr differ for complete graphs, and do so as early as K_8 for which $\mathrm{cr}(K_8) = 18$, as we saw in Chapter 1.

Theorem 6.12 (Barton [49]; Singer [316]). *We have*

$$\overline{\mathrm{cr}}(K_8) = 19.$$

Proof. Let D be a $\overline{\mathrm{cr}}$-minimal drawing of K_8, and let $c_i := E_i(D)$ be the number of i-edges in D. Clearly, $c_0 + c_1 + c_2 + c_3 = \binom{8}{4} = 28$. We know that $c_0 \geq 3$, $c_1 \geq 6$, $c_2 \geq 9$, by Lemma 6.8 and

$$\overline{\mathrm{cr}}(D) = 210 - 5c_1 - 8c_2 - 9c_3,$$

by Theorem 6.7. So we can view this as a constrained minimization problem; the unique minimal solution to this problem is $(c_0, c_1, c_2, c_3) = (3, 6, 9, 10)$, which gives a value of 18; we argue that this vector of values is not realizable in a rectilinear drawing. Note that $c_0 = 3$, so we know that the convex hull of D is a triangle. Each of the vertices of this triangle can be incident to at most one 3-edge, so there must be at least $c_3 - 3 = 7$ many 3-edges that belong to the K_5 lying within the triangle. Every edge on the convex hull of that K_5 must be a 1- or 2-edge in D, and since we have already accounted for all 1-edges, these edges must be 2-edges. As the K_5 has only 10 edges, its convex hull must be a triangle of 2-edges, and the seven edges contained within that triangle are all 3-edges. Let e be the edge within the interior of the 2-edge triangle. At least two of the vertices of the triangle must be on the same side of that edge, so we have the substructure pictured in Figure 6.4.

We argue that this structure cannot be realized. Apart from the solid black vertices shown, there can be only one more vertex below e, so since there have to be at least three vertices below f_2 and f_3, there must be at least one vertex each in the regions labeled I. Since f_1 then has at least 3 vertices on its left, and f_4 at least 3 vertices on its right, there must be two vertices which are both to the left of f_1 and to the right of f_4, which is region II in the figure, but if that is the case, then e is not a 3-edge.

Finally, Figure 6.5 shows a drawing of K_8 with 19 crossings, establishing the upper bound. It shows that the corresponding c-vector is $(3, 6, 10, 9)$. $\qquad\square$

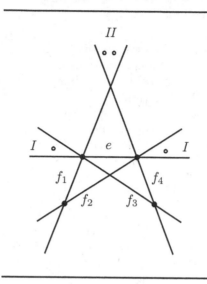

FIGURE 6.4: Non-realizable substructure.

6.1.4 Pseudolinear Drawings and Bishellability

Our work in Section 6.1.3 suggests the question whether we can obtain the lower bound $\mathrm{cr}_D(K_n) \geq Z(n)$ for drawings D which satisfy weaker properties than being rectilinear. We work with the notion of bishellability (simplifying the original work using shellability), introduced in [2].

Definition 6.13. We say a good drawing D of K_n is *s-bishellable* with respect to a reference region R if there are disjoint sequences $a_0, \ldots, a_s$ and $b_0, \ldots, b_s$ of vertices so that a_i is incident on the region of $D - \{a_0, \ldots, a_{i-1}\}$ containing R, and b_i is incident on the region of $D - \{b_0, \ldots, b_{i-1}\}$ containing R, for all $0 \leq i \leq s$.

By definition, $(s+1)$-bishellability implies s-shellability, so bishellability is monotone (this is an advantage over shellability). The reason we care about bishellability is the following result.

Theorem 6.14 (Ábrego et al. [5]; Ábrego et al. [2]). *If K_n has an $(\lfloor n/2 \rfloor - 2)$-bishellable drawing D, then $\mathrm{cr}(D) \geq Z(n)$.*

If we use *bishellable* to mean $(\lfloor n/2 \rfloor - 2)$-bishellable, then the theorem states that every bishellable drawing of K_n has at least $Z(n)$ crossings.

Before we prove the result, let us see some easy consequences.

Lemma 6.15. *If a drawing of K_n contains a cycle $v_0 \ldots v_{t-1}$ so that $v_0 v_{t-1}$ is free of crossings, and any crossing with $v_k v_{k+1}$, $0 \leq k < t-1$ must be with edges $v_i v_j$ where $i < k$ and $j > k+1$, then the drawing is $(\lfloor t/2 \rfloor - 1)$-bishellable.*

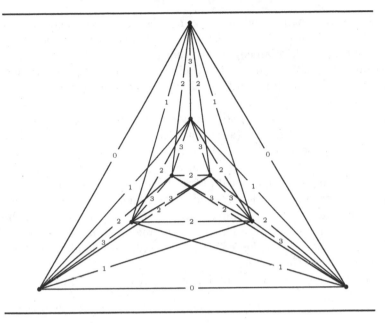

FIGURE 6.5: A $\overline{cr}$-minimal drawing of K_8 with labeled k-edges.

Proof. Let R be a region incident to $v_0 v_{t-1}$, and let $a_i := v_i$ and $b_i := v_{t-1-i}$, for $0 \le i \le \lceil t/2 \rceil - 1$; the two sequences are disjoint, and they witness that the drawing is $(\lfloor t/2 \rfloor - 1)$, since the curve corresponding to the path $a_0 \cdots a_i$ is crossing-free in $D - \{a_0, \ldots, a_i\}$, so R and a_i belong to the same region, and similarly for the b_i. $\qquad\square$

A drawing D of a graph is *x-bounded* if every point on an edge lies between the endpoints of the edge with respect to its x-coordinates.

Lemma 6.16 (Ábrego et al. [5]). *Every x-bounded drawing of K_n has at least $Z(n)$ crossings.*

Proof. Fix a cr-minimal x-bounded drawing of K_n. Label the vertices of the drawing $v_0, \ldots, v_{n-1}$ as they occur, from left to right, in the drawing. Then $v_0 v_{n-1}$ is clear of crossings (if it is not, we could redraw it without crossings by rerouting it below the rest of the drawing), and the cycle $v_0 \cdots v_{n-1}$ satisfies the conditions of Lemma 6.15, which implies the lower bound. $\qquad\square$

Theorem 6.14, together with the lemma, gives us a new proof that $\overline{cr}(K_n) \ge Z(n)$, but it also shows that $\widetilde{cr}(K_n) \ge Z(n)$, since every pseudolinear drawing is isomorphic to a wiring diagram, by Lemma 5.10, which is x-bounded.

We will see further applications in later chapters. It is time to take up the proof of the core result. The proof will work with k-edges again, but since we are dealing with general good drawings now, we need to extend the definition. Actually, we only need to show how to distinguish between points lying on the left and right side of an edge; with that notion as a starting point, k-edge, $E_k(D)$, and $E_{\leq k}(D)$ are defined as before, e.g., a k-edge is an edge which has k vertices on one side, and $n - 2 - k$ vertices on the other side, where $0 \leq k \leq \lfloor n/2 \rfloor$.

Let uv be an edge in a good drawing D of K_n, and let w be a third point. Then uvw is a triangle (adjacent edges do not cross in a good drawing, and there are no self-crossings). To be precise, there are two triangles; when dealing with rectilinear drawings we always picked the bounded triangle, the one not containing the unbounded region. We create some more flexibility now by choosing a reference region R among the regions of $\mathbb{R}^2 - D$ (which could be the unbounded region, of course). When speaking of the region bounded by the triangle uvw, we are referring to the region which is disjoint from R. We then say that w *is on the left of* uv if the region bounded by the triangle uvw touches uv on the left.

With this, we can now restate Theorem 6.7, the proof carries over.

Theorem 6.17 (Ábrego et al. [3]). *If D is a good drawing of K_n, then*

$$\operatorname{cr}(D) = 3\binom{n}{4} - \sum_{k=0}^{\lceil n/2 \rceil - 1} k(n - 2 - k)E_k(D).$$

As a consequence, we can derive an extension of Corollary 6.11, since the proof is an algebraic manipulation of the formula in the theorem.

Corollary 6.18. *If D is a good drawing of K_n, then*

$$\operatorname{cr}(D) = \sum_{k=0}^{\lceil n/2 \rceil - 2} (n - 2k - 3)E_{\leq k}(D) - \frac{1}{2}\binom{n}{2}\left(\left\lfloor \frac{n}{2} \right\rfloor - 1\right).$$

To capture bishellability, we need a smoothed version of $E_{\leq k}(D)$, which we call $E_{\leq\leq k}(D)$ and define as $\sum_{i=0}^{k} E_{\leq k}(D)$. It has the same advantage over $E_{\leq k}(D)$ that $E_{\leq k}(D)$ has over $E_k(D)$: it is less susceptible to small changes in the drawing. We need a version of Corollary 6.18 for $E_{\leq\leq k}$.

Theorem 6.19. *If D is a good drawing of K_n, then*

$$\operatorname{cr}(D) = 2\sum_{k=0}^{\lfloor n/2 \rfloor - 3} E_{\leq\leq k}(D) - \frac{1}{2}\binom{n}{2}\left(\left\lfloor \frac{n}{2} \right\rfloor - 1\right) + c_n E_{\leq\leq \lfloor n/2 \rfloor - 2},$$

where $c_n = 2$ if n is odd, and 1 otherwise.

Proof. By Corollary 6.18 we have

$$\operatorname{cr}(D) = \sum_{k=0}^{\lceil n/2 \rceil - 2} (n - 2k - 3)E_{\leq k}(D) - \frac{1}{2}\binom{n}{2}\left(\left\lfloor \frac{n}{2} \right\rfloor - 1\right))$$

Substituting $E_{\leq k}(D) = E_{\leq\leq k}(D) - E_{\leq\leq k-1}(D)$ into the first term $\sum_{k=0}^{\lceil n/2\rceil-2}(n-2k-3)E_{\leq k}(D)$, gives us

$$2\sum_{k=0}^{\lceil n/2\rceil-3} E_{\leq\leq k}(D) + c_n E_{\leq\leq\lfloor n/2\rfloor-2},$$

where c_n is 2 if n is odd, and 1 otherwise, and the theorem follows. $\qquad\square$

Lemma 6.20 (Ábrego et al. [2]). *If D is a k-bishellable drawing of K_n, where $0 \leq k \leq \lfloor n/2\rfloor - 2$, then*

$$E_{\leq\leq k}(D) \geq 3\binom{k+3}{3}.$$

Before we show the lemma, we use it to complete the proof of the main result.

Proof of Theorem 6.14. Combining Theorem 6.19 and Lemma 6.20 we have

$$\operatorname{cr}(D) \geq 2\sum_{k=0}^{\lfloor n/2\rfloor-3} 3\binom{k+3}{3} - \frac{1}{2}\binom{n}{2}\left(\left\lfloor\frac{n}{2}\right\rfloor - 1\right) + c_n 3\binom{\lfloor\frac{n}{2}\rfloor+1}{3},$$

where c_n is 2 if n is odd, and 1 otherwise. Applying $\sum_{k=0}^{m}\binom{k+3}{3} = \binom{m+4}{4}$ with $m = \lfloor n/2\rfloor - 3$ we get

$$\operatorname{cr}(D) \geq 6\binom{\lfloor\frac{n}{2}\rfloor+1}{4} - \frac{1}{2}\binom{n}{2}\left(\left\lfloor\frac{n}{2}\right\rfloor - 1\right) + c_n 3\binom{\lfloor\frac{n}{2}\rfloor+1}{3}$$

If n is odd, then the right-hand side turns into $\frac{1}{64}(n-1)^2(n-3)^2 = Z(n)$; if n is even, it becomes $\frac{1}{64}n(n-1)^2(n-4) = Z(n)$. So both cases require $Z(n)$ crossings. $\qquad\square$

We return to the proof of the central lemma.

Proof of Lemma 6.20. Let D be a k-bishellable drawing of K_n with respect to a reference region R, and let $a_0,\ldots,a_k$ and $b_0,\ldots,b_k$ be the witnesses of bishellability. Let e be an edge which occurs (possibly partially) on the boundary of R. By definition of being on a side of e—with respect to R—, all other vertices are on the same side of e, so e is a 0-edge, and since R must be incident to at least three different edges, $E_{\leq\leq 0} = E_0(D) \geq 3$, which implies the base case $k = 0$ of the lemma. So we can assume that $k > 0$. The drawing $D' = D - \{a_0\}$ of $K_n - \{a_0\}$ is $(k-1)$-bishellable, as witnessed by $a_1,\ldots,a_k$ and $b_0,\ldots,b_{k-1}$, so by induction $E_{\leq\leq k-1}(D') \geq 3\binom{(k-1)+3}{3}$. By definition, $E_{\leq\leq k-1}(D') = \sum_{i=0}^{k-1} E_{\leq i}(D') = \sum_{i=0}^{k-1}(k-i)E_i(D')$, so

$$\sum_{i=0}^{k-1}(k-i)E_i(D') \geq 3\binom{k+2}{3}.$$

An i-edge e in D' is an i-edge or an $(i+1)$-edge in D, depending on which side of e lies a_0. We name these two types of edges *invariant* and *variant*. Then $(\leq k)$-edges in D fall into three categories: incident to a_0, invariant, and variant. We claim that

(i) an invariant i-edge contributes $k-i$ to both $E_{\leq\leq k-1}(D')$ and $E_{\leq\leq k}(D)$, while a variant i-edge contributes $k-i$ to $E_{\leq\leq k-1}(D')$ and $k-i+1$ to $E_{\leq\leq k}(D)$,

(ii) there are $\binom{k+2}{2}$ invariant $(\leq k)$-edges, and

(iii) edges incident to a_0 contribute at least $2\binom{k+2}{2}$ to $E_{\leq\leq k}(D)$.

Assuming the truth of the claims, we conclude

$$
\begin{aligned}
E_{\leq\leq k}(D) &\geq E_{\leq\leq k-1}(D') + \binom{k+2}{2} + 2\binom{k+2}{2} \\
&\geq 3\binom{k+2}{3} + \binom{k+2}{2} + 2\binom{k+2}{2} \\
&= 3\binom{k+3}{3},
\end{aligned}
$$

which is what we had to show. We are left with the proofs of the three claims. Claim (i) follows immediately from comparing the coefficient of E_i in $E_{\leq\leq k-1}(D') = \sum_{i=0}^{k-1}(k-i)E_i(D')$ to the coefficients of E_i (invariant edges) and E_{i+1} (variant edges) in $E_{\leq\leq k}(D) = \sum_{i=0}^{k}(k+1-i)E_i(D)$. For claim (ii) we show that b_i is incident to at least $k-i$ invariant edges. We start with $i=0$. Let e_0 be one of the two edges at b_0 incident to the reference region R, and let $e_1, \ldots e_k$ be the next k consecutive edges in the rotation away from R. We have two choices for e_0, and, since b_0 is incident to $n-1 > 2(k+1)$ edges, the two resulting groups of edges are disjoint, so we can assume that the group we picked does not contain $b_0 a_0$. On one of its sides, e_i can only have the endpoints of $e_{i'}$ with $i' < i$ other than b_0, and a_0 is on its other side, so e_i is an invariant j-edge for some $j \leq i$, giving us $k+1$ invariant $(\leq k)$-edges. For $i > 0$ we make a similar argument: b_i is incident to at least $k+1-i$ invariant $(\leq k-i)$-edges in $D - \{b_0, \ldots, b_{i-1}\}$. Adding back $b_0, \ldots, b_{i-1}$ does not change that the edges are invariant, but, depending on which side of an edge they go on, may turn $(\leq k-i)$-edges into $(\leq k-i+i)$-edges, so $(\leq k)$-edges, which is what we are counting. In total we then have $\sum_{i=0}^{k} i+1 = \binom{k+2}{2}$.

To see claim (iii), label $2(k+1)$ consecutive edges at a_0 as $f_k, \ldots, f_0, f_0', \ldots, f_k'$, where f_0 and f_0' are both incident to R at a_0. As we argued above (for edges incident to b_0), each f_i is a j-edge for some $j \leq i$, so it contributes $k+1-j \geq k+1-i$ to $E_{\leq\leq k}(D) = \sum_{i=0}^{k}(k+1-i)E_i(D)$. Therefore, $f_0, \ldots, f_k$ contribute at least $\sum_{i=0}^{k}(k+1-i) = \binom{k+2}{2}$ to $E_{\leq\leq k}(D)$, as do the f_i'; since the two groups are disjoint $(n-1 > 2(k+1))$, this completes the proof of claim (iii). $\qquad\square$

Bishellability is strictly stronger than cr-minimality; for example, Figure 6.6 shows a cr-minimal drawing of K_9 which is not 0-bishellable (example from [2]), since there are no crossing-free edges.

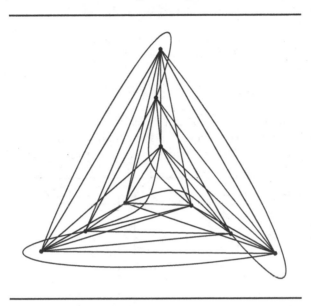

FIGURE 6.6: A cr-minimal drawing of K_9 which is not 0-bishellable (based on [2]).

6.2 Other Graphs

Apart from the deep and extensive work on $\overline{cr}(K_n)$, our knowledge of $\overline{cr}$ remains quite limited. Since the Zarankiewicz drawings of $K_{m,n}$ showing $cr(K_{m,n}) \leq Z(m,n)$ are rectilinear drawings, it is conjectured that $\overline{cr}(K_{m,n}) = cr(K_{m,n})$. Reportedly, there is work by Norin and Zwols using Razborov's flag algebras showing that $\lim_{n \leftrightarrow \infty} \overline{cr}(K_{n,n})/Z(n,n) > 0.905$, but this remains unpublished [329]. Gethner et al. [149] uses the same method to show that $\lim_{n \leftrightarrow \infty} \overline{cr}(K_{n,n,n})/Z(n,n,n) > 0.973$, where $Z(m,n,k)$ is the upper bound on $cr(K_{m,n,k})$ we discussed in Section 2.2.3. The authors' conjecture that $\overline{cr}(K_{m,n,k}) = Z(m,n,k)$ and $\overline{cr}(K_{m,n,k}) = cr(K_{m,n,k})$. While it is tempting to conjecture that $\overline{cr}$ and cr are the same for all complete k-partite graphs with $k \geq 2$ we know this fails for $k = 8$, since K_8 is an 8-partite graph.

Question 6.21. What is the smallest k so that $cr(G) < \overline{cr}(G)$ for a k-partite graph G?

Similarly, the conjecture by Harary, Kainen, and Schwenk that $\mathrm{cr}(C_m \Box C_n) = (m-2)n$ for $m \leq n$ implies that $\overline{\mathrm{cr}}(C_m \Box C_n) = (m-2)n$, since the upper bounds can be achieved by rectilinear drawings. It may be easier to obtain lower bounds on $\overline{\mathrm{cr}}(C_m \Box C_n)$, but we are not aware of any work in that direction.

Theorem 6.22 (Faria, de Figueiredo, Richter, and Vrt'o [130]).

$$\overline{\mathrm{cr}}(Q_n) \leq \frac{47}{256} 4^n + 2^{n-8}(-32n^2 - 32n - 64).$$

Proof. We start with a simpler bound which illustrates the surprisingly simple construction. Start with the drawing D_2 of Q_2 shown on the left in Figure 6.7. Make a copy of D_2 and move it slightly to the right and below the original one

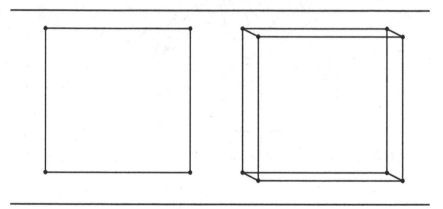

FIGURE 6.7: *Left:* Drawing of Q_2. *Right:* Derived drawing of Q_3.

(so that no vertices come to lie on an edge), and connect duplicate vertices by straight-line segments. The resulting drawing D_3 is shown on the right of the figure (it is suboptimal, having two crossings). There are two types of crossings: crossings corresponding to original crossings (of which there are none in our example), and crossings close to vertices (two in our example). Each crossing in the original drawing results in four crossings in the new drawing. The other type of crossing is harder to analyze; say a vertex is of type a if it is incident to a edges leaving it upwards, and let c be the vector counting how many vertices there are of each type as a ranges from 0 to n. This vector is $(1, 2, 1)$ for D_2 and $(1, 3, 3, 1)$ for D_3. If the initial vector is c, then the vector after the construction is $(0, c) + (c, 0)$, which implies that the entries for D_n are the binomial coefficients, so there are $\binom{n}{a}$ vertices of type a in D_n. With this observation, we can count the crossings close to vertices in D_n: a vertex of type a in D_{n-1} is incident to a edges above and $(n-1) - a$ edges below; by duplicating it in D_n we cause $\binom{a}{2} + \binom{n-a}{2}$ crossings between the edges leaving the duplicate vertices. Hence,

$$
\begin{aligned}
\operatorname{cr}(D_n) &= 4\operatorname{cr}(D_{n-1}) + \sum_{a=0}^{n-1} \binom{n-1}{a}\left(\binom{a}{2} + \binom{(n-1)-a}{2}\right) \\
&= 4\operatorname{cr}(D_{n-1}) + 2\sum_{a=0}^{n-1} \binom{n-1}{a}\binom{a}{2} \\
&= 4\operatorname{cr}(D_{n-1}) + 2^{n-3}(n^2 - 3n + 2),
\end{aligned}
$$

and $\operatorname{cr}(D_2) = 0$. The solution to this recursion is $\operatorname{cr}(D_n) = 2^{n-3}(2^{n+1} - n^2 - n - 2) = \frac{1}{4}4^n + o(2^n)$. Like any sub-optimal construction, this result depends on the choice of the initial drawing; for example, there is a planar rectilinear drawing of Q_3 with $c = (1, 3, 3, 1)$ leading to the same recursion with starting point $\operatorname{cr}(D_3) = 0$. This yields a bound of $\operatorname{cr}(D_n) = 2^{n-5}(72^n - 4n^2 - 4n - 8) = \frac{7}{32}4^n + o(2^n)$. The currently best bound is achieved by a drawing of Q_5 with 60 crossings and $c = (1, 5, 10, 10, 5, 1)$ with solution $\operatorname{cr}(D_n) = 2^{n-8}(472^n - 32n^2 - 32n - 64) = \frac{47}{256}4^n + o(2^n)$. The drawing can be found in [130].   □

6.3   Notes

For an entertaining history of Sylvester's Four-Point problems, mostly focussing on the version for convex regions, see [280]. Pegg and Exoo [274] catalogued small $\overline{\operatorname{cr}}$-critical graphs. For a comprehensive survey of the amazing recent progress on $\overline{\operatorname{cr}}(K_n)$, see the survey by Ábrego, Fernández-Merchant, and Salazar [11].

6.4   Exercises

(6.1) We saw that the rectilinear crossing constant equals $\min q(R)$, where the minimum is over all open regions of area 1. Does the result remain true if we require R to be simply connected (connected and no holes)?

(6.2) Let $\vartheta(G)$ be the number of pairs of independent edges in G (this is the so-called thrackle bound, we will meet it again in Chapter 12). Then $\overline{\operatorname{cr}}(G) \leq \overline{\nu}^*\vartheta(G)/3$, where $\overline{\nu}^*$ is the rectilinear crossing constant.

(6.3) Show that Theorem 6.3 implies that $\overline{\nu}^* \leq 5/13$.

(6.4) Show that K_n has a polygonal drawing with $Z(n)$ crossings in which every edge has at most one bend. *Note:* We do not know whether this is true for cr-minimal drawings.

(6.5) Work out the details of the proof of Theorem 6.5. *Hint:* Keep doubling until you are in the right range.

(6.6) Show that the graph in Figure 6.6 is 0-bishellable, but not 1-bishellable.

(6.7) Call a drawing of a graph *cylindrical* if all vertices of the graph can be placed on two concentric circles so that neither circle is crossed by any edge. The *cylindrical crossing number*, $cr_\odot$, of the graph is the smallest number of crossings in a cylindrical drawing of the graph. Show that $cr_\odot \geq Z(n)$. *Note:* The polygonal drawings of K_n we introduced in Chapter 1 are cylindrical.

(6.8) Show that $cr(Q_n) = \frac{7}{32}4^n + o(2^n)$. *Hint:* Theorem 6.22 contains a hint.

Chapter 7

The Local Crossing Number

7.1 Local Crossings

7.1.1 Simple, or Not?

So far we have counted crossings in a drawing globally; in this chapter we change perspective, and count crossings locally, along each edge.

Definition 7.1. The *local crossing number of a drawing* is the largest number of crossings along any edge of the drawing. The *local crossing number* of G, $\mathrm{lcr}(G)$, is the smallest local crossing number of any drawing of G.

Traditionally, the local crossing number was restricted to good drawings only; to distinguish this variant, we write $\mathrm{lcr}^*(G)$ for the *simple local crossing number* which is the smallest local crossing number of any good drawing of G.[1] Up to three crossings per edge, these two notions agree.

Lemma 7.2 (Pach, Radoičić, Tardos, and Tóth [256]). *We have* $\mathrm{lcr}(G) = \mathrm{lcr}^*(G)$ *for* $\mathrm{lcr}(G) \leq 3$. *For* $\mathrm{lcr}(G) = 4$, *there are examples of graphs for which any* lcr-*minimal drawing requires adjacent edges to cross.*

Proof. Fix a drawing of G realizing $\mathrm{lcr}(G) \leq 3$. By definition $\mathrm{lcr}(G) \leq \mathrm{lcr}^*(G)$, so it is sufficient to show that if the drawing is not good, then it can be made good without increasing its local crossing number. We can remove any self-crossing of an edge without increasing lcr or the drawing. So if the drawing is not good, there must be two edges e and f which intersect at least twice, possibly including a common endpoint. We can swap the arcs of e and f connecting the two intersections as we have done before (Figure 1.4). This redrawing removes at least one crossing between e and f and does not change the number of crossings along any other edge. The only reason the swap could fail is if it increases the local crossing number of the drawing. For that to happen the difference between the number of crossings on the two arcs must have been at least two (if it had been at most one, the loss of a crossing in the swap would have compensated for it). Since each edge is involved in at most three crossings, and e and f had at least one crossing, once arc must have

---

[1]Good drawings are sometimes called simple topological graphs.

157

contained two, and the other no crossings. But then we can reroute the arc with two crossings along the arc with no crossings, again removing at least one crossing between e and f. The redrawing may introduce self-crossings of edges, but those can be removed locally as before (Figure 1.3). All our redrawings reduced the crossing number of the drawing, so if we start with a cr-minimal drawing realizing $\text{lcr}(G) \leq 3$, it satisfies $\text{lcr}(G) = \text{lcr}^*(G)$.

For $\text{lcr}(G) = 4$ a hole opens up in the argument: it is possible that one of the arcs has one crossing, and the other three. This set-up can be turned into an example in which two adjacent edges cross (Exercise (7.1)). Our argument still shows that two *independent edges* cross at most once in this case. A counterexample to this requires $\text{lcr}(G) = 5$ and can be obtained from the counterexample for adjacent edges for $\text{lcr}(G) = 4$ (Exercise (7.2)).   □

It is open, whether lcr^* can be bounded in lcr, even for $\text{lcr}(G) = 4$, or whether $\text{lcr}^*(G)$ can be arbitrarily large in this case.

7.1.2   Density of Graphs with Few Local Crossings

Graphs with small local crossing numbers cannot be very dense; for $\text{lcr}(G) \leq 2$ this is made precise in the following theorem.

Theorem 7.3 (Schumacher [306, 307]). *If G is a graph with $\text{lcr}_\Sigma(G) \leq 2$ on a surface Σ with Euler genus eg, then*

$$m \leq (\text{lcr}_\Sigma(G) + 3)(n + \text{eg} - 2),$$

where $m = |E(G)|$, $n = |V(G)|$.

The bounds are tight; for the sphere, $\text{eg} = 0$, this follows from the following examples: $\text{lcr}(K_4) = 0$ and $m = 6 = 3(4 - 2)$; if G is the hypercube Q_3 with all diagonals in its 4-faces, then $\text{lcr}(G) = 1$, and $m = 24 = 4(8 - 2)$; finally, if G is the dodecahedron with all diagonals in its 5-faces, then $\text{lcr}(G) = 2$, and $m = 30 + 12 \cdot 5 = 90 = 5(20 - 2)$.

Proof. Fix a drawing D of G on Σ with $\text{lcr}(D) \leq 2$. Lemma 7.2 allows us to assume that D is good (the proof of that lemma works on any surface). If we can insert a crossing-free edge into D, we do so, even if this creates multiple edges between two vertices, as long as we do not create two edges bounding an *empty disk* (a disk-homeomorph that contains no part of the graph in its interior). This process terminates after a finite number of steps (view each crossing as a vertex, then the Euler-Poincare formula, Corollary A.17, gives us an upper bound on the number of edges; note that all faces have size at least 3). We use G and D for the resulting multigraph and its drawing. Since we added edges, it is sufficient to prove the upper bound for them.

If $\text{lcr}(D) = 0$, then the Euler-Poincare formula implies that $m \leq 3(n + \text{eg} - 2)$, which is what we had to show.

Next, let $\text{lcr}(D) = 1$. Suppose there are two edges ux and uy, consecutive at u, so that both ux and uy are involved in a crossing. Starting at u between ux and uy, we follow ux up to its crossing with another edge vw, and then continue along vw (without crossing it) to one of its endpoints, say w. See the left half of Figure 7.1. This shows that we can add a crossing-free edge uw

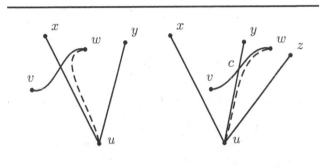

FIGURE 7.1: *Left:* Two consecutive edges ux and uy involved in crossings. *Right:* Three consecutive edges ux, uy, and uz involved in crossings.

to the drawing; the only reason we would not have done so is if uw forms an empty disk with another edge, but that must be ux or uy, both of which have a crossing, so they cannot form an empty disk with uw, which is a contradiction. We conclude that no two consecutive edges at a vertex are both involved in crossings. Let G' be the graph obtained from G by removing all k edges involved in crossings. Then $n' = |V(G')| = n$ and $m' = |E(G')| = m - k$, and G' has an embedding D' in Σ. By our earlier argument, a 3-face contains no crossing edges, a 4-face two crossing edges, and there are no i-faces for $i \geq 5$, since any pair of crossing edges is enclosed in a 4-face, and all other faces were triangulated by adding edges. So letting f_i' be the number of i-faces in D', and $f' = \sum_{i=3}^{\infty} f_i'$, we have

$$
\begin{aligned}
2m &\leq 3f_3' + (4 + 2 \cdot 2)f_4' = 3f_3' + 8f_4' \\
&\leq 4\sum_{i=3}^{\infty} if_i' - 8\sum_{i=3}^{\infty} f_i' \\
&\leq 8m' - 8f'.
\end{aligned}
$$

Using $n - m' + f' = 2 - \text{eg}$, that is $f' = 2 - \text{eg} - n + m'$, we obtain $2m \leq 8(n + \text{eg} - 2)$, so $m \leq 4(n + \text{eg} - 2)$, which is what we had to show.

A similar approach works for $\text{lcr}(D) = 2$. Suppose we can show that no three consecutive edges at a vertex are involved in a crossing. With G' and D' as above, we conclude that a 3-face contains no crossing edges, a 4-face at most two crossing edges, and an i-face, for $i \geq 5$, at most i crossing edges (and this can actually happen, in a convex drawing of K_5, for example, every diagonal crosses two other diagonals; after their removal we are left with a

5-face). Then

$$2m \;\le\; 3f_3' + (4 + 2 \cdot 2)f_4' + \sum_{i=5}^{\infty}(i + 2i)f_i'$$

$$\le\; 3f_3' + 8f_4' + \sum_{i=5}^{\infty} 3if_i'$$

$$\le\; 5\sum_{i=3}^{\infty} if_i' - 10\sum_{i=3}^{\infty} f_i'$$

$$\le\; 10m' - 10f'.$$

Again using $f' = 2 - \text{eg} - n + m'$, we conclude that $m \le 5(n + \text{eg} - 2)$. To complete the proof, suppose ux, uy, and uz are three consecutive edges at u all involved in crossings. Let vw be the closest edge to u that crosses uy in a crossing point c. See the right half of Figure 7.1 for an illustration. Either vc or cw must be free of crossings (since vw has at most two crossings), say there are no crossings on cw. We can then add uw as a crossing-free edge ending between uy and uz. Since we did not add that edge, uw must form an empty disk with uy or uz which is not possible, since both are involved in crossings. □

Theorem 7.3 implies that a sufficiently dense graph must have many edges with many crossings. Pach and Tóth [263] used this observation to improve the crossing lemma constant.

Lemma 7.4. *If G is a graph on $n \ge 3$ vertices and m edges, then $\text{cr}(G) \ge 3m - 12(n - 2)$.*

Proof. If $m \le 3(n - 2)$, the right-hand side is negative, and there is nothing to prove, so let us assume that $m > 3(n - 2)$. Then we have to remove at least $m - 3(n - 2)$ edges before G becomes planar, so $\text{cr}(G) \ge m - 3(n - 2)$, which is larger than $3m - 12(n - 2)$. If $m > 4(n - 2)$, by Theorem 7.3 we have to remove at least $m - 4(n - 2)$ edges, each with two crossings, before G has local crossing number at most 1 (with $4(n - 2)$ edges). So G has at least $2(m - 4(n - 2)) + (4(n - 2) - 3(n - 2)) = 2m - 7(n - 2)$ crossings, which is larger than $3m - 12(n - 2)$ for $m > 4(n - 2)$. If $m > 5(n - 2)$, Theorem 7.3 implies that we have to remove at least $m - 5(n - 2)$ edges with 3 crossings each, before G is 2-planar (with $5(n - 2)$ edges). So G has crossing number at least $3(m - 5(n - 2)) + (2(5(n - 2)) - 7(n - 2) = 3m - 12(n - 2)$. □

This lower bound on $\text{cr}(G)$ combined with the proof of the crossing lemma, Theorem 1.8, increases the crossing lemma constant from $1/64$ to $1/36$.

Theorem 7.5 (Pach and Tóth [263]). *If $m \ge 6n$, then $\text{cr}(G) \ge \frac{1}{36}m^3/n^2$, where $m = |E(G)|$, and $n = |V(G)|$.*

Proof. Fix a drawing of G with $\operatorname{cr}(G)$ crossings, and a random subgraph H of G, where each vertex is chosen with probability p. As before, the expected number of vertices of H is pn, the expected number of edges $p^2 m$, and the expected number of crossings, $p^4 \operatorname{cr}(G)$. By Lemma 7.4 we have

$$p^4 \operatorname{cr}(G) \geq 3mp^2 - 12(pn - 2) \geq 3mp^2 - 12pn.$$

Choosing $p = \lambda n/m$ yields $\operatorname{cr}(G) \geq (3\lambda^{-2} - 12\lambda^{-3})m^3/n^2$. The optimal constant occurs for $\lambda = 6$. $\qquad\square$

To improve the crossing lemma constant further, we need better lower bounds in Lemma 7.4, which, in turn, requires better upper bounds on m for larger values of lcr. Pach and Tóth [263] showed that $m \leq (\operatorname{lcr}(G)+3)(n-2)$ for $\operatorname{lcr}(G) \leq 4$, and this already improves the constant to $1/33.75$, but these bounds can be pushed further. The currently best crossing lemma constant of 0.0345, due to Ackerman [13], is based on two additional bounds:

- $m \leq 5.5(n-2)$, if $\operatorname{lcr}(G) \leq 3$ (Pach, Radoičić, Tardos, and Tóth [256]).

- $m \leq 6(n-2)$, if $\operatorname{lcr}^*(G) \leq 4$ (Ackerman [13]).

For $\operatorname{lcr}(G) \leq 2$, the coefficients of $(n-2)$ seemed to increase proportionally with lcr, but the last two bounds show that this is not the case. Asymptotically, the coefficient of $(n-2)$ increases proportionally with $\operatorname{lcr}(G)^{1/2}$.

Theorem 7.6 (Pach and Tóth [263]). *If G is a graph on n vertices and m edges, then $m \leq c\operatorname{lcr}(G)^{1/2} n$, and the bound is asymptotically tight.*

Proof. By the crossing lemma, $\operatorname{cr}(G) \geq c'm^3/n^2$ while, on the other hand, $\operatorname{cr}(G) \leq m\operatorname{lcr}(G)/2$, so $m \leq c\operatorname{lcr}(G)^{1/2} n$. To see that the bound is asymptotically tight, take a $n^{1/2} \times n^{1/2}$ square grid and perturb its vertices slightly so that they are in general position; let that be the vertex set of G; the edge-set consists of all edges between vertices of G whose Manhattan distance is at most $d = (m/n)^{1/2}$. Then $nd^2 \leq |E(G)| \leq n(2d)^2$, so $|E(G)| = \Theta(m)$. Moreover, $d^4/4 \leq \operatorname{lcr}(G) \leq (3d)^4$, since the endpoints of two edges crossing can have distance at most $3d$ (upper bound), and a horizontal edge of length d is crossed by any edge with endpoints in the $d/2 \times d$ rectangle above and the $d/2 \times d$ rectangle below it (lower bound). So $\operatorname{lcr}(G) = \Theta(d^4)$. Since we defined $m = d^2 n$, this implies that $m = \Theta(\operatorname{lcr}(G)^{1/2}) n$. $\qquad\square$

Using $c' = 1/64$ from the (original) crossing lemma, we get $c = 32^{1/2} \leq 5.66$. Using $c' = 1/33.75$ gives $c \leq 4.108$, while the best-known value of 0.0345 gives $c \leq 3.81$.

Corollary 7.7. *We have*

$$\frac{1}{30} \leq \lim_{n \to \infty} \frac{\operatorname{lcr}(K_n)}{\binom{n}{2}} \leq \frac{2}{9}.$$

The values of lcr(K_n) appear unexplored, though some values are known on the torus [170].

Proof. By Theorem 7.6, we have $\binom{n}{2} \leq 3.81 \, \text{lcr}(K_n)^{1/2} \, n$, which implies the lower bound. The upper bound will follow from Theorem 7.18. □

Another consequence of Theorem 7.6 is that a graph with lcr(G) = k has chromatic number $\chi(G) = O(k^{1/2})$, and that result is asymptotically tight, since $G = K_{k^{1/2}}$ satisfies lcr(G) $\leq k$ and requires $k^{1/2}$ colors.

Theorem 7.6 combined with Theorem 8.5 shows that the book thickness of a graph with lcr(G) = k is at most $100k^{1/4}n^{1/2}$. For $k = 1$ it is known that the book thickness is constant [56].

7.1.3   Local Crossings on Surfaces

It has been conjectured that the local crossing number decreases proportionally with the Euler genus of the underlying surface Σ.

$$\text{lcr}_\Sigma(G) \leq O\left(\frac{m}{\text{eg}(\Sigma) + 1}\right) ?$$

Clearly, the extreme cases are true: on the sphere, lcr(G) $\leq m$, and lcr$_{S_m}$(G) = 0, so it is the midrange that is of interest. Dujmović, Eppstein and Wood [118] were able to show that the conjecture is true up to a polylogarithmic factor.

Theorem 7.8 (Dujmović, Eppstein, and Wood [118]). *If Σ is a surface of Euler genus $k = \text{eg}(\Sigma) > 0$, then*

$$\text{lcr}_\Sigma(G) \leq O\left(\frac{m}{k} \log^2 k\right).$$

To make optimal use of the surface, we group the edges so they are as balanced as possible.

Lemma 7.9 (Dujmović, Eppstein, and Wood [118]). *Suppose we are given a graph G with n vertices and m edges, and an integer $q > 0$. Let $m = qb + r$, where $0 \leq r < q$. We can construct a weighted bipartite graph with $n + q - 1$ edges between vertices $u_1, \ldots, u_n$ and $v_1, \ldots, v_q$ so that the total weight of edges incident to u_i equals the degree of the i-th vertex of G, and the total weight of edges incident to v_i is $b + 1$ for $1 \leq i \leq r$ and b for $r < i \leq q$.*

Proof. Let d_i be the degree of the i-th vertex in G. Starting with the empty graph on $u_1, \ldots, u_n$ and $v_1, \ldots, v_q$, call all vertices *unprocessed*. Vertices u_i have capacity d_i and vertices $v + j$ capacity b or $b + 1$ depending on j (as in the statement of the lemma). Let u_i and v_j be the vertices with smallest index which are unprocessed. Add an edge between u_i and v_j and assign to it the maximum weight compatible with the capacities of u_i (total weight at most d_i) and v_j (total weight at most b or $b + 1$ depending on j). Then at least one, possibly both, of the two constraints at u_i and v_j is at capacity, and we have satisfied one of the weight conditions at the two vertices. Call this vertex (or vertices) *processed*, and continue until all vertices are processed. The algorithm terminates, since the total weight of both partitions is the same, so once we have processed the last vertex on one side, the last vertex on the other side must also have been processed, and we have found the graph we needed. Since we process at least one vertex in each step, we added at most $n + q - 1$ edges. $\qquad\square$

A *subcubic* graph is a graph with maximum degree at most three. Loopless multigraphs can be decomposed into a small number of subcubic graphs as the following lemma shows (we base the proof on [337], which contains stronger decomposition results).

Lemma 7.10. *Every loopless multigraph is the union of at most $\Delta - 2$ subcubic graphs, where Δ is the maximum degree of the graph.*

Proof. We prove the result by induction on $\Delta = \Delta(G)$. If $\Delta \leq 3$, we are done, so we can assume that $\Delta > 3$. Let H be a graph extending G so that H is Δ-regular (if there are two vertices with degree less than Δ, add an edge between them; if all but one vertex has degree Δ, create two copies of the graph, and add edges between the two degree-deficient vertices until their degree is Δ). Let M be a maximal matching in H. If $V(M) = V(H)$, we can use M as one of the subcubic subgraphs, and apply induction to $H - M$ which has maximum degree at most $\Delta - 1$ to find the remaining graphs for the union. Otherwise, $V(H) - V(M)$ must be an independent set, with each vertex of degree Δ, so by Hall's Theorem the graph consisting of edges incident to $V(H) - V(M)$ contains a matching M'. Then we can use $M \cup M'$ as one of the subcubic graph (its maximum degree is at most 2), and apply induction to $V(H) - V(M \cup M')$, which has maximum degree at most $\Delta - 1$. $\qquad\square$

We need one more classical result, due to Leighton and Rao [226] on expander graphs. Recall that we defined the congestion of an embedding of a graph into another graph as the largest number of paths passing through an edge; the *dilation* of an embedding is the length of the longest path in the embedding.

Theorem 7.11 (Leighton and Rao [226]). *If G is a bounded-degree graph on q vertices, and H is a graph on q vertices with positive edge expansion, then G can be embedded into H with congestion and dilation $O(\log q)$.*

There are d-regular graphs with positive edge expansion for any fixed $d \geq 3$ [193] (a random d-regular graph, for example).

Proof of Theorem 7.8. We first prove a weaker bound to illustrate the underlying idea. Let $q = \lfloor k^{1/2} \rfloor$, where $k = \mathrm{eg}(\Sigma)$. Take the weighted bipartite graph constructed from G as in Lemma 7.9, and add a complete graph on vertices $v_1, \ldots, v_q$. Let G' be the resulting graph. Then G' has $n + q$ vertices, and at most $n + q - 1 + \binom{q}{2}$ edges, so as Theorem A.16 shows, G' can be embedded on a surface Σ of Euler genus at most $n + q - 1 + \binom{q}{2} - (n + q) = \binom{q}{2} - 1 \leq k$. We only have to show how to draw G based on the embedding of G' without creating too many local crossings, but this is easy. To draw an edge between the i-th and the j-th vertex of G, we start at u_i, follow the edge in G' to one of the q vertices in the other partition, move from there to the vertex connected to u_j, and then to u_j. Since at most $b + 1$ edges pass through each v-vertex, and each edge passes through at most two v-vertices, each edge has at most $2(b + 1) \leq 2(m/q + 1) = O(m/k^{1/2})$ crossings.

For the stronger bound we need $q = \Omega(k)$, so we need a graph that has connectivity like a complete graph, but with smaller Euler genus. That is not really possible, unless we relax another parameter: the number of steps we take through that graph. So construct G' as above, for $q = 2/3k$, but instead of using K_q on the q-vertices, we use a bounded-degree graph Q with positive edge expansion. For example, we can use a random 3-regular graph [193], so Q has at most $3/2q$ edges.

G' has $n + q$ vertices, and at most $n + q - 1 + 3/2q$ edges, so it has Euler genus at most $3/2q - 1 \leq k$. Hence, we can embed G' on a surface of Euler genus at most k. To draw G we again use G' as a template. As before we want to connect the i-th vertex in G to the j-th vertex, by starting at u_i and u_j, following the edges in G' to Q, and then find a path of length $O(\log q)$ in Q connecting them. To apply Theorem 7.11 we need to work with bounded-degree graphs, and we can do so as follows: map each edge in G to the vertex in Q its endpoints are connected to. This gives us a multigraph $G^{\#}$ on $V(Q)$ with maximum degree at most $m/q + 1$. By Lemma 7.10, we can partition $G^{\#}$ into at most m/q graphs of degree at most 3. To each of these graphs, we can apply Theorem 7.11 so that each edge in the graph maps to a path of length at most $O(\log q)$ in Q, and so that each edge of Q has congestion at most $\log q$, *for each of the subgraphs.* We conclude that for each edge in G we can find a path in Q of length $O(\log q)$ so that each edge in Q has congestion at most $\frac{m}{q} O(\log q)$. So, as we draw an edge of G, we traverse $O(\log q)$ vertices, each of which is traversed by at most $\frac{m}{q} O(\log q)$ other paths, implying that each edge requires at most $O(\frac{m}{q} \log^2 q)$ crossings, which is $O(\frac{m}{q} \log^2 k)$. $\qquad\square$

7.2   1-planarity

7.2.1   A Map Coloring Problem

Ringel [291] posed the following problem: Given a map in the plane (regions bounding each other), how many colors do we need to color the countries and the corner vertices so that no two countries that share a border, no two corners connected by a border, and no corner and a country it is incident to have the same color. Adding a vertex to each region and connecting it to all vertices representing bordering countries as well as all corner vertices of that region yields a 1-*planar* graph, that is, a graph with local crossing number 1. Figure 7.2 shows an example. Ringel's map coloring problem is

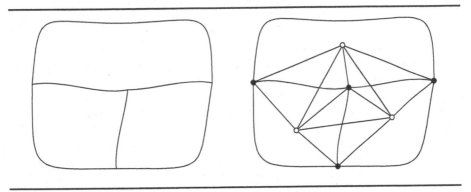

FIGURE 7.2: A map and its induced 1-planar graph.

then generalized by the problem of determining the chromatic number of 1-planar graphs (there are 1-planar graphs which do not correspond to maps). By Theorem 7.3 we know that the average vertex degree of a 1-planar graph is less than 7, so 1-planar graphs are 7-*degenerate* (always contain a vertex of degree at most 7), and are therefore 8-colorable. Ringel showed that 7 colors are sufficient, and conjectured that 6 colors always work, a fact proved later by Borodin [74]. Six colors are optimal, since K_6 has a 1-plane drawing (also see Exercise (7.3)).

Theorem 7.12 (Ringel [291]). *If G is 1-planar, then $\chi(G) \leq 7$.*

The proof yields a polynomial-time algorithm for 7-coloring a 1-plane graph (so a 1-plane embedding needs to be given).

Proof. We prove the result by induction on $n = |V(G)|$. Fix a 1-plane drawing of G. If we can add a crossing-free edge to the drawing without creating two edges forming an empty disk (the interior is empty), we do so until no more edges can be added. The proof of Theorem 7.3 tells us that (i) there is a vertex of degree at most seven in the graph, and (ii), no two consecutive edges in a rotation at a vertex are both involved in a crossing.

If there is a vertex of degree at most six, we can remove it, apply induction, and then use the seventh color different from its neighbors to color it; this, together with (i), implies that there is a vertex v of degree seven. We observe two facts about that vertex:

If v is incident to two crossing-free edges, they form part of a triangle. If v is incident to vu and vw and the graph does not contain uw, contract vu and vw to v, let G' be the resulting graph (after removing multiple edges). Since

uw is absent, G' has no loops, and is 1-plane, so we can apply induction to find a 7-coloring of G'. Color vertices of $G - \{u, v, w\}$ as in G'. Give u and w the color of v in G'. This does not cause a conflict, since uw does not belong to G. Now v has 7 neighbors, but two of them have the same color, so we can color v with the missing seventh color.

Vertex v is incident to at least four crossing-free edges whose endpoints induce a K_5. If edge vu is involved in a crossing, then the edges preceding and succeeding vu in the rotation at v must be free of crossings by (ii). Hence, there must be at least as many crossing-free edges as there are crossing edges incident to v, which implies that v is incident to at least four crossing-free edges, say vu_1, vu_2, vu_3 and vu_4. By the previous observation, there is an edge between any two vertices in $\{u_1, u_2, u_3, u_4\}$, so these four vertices together with v induce a K_5.

With these two observations, we are ready to prove a contradiction. Let $u_1, \ldots u_4$ be the four neighbors of v so that vu_i are crossing-free, $1 \le i \le 4$, labeled according to the order in which they appear in the rotation at v. Then u_1vu_3 is a triangle that is only crossed by u_2u_4, so removing u_2u_4 from G gives us a crossing-free separating triangle u_1vu_3. So $G - \{u_2u_4\} = G_1 \cup G_2$, with $G_1 \cap G_2 = (\{u_1, v, u_3\}, \{u_1v, vu_3, u_1u_3\})$, and $u_2 \in V(G_1)$, $u_4 \in V(G_2)$. Apply induction to G_1 and G_2. We can relabel colors, so they agree on u_1, v, and u_3. Moreover, since both u_2 and u_4 have colors different from the colors of u_1, v, and u_3, since they are part of a K_4 in G_1 and G_2, we can relabel the colors in G_2 so that u_1, v, and u_3 remain unchanged, and the color of u_4 is different from the color of u_2 in G_1. We can then merge the colorings of G_1 and G_2 to get a coloring of G. It is a proper coloring (except for u_2u_4 all edges belong to either G_1 or G_2; and u_2u_4 we ensured is properly colored). □

The class of 1-planar graphs can be refined; Schumacher [306] observed that if the endpoints of two pairs of crossing edges overlap in three vertices, the crossing edges must be the same in both pairs; so if we have two different pairs, the overlap is at most 2. Graphs which have 1-plane drawings in which there is no overlap between any two different pairs of crossing edges are called *IC-planar (independent-crossing planar)*; if there are 1-plane drawings in which the overlap is at most 1, they are called *NIC-planar (near-independent planar)*.

7.2.2 Complexity of 1-Planarity Testing

Testing whether $\mathrm{lcr}(G) = 0$ is the same as planarity testing, so it can be solved in polynomial time. Our goal in this section is to show that the next case, testing whether $\mathrm{lcr}(G) \le 1$, is already **NP**-complete.

Lemma 7.13. *Given a graph G and an edge $e \in E(G)$, let G' be the graph obtained from G by adding a K_6 and identifying one of its edges with e. Then G has a 1-planar drawing in which edge e is free of crossings if and only if G' is 1-planar.*

Proof. If G is 1-planar with e free of crossings, we can identify e with a crossing-free edge in a 1-planar drawing of K_6 (see Figure 7.2 for such a drawing) to get a 1-planar drawing of G'. For the other direction fix a 1-planar drawing of G'. If e is free of crossings, or e crosses another edge of the K_6, we can remove the K_6, excepting e, to get a 1-planar drawing of G with e free of crossings. So e crosses some edge of G, but is free of crossings with edges of the K_6. Let f and g be two edges of the K_6 incident to different endpoints of e. If f and g cross, they cannot be involved in any other crossings. In that case we can reroute e using $f \cup g$ to obtain a crossing-free drawing of e. So none of the K_6-edges incident to e can cross each other (by the argument just given, or because they are adjacent to each other). If we consider the drawing of K_6 on the sphere, the four triangles are nested within each other, but this is impossible, since the edge connecting the apex of the outermost triangle to the innermost triangle has to either cross the other two triangles (which it cannot), or e, which we already excluded. $\qquad\square$

Theorem 7.14 (Grigoriev and Bodlaender [159]; Korzhik and Mohar [214]). *Testing for 1-planarity is* **NP**-*complete.*

Proof. A witness for the 1-planarity of a graph $G = (V, E)$ consists of a set $W \subseteq E^2$ of pairs of edges of the graph. In polynomial time we can check that no edge occurs more than once in R, and that G can be drawn so that only the pairs of edges in W cross. We simply build the planarization of G using W and test that it is planar in polynomial time.

To show **NP**-hardness, we use a slight modification of the construction in Theorem 4.1. Again we start with a graph G and three terminals (t_1, t_2, t_3) and encode the question whether G has a multiway cut of size at most k. We construct a modified frame: take a $K_5 - e$, and let a_1, a_2, a_3 be the triangle which separates the two independent vertices x and x' (endpoints of the removed edge). Subdivide each edge of the triangle, creating vertices t_1, t_2, t_3, where t_i is the vertex not adjacent to a_i. Identify every edge not incident to x with an edge from a K_6 (a new K_6 for each edge), and replace each xa_i with paths of length k. Finally, add k paths of length ℓ, to be determined later, between t_i and a_i for $i \in \{1, 2\}$. This completes the construction of the frame, see Figure 7.3. We now take G, subdivide each of its edges ℓ times, and identify vertices t_1, t_2, t_3 in G with their namesakes in the frame. For the resulting graph H we claim that H is 1-planar if and only if G has a (t_1, t_2, t_3) multiway cut of size at most k.

Suppose H has a 1-planar drawing. By repeatedly applying Lemma 7.13 we can ensure that all edges of the frame which share an edge with one of the K_6 gadgets we added are free of crossings. Each of the paths xa_i, $i \in \{1, 2, 3\}$ is involved in at most k crossings, so $3k$ crossings overall. However, the $2k$ paths connecting t_i to a_i, $i \in \{1, 2\}$ require at least $2k$ of these crossings, leaving at most k crossings of the paths xa_i with subdivided edges of G. Let E' be the set of those edges, so $|E'| \leq k$. Then E' is a (t_1, t_2, t_3)-multiway cut

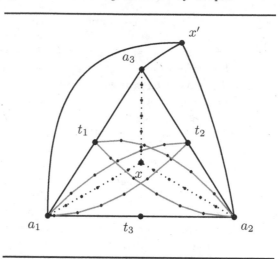

FIGURE 7.3: Drawing of the framework. Edges identified with K_6-gadgets in black; $t_i a_i$-paths in gray, and $x a_i$-paths dotted.

of the subdivided G, which corresponds to a (t_1, t_2, t_3)-multiway cut of size at most k in G.

For the other direction, suppose G has a cut E' of size at most k. We draw the crossing-free edges of the frame as shown in Figure 7.3, and add 1-planar drawings of the K_6 gadgets. We draw $G - E'$ as in the earlier proof (temporarily violating 1-planarity), and add the edges of E'. Let us say the edges of E' have k_i crossings with paths $x a_i$, $i \in \{1, 2, 3\}$, so $k_1 + k_2 + k_3 \leq k$. Add $k - k_1$ edges from t_2 to a_2 crossing $x a_1$, and $k - k_2$ edges from t_1 to a_1 crossings $x a_2$. The remaining $t_1 a_1$ and $t_2 a_2$ edges we route through $x a_3$. Since there are at most $2k - (k - k_1) - k - (k_2) = k_1 + k_2 \leq k - k_3$ of those, there are sufficiently many subdivided edges left along $x a_3$ to accommodate them. The edges we added to the frame, namely edges of G and edges connecting t_i and a_i violate 1-planarity and simplicity of the graph. However, every such edge is involved in at most $|E(G)| + 2k$ crossings (with other edges of G with $x a_i$ or with $t_i a_i$), so subdividing all these edges $\ell := |E(G)| + 2k$ times is sufficient to make the drawing 1-planar (and remove multiple edges). $\qquad\square$

It is easy to adjust the proof to show that 1-planarity remains **NP**-hard if the rotation system is given and the graph is 3-connected (see Exercise (7.11)).

7.3 Rectilinear Drawings

7.3.1 Rectilinear 1-planarity

We saw that testing 1-planarity is **NP**-complete, even if the rotation system is given, see Exercise (7.11). What happens if we consider rectilinear drawings instead? How hard is it to tell whether a given 1-plane drawing of a graph is *stretchable* (also known as *rectifiable*), that is, isomorphic to a rectilinear drawing (necessarily 1-planar). Thomassen identified two obvious obstructions to stretchability which he called the *B*- and *W*-configurations, see Figure 7.4.

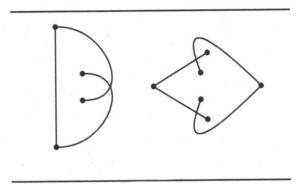

FIGURE 7.4: The *B*- and *W*-configuration.

He was then able to show that these are the only obstructions to stretchability of a 1-plane drawing.

Theorem 7.15 (Thomassen [332]). *A 1-plane graph is stretchable if and only if it does not contain a B- or a W-configuration.*

As a consequence, we can check in polynomial time whether a given 1-plane drawing is stretchable. Thomassen's theorem is an analogue of Fary's theorem for planar graphs, and we will use a similar strategy; the simple idea is to remove the crossing edges, find a convex embedding of the resulting graph, that is, an embedding in which every face but the outer one is convex, and then reintroduce the crossing edges. Fleshing out this simple idea takes some work.

Hong and Nagamochi [192] replaced isomorphism with weak isomorphism, and were able to give a characterization of this case based on forbidden substructures, which also leads to a polynomial-time algorithm.

Proof. Since *B*- and *W*-configurations are not stretchable, we only need to show that a 1-plane graph without these obstructions has an isomorphic, rectilinear drawing. Start with a 1-plane drawing without B- and W-configurations.

We want to extend the graph so that every pair of crossing edges lies in a 4-face. While that is not always possible, we can deal with the cases of failure. Let f and g be a pair of crossing edges in the drawing, and let u and v be two endpoints of f and g from different edges. In the drawing we can connect u to v by a crossing-free curve γ_{uv} close to $f \cup g$. Figure 7.5 illustrates the set-up.

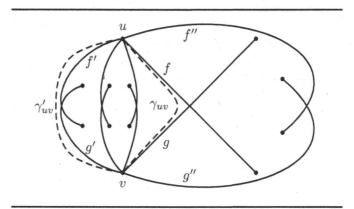

FIGURE 7.5: Enclosing a crossing pair in a 4-face.

Suppose the drawing does not already contain an edge e connecting u and v. We can then add e along the curve γ_{uv}, unless this creates a B- or W-configuration. Since γ_{uv} is crossing-free and all edges in the W-configuration are involved in crossings, it must be that adding e as γ_{uv} creates a B-configuration. Let f' and g' be two crossing edges so that γ_{uv}, f', and g' form a B-configuration (with γ_{uv} as the stem of the B, since it is crossing-free). Choose f' and g' so that the region bounded by f', g', and γ_{uv} is maximal (with respect to containment), see Figure 7.5. This region cannot contain f and g, since otherwise f, g, f', and g' form a W-configuration (f'', g'' and γ_{uv} in Figure 7.5 bound such a region, and f'', g'', f and g form a W-configuration). Let γ'_{uv} be the curve connecting u and v along f' and g'. We claim that we can add e as γ'_{uv} to the drawing without creating B- or W-configurations. This is obvious for W-configurations, since γ'_{uv} is crossing-free. Suppose γ'_{uv} forms a B-configuration with two edges f'' and g'', necessarily incident to u and v. By choice of f' and g', they form a W-configuration with f'' and g'' which is a contradiction (see Figure 7.5).

By repeating this process as often as possible, we obtain a 1-plane drawing without B- or W-configurations in which the endpoints of every pair of crossing edges induce a K_4.

We conclude that the drawing already contains an edge e connecting u to v. Let us choose u and v so that $e \cup \gamma_{uv}$ is minimal (w.r.t. containment of the enclosed region; since $e \cup \gamma_{uv}$ is involved in at most one crossing, this is a partial order, so has a minimal element). If $e \cup \gamma_{uv}$ bounds an empty disk, then there is no need to add γ_{uv}, since there is a homeomorphism of the plane that takes e to γ_{uv} and leaves the remainder of the drawing unchanged. Hence $e \cup \gamma_{uv}$ must contain parts of the graph. There are two cases, depending on whether e is involved in a crossing or not.

Let us first consider the case that e is crossing-free. In that case e and its endpoints separate the graph: $G = G_1 \cup G_2$, where $G_1 \cap G_2 = (\{u, v\}, \{e\})$. We can apply induction to G_1 and G_2 and merge their rectilinear embeddings along e using linear transformations to make the graph enclosed by $e \cup \gamma_{uv}$ fit close to e.

We are left with the case that e crosses some edge xy. Let y lie in the region enclosed by $e \cup \gamma_{uv}$. Then we know that the drawing contains edges yu and yv, see Figure 7.6.

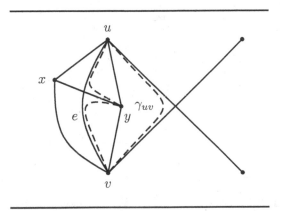

FIGURE 7.6: Edge $e = uv$ crossed by edge xy.

Let γ_{yu} and γ_{yv} be curves close to $xy \cup e$ connecting y to u and y to v. By minimality of $e \cup \gamma_{uv}$, $\gamma_{yu} \cup yu$ and $\gamma_{yv} \cup yv$ bound empty disks. We can then write $G - \{xy\} = G_1 \cup G_2$, where $G_1 \cap G_2 = (\{u, v\}, e)$, where G_2 contains a face uvy. Apply induction to G_1 and G_2 to find rectilinear, 1-plane drawings of both graphs. Merge G_1 and G_2 along e, G_2 can be transformed to fit close to e, but one problem remains: we need to reinsert xy. First consider the triangle xuv. Anything drawn within that triangle can be drawn arbitrarily close to xu or xv, since the edge xy separated that triangle, and the new drawing is isomorphic. In particular, we can make room to draw a straight line from x to some point on e. We linearly transform G_2 so that this straight line can be extended to end in y (this cannot lead to crossings with uvy since that is a face in the drawing of G_2).

At this point we know that every pair of crossings edges not only induces a K_4 in the graph, it is contained in a 4-face of the graph consisting of crossing-free edges only. Triangulate that plane graph, with the exception of the 4-faces, and let H be the resulting graph. We can triangulate without creating edges parallel to the (absent) crossing edges: suppose some 4-face $abcd$ is incident to an edge ac (without loss of generality) in H. It cannot have been before, since ac would have duplicated a diagonal edge. So ac is an edge we added for the triangulation, and we can exchange it with the other diagonal $b'd'$ in its 4-face $ab'cd'$, see Figure 7.7. The only obstacle would be if there already is an edge $b'd'$. Since $b'd'$ cannot lie inside $ab'cd'$ (otherwise it would have blocked ac), it must pass through the 4-face $abcd$, and this is only possible if $b'd' = bd$. In that case, the graph we are dealing with is K_4, for which the result we are trying to establish is immediate.

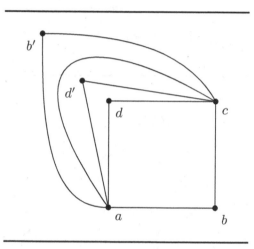

FIGURE 7.7: Triangulating a 4-face without creating parallels.

Graph H is 3-connected; suppose H has a cutset S with $|S| \leq 2$. Add to each 4-face of H one of the diagonal edges so that it is incident with as many vertices of S as possible. The resulting graph H' has no parallel edges by the argument in the previous paragraph, and, as a triangulation, is 3-connected (Lemma A.8). Now $H - S$ and $H' - S$ only differ in faces not incident to vertices of S, so, since $H' - S$ is connected, so is $H - S$, and we conclude that H' is 3-connected. By Theorem 5.2 we can fix a strictly convex embedding of H, and reintroduce crossing edges as straight lines into the 4-faces. □

For an application of Theorem 7.15 to simultaneously drawing both a graph and its dual, see Exercise (7.14).

While the proof of Theorem 7.15 is constructive, the repeated use of linear transformations to squeeze drawings into available space can lead to very dense drawings. Unfortunately, this is unavoidable. If we require that every pair of vertices has distance at least 1 from each other, then there are 1-plane graphs whose isomorphic rectilinear drawings require exponential area.

Theorem 7.16 (Hong, Eades, Liotta, and Poon [191]). *There are 1-plane graphs on n vertices which require area at least $2^{n/2-3}$ in any isomorphic rectilinear drawing if every pair of vertices has distance at least 1.*

Proof. Let $n = 2(k + 1)$ and consider the n-vertex graph consisting of two vertex-disjoint paths of length k shown in Figure 7.8. In an isomorphic

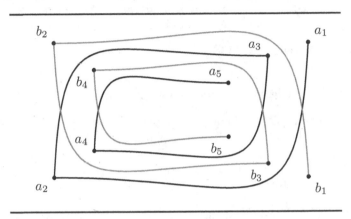

FIGURE 7.8: Two paths, intertwined (adapted from [191, Fig. 7].

straight-line realization, the initial parts of the paths must look similar to the illustration in Figure 7.9. Let c_i be the crossing between $a_i a_{i+i}$ and $b_i b_{i+1}$.

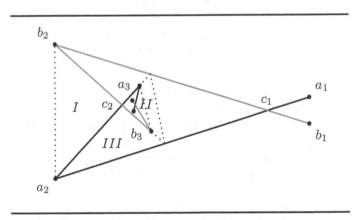

FIGURE 7.9: Two paths, stretched.

Consider the triangle $\Delta_i := \Delta(c_i, a_{i+1}, b_{i+1})$. We claim that the area of Δ_{i+1} is at most half of the area of Δ_i. It is sufficient to prove this for Δ_1 and Δ_2, which are shown in Figure 7.9. Triangle Δ_2 is contained in triangle II. The triangle consisting of $I + III$ is larger than the triangle consisting of $II + III$, since they have the same base, and the second triangle has shorter height. But then II, and therefore, Δ_2, must be smaller than I. Since both are disjoint, and part of Δ_1, the claim follows. The final triangle, Δ_k, must have area at least $1/2$, since its vertices have distance at least 1 from each other. Hence, Δ_1 has area at least $2^{k-2} = 2^{n/2-3}$.   □

Call a graph *rectilinear* 1-*planar* if it has a rectilinear, 1-plane drawing. By Theorem 7.3, a 1-planar graph on n vertices has at most $4(n-2)$ edges. For rectilinear 1-planar graphs, that bound can be improved.

Theorem 7.17 (Didimo [112]). *Rectilinear* 1-*planar graphs have at most* $4(n-2)-1$ *edges.*

This bound is tight for infinitely many n, see Exercise (7.15).

Proof. By Theorem 7.3 we know that 1-planar graphs have at most $4n-8$ edges. Suppose there is a rectilinear 1-plane graph with $4n-8$ edges. Fix a vertex-minimal such graph G. Then G contains no edge uv so that $\{u,v\}$ disconnects G: if $G = G_1 \cup G_2$, and $G_1 \cap G_2 = (\{u,v\},\{e\})$, we know that G_i contains at most $4|V(G_i)| - 9$ edges, $i \in \{1,2\}$, so G contains at most $4(|V(G_1)| + |V(G_2)|) - 18 - 1 = 4(|V(G)| + 2) - 19 = 4|V(G)| - 11$ edges. In other words, every crossing pair lies in a 4-face of the subgraph H of crossing-free edges of G. Adding one of the crossing edges to each 4-face yields a plane graph with at most $3n - 6$ edges; since we are assuming that there are $4n - 8$ edges overall, there must be $n - 2$ crossing pairs, so just as many 4-faces in H. Let f be the number of faces of H. By Euler's formula, $|V(H)| - |E(H)| + f = 2$, so $n - 2 \le f = 2 - n + |E(H)|$, so $|E(H)| \ge 2(n-2)$. So when we add diagonals to the $n - 2$ faces which contained crossings, we have a graph with $3(n - 2)$ edges, which must be a triangulation. Hence H has 4-faces only. Now one of these must be the outer face. Adding the crossing pair to that face forces an illegal B-configuration.   □

We conclude that not every 1-planar graph has a rectilinear 1-planar drawing (not necessarily isomorphic or weakly isomorphic), and the theorem shows us how to construct explicit counterexamples from quadrangulations of the plane. For example, for $n = 8$, we can take a cube (octahedron), and add a crossing pair to each face. We are not aware of any characterization of 1-planar graphs that have a rectilinear 1-planar drawing, but see Exercise (7.18) for a partial result.

7.3.2   Rectilinear Local Crossing Number

As we mentioned earlier, little is known about values of the local crossing number, even $\mathrm{lcr}(K_n)$ has not yet been determined. Somewhat surprisingly then, we do know $\overline{\mathrm{lcr}}(K_n)$, the *rectilinear local crossing number*, in which lcr is restricted to rectilinear drawings.

Theorem 7.18 (Ábrego and Fernández-Merchant [10]). *For $n \neq 8, 14$, we have*

$$\overline{\mathrm{lcr}}(K_n) = \left\lceil \frac{1}{2} \left(n - 3 - \left\lceil \frac{n-3}{3} \right\rceil \right) \left\lceil \frac{n-3}{3} \right\rceil \right\rceil,$$

and $\overline{\mathrm{lcr}}(K_8) = 4$ and $\overline{\mathrm{lcr}}(K_{14}) = 15$.

The case $n \equiv 2 \bmod 3$ turns out to be somewhat special, and we do not show the construction which achieves the upper bound in this case; the lower bounds on $\overline{\mathrm{lcr}}(K_8)$ and $\overline{\mathrm{lcr}}(K_{14})$ require computer search.

Proof. Fix a rectilinear drawing of K_n minimizing the local crossing number. Let $u_1, \ldots, u_k$ be the vertices of the convex hull. If there are two vertices u_i, u_j on the convex hull and a third vertex v not on the convex hull, so that the number of vertices in the hull on the two sides of $u_i v u_j$ differs by at most $(n-3)/3$, then neither side can have fewer than $(n-3)/3$ vertices, so both sides must have at least $\lceil (n-3)/3 \rceil$ vertices. These vertices cause at least $(n-3-\lceil (n-3)/3 \rceil)\lceil (n-3)/3 \rceil$ crossings with $u_i v u_j$—since $f(x) = (c-x)x$ is concave, its local minimum is on the boundary, so either $u_i v$ or $v u_j$ contains at least half of those crossings, proving the bound in the lemma.

Suppose there is a triangle $u_i u_j u_k$, with vertices on the boundary of the hull, so that the triangle contains at least $(n-3)/3$ vertices. Let v be any of the remaining $n-3$ vertices; then $v u_i$, $v u_j$, and $v u_k$ split the convex hull into three (disjoint) sectors. If any of the three sectors contains at least $(n-3)/3$ vertices, and at most $2(n-3)/3$ vertices, the argument in the first paragraph establishes the lower bound, so each sector contains either less than $(n-3)/3$ or more than $2(n-3)/3$ vertices. Since there are only $n-3$ vertices in the sectors, it must be that exactly one of them contains more than $2(n-3)/3$ vertices, and the other two less than $(n-3)/3$ vertices. If this is the case, we assign v to the two vertices (among u_i, u_j, u_k) on the boundary of that sector. Since there are $n-3$ choices for v, there must be one pair, let us say $\{u_i, u_j\}$, which is assigned to at least $(n-3)/3$ vertices. Among those vertices pick a vertex v' which minimizes the number of vertices in its largest sector $u_i v' u_j$. That number is larger than $2(n-3)/3$. The boundary of that sector is convex: if v' and u_k do not lie on the same side of $u_i u_j$, then the sector and the triangle are disjoint, so would contain more than $(n-3)/3 + 2(n-3)/3 = n-3$ vertices, which is not possible. Consider any vertex v inside that sector. Since the sector $u_i v u_j$ has fewer vertices than $u_i v' u_j$, v is not assigned to $\{u_i, u_j\}$. But this means more than $2(n-3)/3$ vertices are not assigned to $\{u_i, u_j\}$, which is a contradiction.

To find a triangle containing at least $(n-3)/3$ vertices, consider triangles $u_1 u_i u_{i+1}$ for $2 \leq i < k$. If any one of these has at least $(n-3)/3$ vertices, we are done, so we can assume that they all contain less than $(n-3)/3$ vertices. There is an i so that the number of vertices on the left of $u_1 u_i$ is less than $(n-3)/3$ and the number of vertices on the left of $u_1 u_{i+1}$ is not. Then the number of vertices to the left of $u_1 u_{i+1}$ is at most $2(n-3)/3$ and at least

$(n-3)/3$. But then u_1u_{i+1} has at least $2((n-3)/3)^2$ crossings, which is larger than the lower bound we need.

For the upper bound we consider three (very flat) arcs of $\lfloor n/3 \rfloor$ vertices each; if $n \not\equiv 0 \bmod 3$ add the remaining one or two vertices to different arcs. We place the arcs arching in the same direction, on three rays at an angle of $2\pi/3$ around the origin, as shown in Figure 7.10.

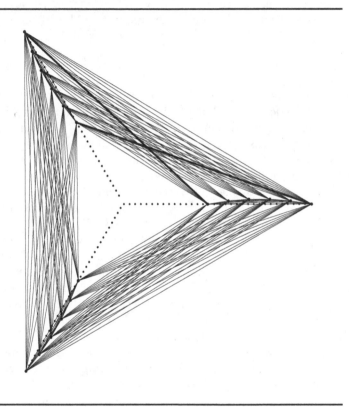

FIGURE 7.10: An $\overline{\text{lcr}}$-minimal drawing of K_n, if $n \not\equiv 2 \bmod 3$.

If the arcs are sufficiently flat, crossings only occur between two groups. The largest number of crossings is incurred by an edge connecting a farthest point on a ray to a closest point on another ray (two of these edges are drawn more heavily in Figure 7.10). If $n \equiv 0 \bmod 3$, such an edge has at most $((n/3)-1)^2 = 1/9(n-3)^2$ crossings. If $n \equiv 1 \bmod 3$, such an edge has at most $((n-1)/3-1)(n-1)/3 = 1/9(n-1)(n-4)$ crossings. Both bounds agree with our lower bound. If $n \equiv 2 \bmod 3$ we get an upper bound of $((n-2)/3^2 = 1/9(n-2)^2$, which does not agree with the lower bound. This case requires a different construction, which can be found in [10]. $\qquad\square$

7.4 Quasi-k-planarity

A drawing of a graph is k-*quasi-plane* if no k edges cross pairwise; a graph with a k-quasi-plane drawing is called k-*quasi-planar*. Trivially, every k-planar graph (a graph with local crossing number at most k) is $k+2$-quasi-planar, and it may not be k-quasi-planar, leaving open the case of $k + 1$-quasi-planarity. Somewhat surprisingly, the answer is not: it depends. The following theorem has been announced.

Theorem 7.19 (Angelini et al. [31]; Hoffmann and Tóth [190]). *Every k-planar graph is $k + 1$-quasi-planar for $k \geq 2$.*

Angelini et al. [31] prove the case $k \geq 3$, and Hoffmann and Tóth [190] the case $k = 2$. Theorem 7.19 does not have a reverse: it is easy to construct 3-quasi-planar graphs with unbounded local crossing number (Exercise (7.22)).

Still, k-quasi-planar graphs cannot be too dense as the following application of the bisection-width method shows.

Theorem 7.20 (Pach, Shahrokhi, and Szegedy [258]). *If a graph G on n vertices has a good, k-quasi-plane drawing, then $|E(G)| \leq 3n(6 \log n)^{2k-4}$.*

We will later see that we can drop the assumption that the drawing is good (Exercise (9.6)). It is conjectured that the upper bound is linear for every fixed k (and drawings which are not necessarily good). This is known to be true for $k \leq 4$ (for $k = 2$ this follows from Euler's formula, for $k = 3$ it was proved by Pach, Radoičić, and Tóth [257], building on [16], which proved the result for good drawings; for $k = 4$, the linear bound was established by Ackerman [12]). For upper bounds improving Theorem 7.20, in particular if the local crossing number is bounded, see [326]. For convex drawings, a linear upper bound is known, see Theorem 8.14.

Proof. We show inductively, on k and n, that if a graph can be drawn so that no k edges cross pairwise, and the drawing is good, then $|E(G)| \leq c_k n \log^{2k-4} n$, where c_k will be determined later. For $k = 2$ the result is true for all n as long as $c_2 \geq 3$, using Euler's formula.

Suppose that G has a good drawing D in which no $k \geq 3$ edges cross pairwise, and that the result has been established inductively for $k' \leq k$ and $n' \leq n = |V(G)|$, where at least one of the inequalities is strict.

An edge e of G cannot cross $k - 1$ edges which cross pairwise; letting G_e be the graph induced by all edges crossing e, applying induction tells us that $|E(G_e)| \leq c_{k-1} n \log^{2k-6} n$. Since each crossing is counted twice, along two edges (here we use that D is good), we have

$$
\begin{aligned}
\mathrm{cr}(D) &\leq \frac{1}{2} \sum_{e \in E(G)} |E(G_e)| \\
&\leq \frac{c_{k-1}}{2} |E(G)| n \log^{2k-6} n.
\end{aligned}
$$

By Theorem 3.1, $16\,\mathrm{cr}(G) + \mathrm{ssqd}(G) \geq (\mathrm{bw}(G)/1.58)^2$. Together with $\mathrm{cr}(G) \leq \mathrm{cr}(D)$ and $\mathrm{ssqd}(G) \leq |E(G)|n$, we get

$$
\begin{aligned}
\mathrm{bw}(G) &\leq 1.58(16\,\mathrm{cr}(G) + \mathrm{ssqd}(G))^{1/2} \\
&\leq 1.58(|E(G)|n)^{1/2}(8c_{k-1}\log^{2k-6}(n) + 1)^{1/2} \\
&\leq 1.58(|E(G)|n)^{1/2}(9c_{k-1}\log^{2k-6} n)^{1/2} \\
&\leq 6c_{k-1}^{1/2}(|E(G)|n)^{1/2}\log^{k-3} n.
\end{aligned}
\tag{7.1}
$$

A bisection (V_1, V_2) of G realizing $\mathrm{bw}(G)$ results in two graphs $G_i = G[V_i]$, $i = 1, 2$, which are smaller than G; we apply induction, and get $\mathrm{bw}(G) = |E(G)| - (|E(G_1)| + |E(G_2)|)$. We bound the second term as follows:

$$
\begin{aligned}
|E(G_1)| + |E(G_2)| &\leq c_k \left(\frac{n}{3}\log^{2k-4}\left(\frac{n}{3}\right) + 2n/3\log^{2k-4}(2n/3)\right) \\
&\leq c_k n \log^{2k-4}(n)\left(\frac{1}{3}\left(\frac{\log(\frac{n}{3})}{\log(n)}\right)^{2k-4} + \frac{2}{3}\left(\frac{\log(2n/3)}{\log(n)}\right)^{2k-4}\right) \\
&\leq c_k n \log^{2k-4}(n)\left(\frac{1}{3}\left(1 - \frac{\log(3)}{\log(n)}\right)^{2k-4} + \frac{2}{3}\left(1 - \frac{\log(\frac{3}{2})}{\log(n)}\right)^{2k-4}\right) \\
&\leq c_k n \log^{2k-4}(n)\left(1 - \frac{k}{\log(n)}\right),
\end{aligned}
\tag{7.2}
$$

using $(1 - a)^x \leq 1 - ax$ and $(2k - 4)\log 3 \geq k$ for $k \geq 3$. Using bounds 7.1 and 7.2 in $\mathrm{bw}(G) = |E(G)| - (|E(G_1)| + |E(G_2)|)$ we obtain

$$
|E(G)| - 6c_{k-1}^{1/2}(|E(G)|n)^{1/2}\log^{k-3} n \leq c_k n \log^{2k-4}(n)(1 - \frac{k}{\log n}).
$$

If we consider this inequality as an inequality in $x = |E(G)|$, it follows that the inequality holds for $x = 0$, since we can assume that $k < \log n$—otherwise the bound of the theorem is trivially true—so there is a point $x_0 \geq 0$ for which the inequality holds between 0 and x_0 and fails for $x > x_0$. If we try $x = c_k n \log^{2k-4} n$, we get

$$
\frac{k}{\log n} c_k n \log^{2k-4} n \leq 6(c_{k-1}c_k)^{1/2}n\log^{2k-5} n
$$

which is equivalent to $k \leq 6(c_{k-1}/c_k)^{1/2}$, so fails if $c_k \geq 36c_{k-1}$. Hence, $x_0 < c_k n \log^{2k-4} n$ which implies that $|E(G)| < c_k n \log^{2k-4} n$.

It remains to show that we can choose c_k to satisfy all requirements, namely $c_2 \geq 3$, and $6(c_{k-1}/c_k)^{1/2} \leq k$. Setting $c_k = 3 \cdot 6^{2k-4}$ satisfies all requirements. $\qquad\square$

Ackerman, Keszegh, and Vizer [14] introduced an interesting generalization of 1-planarity. Call a drawing of a graph *planarly connected*, if for any two independent crossing edges there is a third edge connecting their end-points. They could show that planarly connected graphs have at most a linear number of edges. Every 1-planar graph has a planarly connected drawing, see Exercise (7.19), giving us an alternative, and more complicated, proof that 1-planar graphs are sparse.

7.5 Notes

The local crossing number can be traced back to Ringel's map coloring problem [291]; this problem led to studies of 1-planar and, less so, 2-planar graphs, as well as local crossing numbers of graphs on surfaces. The bound of $m \leq 4(n-2)$ for 1-planar graphs was first shown by Bodendiek, Schumacher, Wagner [68], and rediscovered several times. We presented Schumacher's generalization for arbitrary surfaces. The topic lay mostly dormant, until Pach and Tóth [263] observed the usefulness of density bounds for improving the crossing lemma, leading to the currently best bounds. Recently there has been an explosion of research on 1-planarity; for an annotated bibliography, see [210].

7.6 Exercises

(7.1) Construct a graph G with $\mathrm{lcr}(G) = 4$, and $\mathrm{lcr}^*(G) > 4$.

(7.2) Show that if $\mathrm{lcr}(G) = 4$, then G has a drawing realizing lcr in which no two independent edges cross more than once. Show that there is a graph G with $\mathrm{lcr}(G) = 5$ in which every drawing realizing lcr requires two independent edges to cross more than once. *Hint:* For the second part, build on Exercise (7.1).

(7.3) Show that there are maps which require at least 6 colors in a proper map-coloring à la Ringel. *Hint:* K_6 will not do, it is 1-planar, but it does not correspond to a map.

(7.4) Show that a 1-plane graph without a B-configuration, and which has all its vertices on the outer face can be drawn with straight lines realizing the same rotation system.

(7.5) Show that $\mathrm{lcr}(K_7) = 2$.

(7.6) Show that there is a graph with crossing number 2 so that any 1-plane drawing of the graph requires at least 3 crossings.

(7.7) Improve the constant in Theorem 3.29. *Hint:* Use Theorem 7.3.

(7.8) Show that for every $n \geq 8$, $n \neq 9$, there is a 1-planar graph on n vertices so that $|E(G)| = 4(n-2)$. (Such graphs are called 1-*optimal.*)

(7.9) Show that $\Theta(G) \leq \mathrm{lcr}(G) + 1$.

(7.10) Show that $m \leq (\mathrm{lcr}(G)+3)(n-2)$ for $\mathrm{lcr}(G) \leq 4$ implies that the crossing lemma holds with a constant of $1/33.75$. *Hint:* Show that $\mathrm{cr}(G) \geq 5m - 25(n-2)$.

(7.11) Show that testing 1-planarity remains **NP**-complete even if the graph is 3-connected. *Hint:* Do this in two steps: show that the construction in Theorem 7.14 works even if the rotation system is fixed, and then replace vertices and edges with hexagonal grids.

(7.12) Show that testing 1-planarity remains **NP**-hard for graphs of bounded degree. *Hint:* Break apart vertices of high-degree in the construction of Theorem 7.14. The K_6-gadget will be a useful tool.

(7.13) Show that Theorem 7.15 implies that if $\mathrm{cr}(G) = 1$, then $\overline{\mathrm{cr}}(G) = 1$ (one of the cases of Theorem 5.3).

(7.14) Call a drawing of a graph *almost rectilinear*, if it has a rectilinear drawing in which at most one edge has a bend. Then $G \cup G^*$ is almost rectilinear if and only if G is 2-connected and 3-edge connected. Here G^* is the topological dual of G. *Hint:* Use Theorem 7.15.

(7.15) Show that for infinitely many n there are rectilinear 1-plane graphs with $4(n-2) - 1$ edges. *Hint:* $n \equiv 3 \bmod 8$.

(7.16) Show that every 1-planar graph is the union of a forest and a planar graph. *Hint:* From each crossing pair pick one edge so as not to create a cycle.

(7.17) Show that testing whether a graph has a rectilinear 1-planar drawing is **NP**-complete.

(7.18) Show that if G is a 3-connected graph that has a 1-plane drawing in which the outer region is a face (so not incident to any crossings), then G has a rectilinear 1-plane drawing.

(7.19) Show that any 1-planar graph has a planarly connected drawing.

(7.20) Construct an n-vertex graph with a planarly connected drawing, and at least $9n - O(1)$ edges.

(7.21) Show that any graph with a planarly connected drawing is 9-plane.

(7.22) Show that $\mathrm{lcr}(G)$ is unbounded for 3-quasi-planar graphs.

Chapter 8

Book and Monotone Crossing Numbers

8.1 Embeddings and Drawings in Books

A *book with k pages* is a set of k half-planes, the *pages*, whose boundaries have been identified to form the *spine* of the book. To draw a graph in a book, we place vertices on the spine, and draw each edge within a single page (so edges do not cross the spine).

Definition 8.1. The *k-page book crossing number*, $\mathrm{bkcr}_k(G)$ is the smallest number of crossings in a drawing of G in a book with k pages.

We study the cases $k = 1$ and $k = 2$ separately in Sections 8.1.2 and 8.1.3, before tackling the general case in Section 8.1.4. We begin with *book embeddings*, book drawings without crossings.

8.1.1 Book Embeddings and Their Thickness

The *book thickness* (also known as *pagenumber* and *stack number*), $\mathrm{bt}(G)$, of a graph G is the smallest k so that G can be embedded in a book with k pages.

Remark 8.2. If we allow edges to cross the spine (sometimes called a *topological book embedding*), every graph can be embedded in a book with 3 pages [40]: take a drawing of the graph in the plane in which all crossings lie on a line. Make this line a spine of a 2-page book, and use the third page to eliminate every crossing locally (this proof is attributed to Babai). This result may explain why topological book embeddings have not been investigated very much, though there is some literature on the *spine crossing number*, the smallest number of crossings of edges with the spine in a topological book embedding. See Exercises (8.1) and (8.2). One reason topological book embeddings deserve more attention is that they can be turned into book embeddings by subdividing edges; since the crossing number is invariant under subdivision, topological book embeddings could lead to improved crossing number bounds.

Graphs with book thickness one are just the outerplanar graphs. The following theorem is a useful characterization of graphs with $\mathrm{bt}(G) \leq 2$.

Theorem 8.3 (Bernhart and Kainen [60]). *A graph is embeddable in a book with two pages if and only if it is a subgraph of a planar Hamiltonian graph.*

Proof. If G embeds in a book (with any number of pages), we can add a Hamiltonian cycle to it: follow the spine closely, adding an edge between two consecutive vertices if it does not exist yet, and finally connect the first and last vertex of the graph with an edge avoiding all other edges on one of the pages. Since two pages form a plane, this settles the forward direction. In the other direction, we can extend the graph so it is planar and Hamiltonian. The Hamiltonian cycle partitions the edges of the graph into two sets: those inside, and those outside the cycle (where we can add the edges of the cycle to either one of these two sets). We place the vertices along the spine in the order that they occur on the Hamiltonian cycle, and then draw the edges in each set on a separate page. ☐

Since we can separate the pages of a book and merge pairs of pages into planes, we have, $\theta(G) \leq \lceil \mathrm{bt}(G)/2 \rceil$, which means that lower bounds on thickness or upper bounds on book thickness have implications for the other parameter, either directly, or via similar arguments.

Lemma 8.4. *A graph embeddable in a book with k pages has at most $k(n - 3) + n$ edges, where $n = |V(G)|$.*

We conclude that $\mathrm{bt}(G) \geq \lceil (m - n)/(n - 3) \rceil$, where $m = |E(G)|$.

Proof. Every page of a book embedding contains an outerplanar subgraph G_i of G; if we remove from G_i all edges between consecutive vertices and the first and last vertex along the spine, then there remain at most $2n - 3 - n = n - 3$ edges. So G has at most $n - 3$ edges for each page plus the at most n edges we removed, for a total of at most $\mathrm{bt}(G)(n - 3) + n$ edges, which yields the bound of the lemma. ☐

Imagine the pages of the book are transparent, and we open the book so that $\lfloor k/2 \rfloor$ pages are stacked on the top and $\lceil k/2 \rceil$ pages are stacked at the bottom (do not try this with real books at home). Then we have a drawing of G in the plane which, after making it good, contains at most $\binom{\lfloor k/2 \rfloor n}{2} + \binom{\lceil k/2 \rceil n}{2} \leq \frac{1}{2}\binom{kn}{2}$ crossings, so

$$\mathrm{cr}(G) \leq \frac{1}{2}\binom{\mathrm{bt}(G)n}{2}.$$

Using a probabilistic argument, Malitz [235] was able to give a bound on the book thickness in terms of edges (the analogue result for thickness is easier to obtain, see Theorem 3.19).

Theorem 8.5 (Malitz [235]). *If G is a graph on m edges, then*

$$\mathrm{bt}(G) = O(m^{1/2}).$$

Two edges in a (vertex)-ordered graph can be *twisted* (also known as *crossing*), *nested*, or *non-overlapping* depending on whether their endpoints occur in the order *efef*, *effe*, or *eeff*. A k-*twist* in a vertex-ordered graph is a set of k disjoint edges $u_i v_i$ whose endpoints have order $u_1 < \cdots < u_n < v_1 \cdots v_n$. A k-twist in a book embedding, with vertices ordered as they occur along the spine, requires at least k pages, since no two edges in a k-twist can lie in the same page. If there are no non-overlapping pairs of edges, then the absence of a $(k + 1)$-twist is sufficient for embeddability in a k-page book.

Lemma 8.6. *If a graph has an ordering without $(k + 1)$-twists and non-overlapping edges, then the graph can be embedded in a book with k pages.*

The proof uses Dilworth theorem (see the discussion after Theorem 5.35 for a statement and a definition of chains and antichains).

Proof. Fix the ordering of the graph in which every two edges overlap, and there are no $(k + 1)$-twists. Define a partial order of the edges of the graph: $uv \preceq xy$ if $u \le x < y \le v$, that is, xy is nested within uv. By assumption there is no antichain of size k, so by Dilworth theorem, the graph is the union of at most k chains. A chain consists of pairwise nested edges (since we do not have any non-overlapping edges), so it can be embedded on a single page. $\square$

So in the absence of non-overlapping edges, we can find book embeddings whose pagenumber is bounded by the largest twists; by bipartioning the vertex set recursively, we can find large sets of non-overlapping edges, which we can exploit to construct a book embedding with few pages. We control the size of the twists by starting with a random ordering of the vertices.

Proof of Theorem 8.5. By at most doubling the order $n = |V(G)|$ of the graph, we can ensure that n is a power of 2. Fix a random linear order of the vertices of G. Split this order into two halves of equal size. Let us calculate the probability that there is a k-twist among the edges connecting the two halves. This probability is at most

$$2^k \binom{m}{k} \binom{n/2}{k}^2 \frac{k!(n - 2k)!}{n!},$$

as we need to choose k out of the m edges, $\binom{m}{k}$, determine their left and right endpoints in the order, 2^k, identify those endpoints with vertices in the opposite halves, $\binom{n/2}{k}^2$, ensure they are in reverse order, $(k!)^2/(k!) = k!$, and assign the remaining vertices of the graph, $(n-2k)!$. Stirling's approximation of $n!$ yields the bounds $(2\pi n)^{1/2}(n/e)^n < n! < 2(2\pi n)^{1/2}$, which implies that $(n - 2k)!/n! \le 2((n-2k)/e)^{n-2k}/(n/e)^n = 2e^{2k}/n^{2k}$. Then the original probability can be bounded by

$$2^k \binom{m}{k} \binom{n/2}{k}^2 \frac{k!(n-2k)!}{n!} \quad \leq \quad \frac{(2m)^k}{k!} k! \frac{n^{2k}}{(k!)^2 2^{2k}} \frac{(n-2k)!}{n!}$$

$$\leq \quad 2(2m)^k \left(\frac{e^2}{k}\right)^{2k}.$$

In the second stage, we halve each of the original halves and consider the edges connecting the halves within each half. What is the probability that there is a k-twist in either half? This probability is at most

$$2^1 2^k \binom{m}{k} \binom{n/2^2}{k}^2 \frac{k!(n-2k)!}{n!}.$$

As we keep halving, in the i-th stage, the probability is at most

$$2^{i-1} 2^k \binom{m}{k} \binom{n/2^i}{k}^2 \frac{k!(n-2k)!}{n!},$$

and this probability can be bounded above by

$$e 2^{i-1} 2 \left(\frac{2^{1/2} m^{1/2} e^2}{2^i k}\right)^{2k} \quad \leq \quad \left(\frac{2^{1/2} m^{1/2} e^2}{2^{1/2} k}\right)^{2k}.$$

For edges in stage i we set $k = k_i = 2^{1/2} m^{1/2} e^2 / 2^{i/2-1}$. Then the probability that there is a k_i-twist in stage i for any $1 \leq i \leq \log n$ is at most

$$\sum_{i=1}^{\log n} \left(\frac{2^{1/2} m^{1/2} e^{2^{i/2}}}{2} k_i\right)^{2k_i} \quad \leq \quad \sum_{i=1}^{\log n} 2^{-2k_i}$$

$$= \quad \sum_{i=1}^{\log n} 2^{-4(2m/2^i)^{1/2} e^2}$$

For $i = \log n$, we have $4(2m/2^i)^{1/2} e^2 > 41$ (if $m < n$, then the graph is not connected), so the last term of the sum is less than $1/2^{41}$. Since the exponents differ by at least $4e^2$, we can bound the sum by the value of an infinite geometric series starting with $1/2^{41}$ and a ratio of at least $1/2$, so the sum is less than $1/2^{40}$. Hence, a randomly chosen permutation of the vertices has no k_i-twists at stage i (for any i) with probability close to 1. By Lemma 8.6, we can then embed the edges at stage i in k_i pages. Since every edge belongs to some stage, we can embed the whole graph in a book with

$$\sum_{i=1}^{\log n} k_i \leq 2e^2 2^{1/2} m^{1/2} \sum_{i=1}^{\log n} 2^{-i/2} \leq 2e^2 2^{1/2}(1 + 2^{1/2}) m^{1/2} < 50 m^{1/2}.$$

$\square$

For planar graphs Malitz's theorem implies that $O(n^{1/2})$ pages are sufficient. Somewhat surprisingly, a fixed number of pages suffice.

Theorem 8.7 (Yannakakis [353]). *Every planar graph has book thickness at most 4.*

Using Theorem 8.3 it is easy to construct planar graphs of book thickness 3; it is open whether 3 pages will always work.

There are a few book thickness results for specific families of graphs, but they get intricate very fast.

Theorem 8.8 (Bernhart and Kainen [60]). *The following equalities hold.*

(i) $\mathrm{bt}(K_n) = \lceil n/2 \rceil$.

(ii) $\mathrm{bt}(K_{m,n}) = m$ *for* $n \geq \max(m^2 - m + 1, m)$.

(ii') $\mathrm{bt}(K_{m,m}) \leq m - 1$ *for* $m \geq 4$.

(iii) $\mathrm{bt}(Q_d) \leq d - 1$.

Proof. For (i), the lower bound follows from Lemma 8.4, and it is sufficient to show the upper bound for even n (since it will give the right upper bound for $n - 1$). We start with the plane triangulation shown on the left in Figure 8.1. If vertices are labelled 0 to $n-1$ (in counterclockwise order, say), it consists of $n-3$ chords xy which satisfy $x + y \equiv 0 \bmod n$ or $x + y \equiv 1 \bmod n$. If we rotate the triangulation by i vertices, that is, move the endpoints of each chord by i vertices counterclockwise, we obtain chords xy which satisfy $x + y \equiv 2i \bmod n$ or $x + y \equiv 2i + 1 \bmod n$. This implies that the $n/2$ sets of triangulating chords we get by letting i range from 0 to $n/2 - 1$ are pairwise disjoint, so there are $n/2(n - 3)$ chords; together with the n edges of the C_n this gives us $n + n/2(n - 3) = \binom{n}{2}$ edges, so we have a partition of K_n into $n/2$ plane graph; we include the C_n with each one of them.

The upper bound in (ii) follows, since $K_{m,n}$ is the union of m stars. For the lower bound, suppose we have an embedding of $K_{m,n}$ in a book with k pages. The m vertices of the left partition split the vertices on the spine into m blocks of consecutive vertices (we consider the first and last vertex consecutive, so they belong to the same block); since $n \geq m^2 - m + 1$ at least one of these blocks must contain m vertices, so there are m consecutive vertices from the right partition. Match these with the m vertices of the left partition in reverse order, so that any two edges of the matching have to cross. Then each of the m matching edges needs its own page (they form an m-twist).

For (ii') see Figure 8.2 on how to add two vertices to each partition of a $K_{m-2,m-2}$ using only two pages. The vertices in $K_{m-2,m-2}$ are in four blocks, alternating between the two partitions. We leave the base cases to Exercise (8.3).

The upper-bound construction for (iii) is rather involved, it can be found in [123].

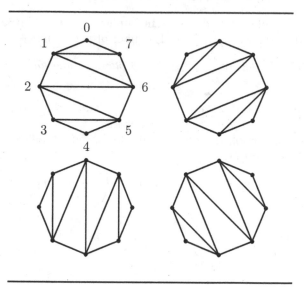

FIGURE 8.1: A plane triangulation of K_8.

For (v) observe that Q_2 can be drawn on a single page; to go from Q_{d-1} to Q_d, embed a Q_{d-1} in a $(d-2)$-page book, and create a mirror copy of this Q_{d-1} using the same $(d-2)$-pages but so that corresponding vertices are in reverse order along the spine; one additional page is then sufficient to connect the corresponding vertices. $\qquad\square$

The bounds for $K_{m,n}$ from the last theorem can be sharpened.

Theorem 8.9 (Enomoto, Nakamigawa, and Ota [123]). *For $m \leq n$ we have*

(i) $\mathrm{bt}(K_{m,n}) \leq \min_{k \in \mathbb{N}} \lceil \frac{mk^2 + k + n}{k^2 + k + 1} \rceil$.

(ii) $\mathrm{bt}(K_{m,n}) = m$ *for* $n \geq m^2/4 + 5/2m^{7/4} + 11m^{3/2} + 8m$.

For the proofs of both results we refer to [123] (and note that to get the expression in the upper bound in (ii) we use $r' = r = n/2$, as observed in [104]). For $n = \lfloor m^2/4 \rfloor$ and $k = \lceil m/2 \rceil - 1$ the bound in (i) implies that $\mathrm{bt}(K_{m,n}) \leq m - 1$, so the transition from $m - 1$ to m pages occurs between $n = \lfloor m^2/4 \rfloor$ and $n = m^2/4 + 5/2m^{7/4} + 11m^{3/2} + 8m$. DeKlerk, Pasechnik, and Salazar [104] showed that for $4 \leq m \leq 7$ the transition occurs exactly at $n = \lfloor m^2/4 \rfloor$, that is, $\mathrm{cr}(K_{m,n}) = m - 1$ and $\mathrm{cr}(K_{m,n+1}) = m$ for $n = \lfloor m^2/4 \rfloor$. Their proof rephrases the question as a coloring problem of the intersection graph of the chords, and then uses a computer to check that for all possible drawings, the chromatic number of that graph is greater than 0. To make this computationally feasible, the chromatic number is replaced by the Lovasz θ-function, which is a lower bound on the chromatic number, and which can be calculated using semidefinite programming.

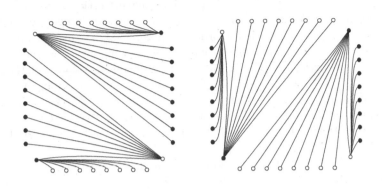

FIGURE 8.2: From $K_{m-2,m-2}$ to $K_{m,m}$. Vertices are colored by partition. The additional four vertices are placed into the corners of the drawing.

Theorem 8.10 (DeKlerk, Pasechnik, and Salazar [104]). $\mathrm{bt}(K_{m,n}) = m - 1$ and $\mathrm{bt}(K_{m,n+1}) = m$ for $n = \lfloor m^2/4 \rfloor$.

8.1.2 A Single Page

8.1.2.1 One-Page Drawings

The 1-page drawing model is equivalent to several other models, and has thus achieved separate attention. If we extend the spine of a 1-page drawing to a closed curve, the drawing lies entirely within that curve, and all vertices lie on this boundary curve, explaining the name *outerplanar* drawing/crossing number. We can go further and find an isomorphic drawing in which the outer curve is convex (a circle), explaining the name *convex* (sometimes *circle*) crossing number for bkcr_1, and this is the name we shall use.

If we fix the cycle order of the vertices along the convex hull, how many ways are there to draw the graph? Or, in Sam Loyd's words: how many ways can the drummers march? (Assuming each drummer crosses every other drummer's path at most once.)

Theorem 8.11 (Felsner [135]; Kynčl [221]). *For a fixed ordering of the endpoints of a graph on the boundary of a circle, there are at most 2^k non-isomorphic, good, convex drawings of the graph, where k is the (unique) number of crossings.*

Proof. All good, convex drawings of the graph are weakly isomorphic, since the order of the endpoints, and the restriction to good drawings, determines which pairs of edges cross, and which do not. Adjacent edges cannot cross, so the rotation at a vertex is determined by the order of the vertices on the boundary, and we can split each vertex v into $\deg(v)$ vertices, so that all edges are independent. The number of crossings k remains unchanged and unique.

What is not determined is the order of crossings along an edge. Arbitrarily orient all edges. For every edge e, let $s(e) \in \{-1, 1\}^{k_e}$ be the vector which encodes the direction of all crossings along e, using a $+1$ for a left-to-right crossings, and a -1 for a right-to-left crossing; k_e is the number of crossings along e. We claim that these vectors determine the drawing up to isomorphism. This already implies the result, since there are at most 2^k ways to choose these vectors (note that a $+1$ crossing between e and f is a -1 crossing between f and e).

We prove the claim by induction on the number of edges and the number of crossings (in this order).

To every edge e correspond two arcs of the circle, let γ_e be the shorter one of the two, where we use the number of vertices along the arc as its length (if both arcs have the same length, we choose one arbitrarily). Let e be an edge with a shortest arc γ_e in this sense. Then no other edge starts and ends on γ_e, otherwise we would have an edge with a shorter arc. If γ_e contains no endpoints, we can remove e, without changing k, and apply induction. So we can assume that γ_e contains at least one endpoint. Orient e and then the edges starting on γ_e so that $s(e) = (+1, \dots, +1)$, which means all edges cross e from left to right. Let f be the edge whose starting point is next to the starting vertex of e and on γ_e. Then $s(f)$ contains a -1 corresponding to it being crossed from right to left by e. Moreover, the entries before that -1 are all $+1$, since before f crosses e, it can only cross other edges starting on γ_e, and those have to cross from the left. In other words, looking at $s(f)$, we can identify the crossing with e by looking for the first -1 entry. All entries before that correspond to edges crossing f, and these edges must next cross e in the triangle formed by the crossing of e and f and its start-vertices (since they cannot end on γ_e). See Figure 8.3. So we can remove the crossing between e and f by swapping their starting points, and removing the -1 entry from f, and the corresponding $+1$ entry from e (they have the same number of crossings before them). This gives us a good, convex drawing with fewer crossings, and we can apply induction to show that the drawing is determined up to isomorphism. □

Theorem 8.11 is a pleasingly precise result, and it gives us good bounds for the complete graph. We essentially proved the following theorem earlier.

Theorem 8.12 (Kynčl [221]). *The number of non-isomorphic good (or pseudolinear) convex drawings of K_n is of order $2^{\Theta(n^4)}$.*

Proof. Review the proof of Theorem 2.4; the lower bound is achieved by convex drawings weakly isomorphic to the initial rectilinear drawing, so all of the drawings are pseudolinear. The upper bound is true for good drawings of K_n. □

In comparison, the number of rectilinear convex drawings is of order

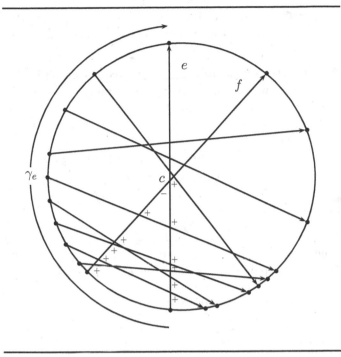

FIGURE 8.3: Two chords e and f, intersections along edges labeled by direction of crossing.

$2^{\Theta(n \log n)}$; the upper bound follows from Theorem 5.26, the lower bound from the fact that we can order the vertices in $n!$ different ways along the convex hull.

We round off this section with some results on substructures of convex drawings. We start with a result on pairwise disjoint edges in convex drawings, commonly attributed to Kupitz [220].

Theorem 8.13. *A good, convex drawing of an n-vertex graph without $k + 1$ pairwise disjoint edges has at most kn edges.*

In other words, a good convex drawing with more than kn edges must contain $k + 1$ pairwise disjoint edges. For a matching lower bound, see Exercise (8.4).

Proof. We can assume that the drawing is convex and rectilinear, with the vertices located at the corners of a regular n-gon. Any two edges that have the same slope are clearly disjoint. Since we assumed that there are no $k + 1$ pairwise disjoint edges, there are at most k edges of each slope, and, since there are at most n slopes, this proves the result. $\square$

For pairwise crossing edges, there also is a sharp bound.

Theorem 8.14 (Capoyleas and Pach [84]). *Every convex drawing of an n-vertex graph in which no $k + 1$ edges cross pairwise, has at most $2kn - \binom{2k+1}{2}$ edges if $n \geq 2k + 1$.*

If $n < 2k + 1$, every convex drawing of an n-vertex graph is $(k + 1)$-quasi-plane, so in that case we only have the trivial bound of $\binom{n}{2}$.

For complete bipartite graphs, more is known. We saw in the proof of Theorem 8.8(ii') that a convex drawing of $K_{m,n}$ contains m pairwise crossing edges if $n \geq \max(m^2 - m + 1, m)$. The proof of Theorem 8.9, which we did not see, establishes the same conclusion under a weaker bound.

Theorem 8.15 (Enomoto, Nakamigawa, and Ota [123]). *Every convex drawing of $K_{m,n}$ contains m pairwise crossing edges if $n \geq m^2/4 + 5/2m^{7/4} + 11m^{3/2} + 8m$.*

Since the authors also showed that $K_{m, \lfloor m/2 \rfloor^2}$ can be embedded in $m - 1$ pages (see the discussion after Theorem 8.9), there are convex drawings of $K_{m, \lfloor m/2 \rfloor^2}$ without m pairwise crossing edges.

8.1.2.2   The Convex Crossing Number

We can get lower bounds on bkcr_1 from $\overline{\mathrm{cr}}(G) \leq \mathrm{bkcr}_1(G)$, but the two parameters may be arbitrarily far apart: $\overline{\mathrm{cr}}(W_n) = 0$, while $\mathrm{bkcr}_1(W_n) = n - 2$. For convex drawings we can easily sharpen the constant of the crossing lemma.

Theorem 8.16 (Crossing Lemma. Shahrokhi, Sýkora, Székely, and Vrt'o [308, 311]). *If G is a graph on n vertices and $m \geq 3n$ edges, then*

$$\mathrm{bkcr}_1(G) \geq \frac{1}{27} \frac{m^3}{n^2}.$$

Proof. If G can be embedded on a single page, then G is outerplanar, so $m \leq 2n - 3$; hence $\mathrm{bkcr}_1(G) \geq m - (2n - 2)$. Using this formula with the standard probabilistic proof of the crossing lemma, see Theorem 1.8, we get

$$\mathrm{bkcr}_1(G) \geq (\lambda^{-2} - 2\lambda^{-3}) \frac{m^3}{n^2},$$

where $p = \lambda n/m$. To maximize the constant, we choose $\lambda = 3$, so we need $m \geq 3n$ for p to be a probability. $\square$

Theorems 3.2 and 5.6 already give us upper bounds on bkcr_1.

Theorem 8.17 (Shahrokhi, Sýkora, Székely, and Vrt'o [308]). *If G is a graph on n vertices and m edges, then*

$$\mathrm{bkcr}_1(G) = O(\log(n)(\mathrm{cr}(G) + \mathrm{ssqd}(G))),$$

and, assuming $m \geq 4n$

$$\mathrm{bkcr}_1(G) = O(\Delta \, \mathrm{cr}(G) \log n).$$

Only a few specific values of the convex crossing number have been studied; the case of the complete graph is easy: $\mathrm{bkcr}_1(K_n) = \binom{n}{4}$ [94, p.34]. The bipartite case is more challenging, and not settled completely.

Theorem 8.18 (Riskin [292]). *If $m|n$, then $\mathrm{bkcr}_1(K_{m,n}) = \frac{1}{12}n(m-1)(2mn-3m-n)$.*

Proof. Color the graph so that the vertices in the left partition are red, and the vertices in the right partition are blue, and assume that $m|n$. We will assume that m is odd (the even case is similar, see Exercise (8.5)).

(i) If D is a convex rectilinear drawing of $K_{m,n}$, and $n_1, \ldots, n_m$ are the number of blue vertices between the i-th and the $i+1$-st red vertex (arbitrarily selecting a first red vertex), then $n_1 + \cdots n_m = n$, and

$$\mathrm{cr}(D) = \binom{m}{2}\binom{n}{2} - \sum_{1 \leq i < j \leq m} (n_i + \cdots + n_{j-1})(n - (n_i + \cdots + n_{j-1})).$$

(ii) $\mathrm{bkcr}_1(K_{m,n})$ is minimized exactly by layouts for which single red vertices alternate with blocks of size n/m of blue vertices.

To see (i), let u be the i-th and v the j-th red vertex. Then the two arcs between u and v contain $(n_i + \cdots + n_{j-1})$ and $n - (n_i + \cdots + n_{j-1})$ blue vertices, and the induced $K_{2,n}$ has $\binom{n}{2} - (n_i + \cdots + n_{j-1})(n - (n_i + \cdots + n_{j-1}))$ crossings; adding this up over all choices of i and j yields (i). For (ii) let us define $n_{i,j} := n_i + \cdots + n_{j-1}$, we use indices modulo m, so $n_{i,m+1}$ is the same as $n_{i,1}$, for example.

$$\begin{aligned}
\binom{m}{2}\binom{n}{2} - \mathrm{cr}(D) &= \sum_{1 \leq i < j \leq m} n_{i,j}(n - n_{i,j}) \\
&= \sum_{i=1}^{m} \sum_{j=1}^{\lfloor m/2 \rfloor} n_{i,i+j}(n - n_{i,i+j}) \\
&= \sum_{j=1}^{\lfloor m/2 \rfloor} \sum_{i=1}^{m} n_{i,i+j}(n - n_{i,i+j}) \\
&= n \sum_{j=1}^{\lfloor m/2 \rfloor} \sum_{i=1}^{m} n_{i,i+j} - \sum_{j=1}^{\lfloor m/2 \rfloor} \sum_{i=1}^{m} n_{i,i+j}^2 \\
&= n^2 \binom{\lfloor m/2 \rfloor + 1}{2} - \sum_{j=1}^{\lfloor m/2 \rfloor} \sum_{i=1}^{m} n_{i,i+j}^2.
\end{aligned}$$

We used the assumption that m is odd (second equality), and that $\sum_{i=1}^{m} n_{i,i+j} = jn$ (last equality). Minimizing $\mathrm{cr}(D)$ is thus equivalent to minimizing $\sum_{j=1}^{\lfloor m/2 \rfloor} \sum_{i=1}^{m} n_{i,i+j}^2$. Which is minimal if we minimize $\sum_{i=1}^{m} n_{i,i+j}^2$ for each $1 \leq j \leq \lfloor m/2 \rfloor$. For $j = 1$ this will be the case if the n_i differ as little as possible, for $j = 2$ if the values of $(n_i + n_{i+1})$ differ as little as possible, and so on. We can achieve this simultaneously for all j, by letting $n_i = n/m$; in that case all the values, for each j, are the same. With this, we have $n_{i,i+j} = jn/m$, and

$$
\begin{aligned}
\mathrm{cr}(D) &= \binom{m}{2}\binom{n}{2} - n^2\binom{\lfloor m/2 \rfloor + 1}{2} + \sum_{j=1}^{\lfloor m/2 \rfloor} \sum_{i=1}^{m} n_{i,i+j}^2 \\
&= \binom{m}{2}\binom{n}{2} - n^2\binom{\lfloor m/2 \rfloor + 1}{2} + \frac{n^2}{24}(m-1)(m+1) \\
&= \frac{n}{12}(m-1)(2mn - 3m - n),
\end{aligned}
$$

as we can verify using a math algebra system, and replacing $\lfloor m/2 \rfloor$ by $(m - 1)/2$. $\qquad\square$

We also know something about the convex crossing number of grid graphs, Exercises (8.6) and (8.7) deal with the cases $P_3 \square P_n$ and $P_4 \square P_n$. For general grid graphs, the next theorem supplies a powerful tool. Let us say that G is $f(k)$-*isoperimetric*, if $|E(U, V(G) - U)| \geq f(k)$ for every $U \subseteq V(G)$ of size k, and $1 \leq k \leq n/2$. Define $\Delta f(i) = f(i+1) - f(i)$, and $\Delta^2 f(i) = \Delta(\Delta f(i))$, the first- and second-order differences of f.

Theorem 8.19 (Shahrokhi, Sýkora, Székely, and Vrt'o [311]). *If G is $f(k)$-isoperimetric and $\Delta f \geq 0$ and decreasing for $1 \leq k < n/2$, where $n = |V(G)| \geq 4$, then*

$$
\mathrm{bkcr}_1(G) \geq -\frac{n}{8} \sum_{k=0}^{\lfloor n/2 \rfloor - 2} f(k)\Delta^2 f(k) - \frac{1}{2}\,\mathrm{ssqd}(G).
$$

Since it is known that grid graphs are $(2k)^{1/2}$-isoperimetric, it follows that

$$
\mathrm{bkcr}_1(P_n \square P_n) = \Omega(n \log n).
$$

A matching upper bound follows from Theorem 3.2.

8.1.3   Two Pages

Using bishellability, it is easy to show that for 2-page drawings of K_n, $Z(n)$ crossings are always necessary.

Theorem 8.20 (Ábrego et al. [3]). *Any good 2-page drawing of K_n has at least $Z(n)$ crossings.*

Proof. We can assume that the 2-page drawing contains a crossing-free Hamiltonian cycle (following the spine closely). Then Lemma 6.15, together with Theorem 6.14, implies the result. $\qquad\square$

Since the Harary-Hill drawings of K_n are 2-page drawings, we have determined the 2-page crossing number of K_n.

Corollary 8.21 (Ábrego et al. [3]).

$$\mathrm{bkcr}_2(K_n) = Z(n).$$

Somewhat more surprisingly (and requiring significantly more analysis) is that for even n the drawing is unique up to isomorphism.

Theorem 8.22 (Ábrego et al. [3]). *For even n there is only one* cr-*minimal 2-page drawing of K_n up to isomorphism.*

As the same paper shows, the result is not true for odd n in which case there can be an exponential number of cr-minimal non-isomorphic 2-page drawings of K_n. Harborth [180] showed that not every good drawing of K_n is isomorphic or even weakly isomorphic to a 2-page drawing: there is a drawing of a K_{14} with a number of crossings which cannot be attained by any 2-page drawing of K_{14}.

For $K_{m,n}$ we first note that the Zarankiewicz drawings we saw in Section 1.2 can be viewed as 2-page drawings (it is easy to add a crossing-free Hamiltonian cycle), so

$$\mathrm{bkcr}_2(K_{m,n}) \le Z(m,n),$$

and we can ask whether equality holds in this model. The best-known lower bounds in general follow from $\mathrm{cr}(G) \le \mathrm{bkcr}_2(G)$, except for the cases $m = 7$ and $m = 8$, where deKlerk and Pasechnik [102] were able to show that $\lim_{n\to\infty} \mathrm{bkcr}_2(K_{m,n})/Z(m,n) = 1$ using semidefinite programming. This implies (using a similar counting argument as the one in Lemma 1.21), that $\lim_{m,n\to\infty} \mathrm{bkcr}_2(K_{m,n})/Z(m,n) \ge 6/7$, which is not as good as the lower bound following from Theorem 1.22.

For the hypercube Q_n we already saw Madej's construction in Section 2.2.2.1 which shows that $\mathrm{bkcr}_2(Q_n) \le \frac{1}{6}4^n - 2^{n-4}(2n^2 + 3 - (-1)^n/3)$. This is asymptotically tight, since we know that $\mathrm{bkcr}_2(Q_n) \ge \mathrm{cr}(Q_n) \ge \frac{1}{20}4^n - (n^2 + 1)2^n$ (Corollary 3.9). The factor in the upper bound can be improved to $125/768$ [130].

8.1.4 The Book Crossing Number

The crossing lemma for convex drawings generalizes to k pages.

Theorem 8.23 (Crossing Lemma. Shahrokhi, Sýkora, Székely, and Vrt'o [308]). *If G is a graph on n vertices and $m \ge 3kn$ edges,*

$$\mathrm{bkcr}_k(G) \ge \frac{1}{27k^2}\frac{m^3}{n^2} - kn.$$

Proof. Fix a crossing-minimal drawing of G in a book with k-pages, and let G_i be the subgraph of G induced by the m_i edges drawn in the i-th page. So

$$\mathrm{bkcr}_k(G) = \sum_{i=1}^{k} \mathrm{bkcr}_1(G_i).$$

Since the function

$$f(x) := \begin{cases} \frac{1}{27}\frac{x^3}{n^2} - n & \text{if } x \geq 3n \\ 0 & \text{otherwise.} \end{cases}$$

is convex, we can use Jensen's inequality[1] to bound $\sum_{i=1}^{k}\mathrm{bkcr}_1(G_i)$ from below; note that $bkcr_1(G_i) \geq f(m_i)$ by Theorem 8.16.

$$\begin{aligned} \mathrm{bkcr}_k(G) &= \sum_{i=1}^{k} \mathrm{bkcr}_1(G_i) \\ &\geq \sum_{i=1}^{k} f(m_i) \\ &\geq k\sum_{i=1}^{k} f\left(\frac{m}{k}\right) \\ &\geq \frac{1}{27k^2}\frac{m^3}{n^2} - kn. \end{aligned}$$

$\square$

Adding pages reduces the book crossing number proportionally.

Theorem 8.24 (Shahrokhi, Sýkora, Székely, and Vrt'o [308])**.** *If $k \geq 1$, then*

$$\mathrm{bkcr}_k(G) \leq \frac{1}{k}\,\mathrm{bkcr}_1(G).$$

Proof. Start with an empty book with k pages. Let D be a cr-minimal convex drawing of G. We will distribute the edges of G among the pages of the empty book using a greedy strategy. At the end, each page will contain a subdrawing of D. Consider each edge $e \in E$, one at a time. If e is involved in $c(e)$ crossings in D, then there must be a page of the book in which e can be inserted (drawn just as it is in D), so that there are at most $c(e)/k$ additional crossings. Repeating this for all edges gives us a k-page drawing of G with at most $\sum_{e \in E} c(e)/k = 1/k\,\mathrm{cr}(D) = 1/k\,\mathrm{bkcr}_1(G)$ crossings. $\square$

Theorem 8.24 allows us to carry over upper bounds from bkcr_1 to bkcr_k at a loss of a factor of $1/k$.

---

[1] If a function f is convex, then $f(\sum_{i=1}^{n} x_j/n) \leq f(\sum_{i=1}^{n} x_i)/n$.

Corollary 8.25 (Shahrokhi, Sýkora, Székely, and Vrt'o [308]). *If G is a graph on n vertices and m edges, then*

$$\operatorname{bkcr}_k(G) = O\left(\frac{1}{k}\log^2(n)(\operatorname{cr}(G) + \operatorname{ssqd}(G))\right),$$

and, assuming $m \geq 4n$

$$\operatorname{bkcr}_k(G) = O(\frac{\Delta}{k}\operatorname{cr}(G)\log^2 n).$$

Corollary 8.25 implies that $\operatorname{bkcr}_k(K_n) \leq 1/k\binom{n}{4}$, but a stronger bound is known.

Theorem 8.26 (Shahrokhi, Sýkora, Székely, and Vrt'o [308]). *If $0 < k < \lceil n/2 \rceil$, then*

$$\operatorname{bkcr}_k(K_n) \leq \frac{1}{12}\frac{n^4}{k^2} - \frac{1}{4}\frac{n^3}{k^2}.$$

The bound is asymptotically sharp by the book crossing lemma, Theorem 8.23. There is a conjectured value for $\operatorname{bkcr}_k(K_n)$, named $Z_k(n)$ in [103]; for a detailed discussion, and exact values for small k and n, see that paper.

Proof. In the proof of Theorem 8.8(*ii*), we saw that K_n can be written as the union of C_n with $\lceil n/2 \rceil$ triangulations $G_1, \ldots, G_{\lceil n/2 \rceil}$ of C_n (without the edges of the C_n), by rotating G_1 one vertex (counterclockwise) at a time (see Figure 8.1). G_1 has $n-3$ edges of which $\lfloor n/2 \rfloor - 1$ are vertical (their endpoints satisfy $x + y \equiv 0 \bmod n$), and at most $\lfloor n/2 \rfloor - 1$ are slanted (their endpoints satisfy $x + y \equiv -1 \bmod n$; if n is even, there are $n/2 - 2$ of these). When we rotate G_1 counterclockwise to get G_2 each vertical edge picks up one crossing (with its former self), and each slanted edge picks up three crossings (with itself, and the two edges it is incident to). After this, each edge picks up at most four crossings with each rotation, two at each end (some edges lose crossing again, we ignore this). So G_1 and G_t, have at most $\lfloor n/2 \rfloor(1 + 3) + 4(n - 3)(t - 2) \leq (n - 3)(4t - 6)$ crossings, where $2 \leq t \leq \lceil n/2 \rceil$. Let b be minimal so that $bk \geq \lceil n/2 \rceil$. To count the crossings of $G_1 \cup \cdots \cup G_b$, we use that $G_i \cup G_j$ has the same number of crossings as $G_1 \cup G_{j-i+1}$, by symmetry:

$$\sum_{1 \leq i < j \leq b} 4(n - 3)(j - i + 1 - 2) = 4(n - 3)\sum_{1 \leq i < j \leq b}(j - i - 1)$$

$$= \frac{2}{3}(n - 3)b(b - 1)(b - 2)$$

$$\leq \frac{n^4 - 3n^3 - 4k^2n^2 + 12k^2n}{12k^3},$$

where we use that $b \leq n/(2k) + 1$. Since k rotates copies of $G_1 \cup \cdots G_b$ are sufficient to cover K_n (we add C_n to one of the pages), this implies that $\operatorname{bkcr}_k(G) \leq 1/12\, n^4/k^2 - 1/4\, n^3/k^2 - 1/3\, n^2 + n \leq 1/12\, n^4/k^2 - 1/4\, n^3/k^2$ (for $n \geq 3$). □

We can use Theorem 8.26 to bound bkcr_k of general graphs in their number of edges.

Corollary 8.27 (Shahrokhi, Sýkora, Székely, and Vrt'o [308]). *If G is a graph on n vertices and m edges, and $0 < k < \lceil n/2 \rceil$, then*

$$\mathrm{bkcr}_k(G) \le \frac{m^2}{3k^2}\left(1 + O\left(\frac{1}{n}\right)\right).$$

Proof. Let D be a cr-minimal drawing of K_n in a book with k pages, where $n = |V(G)|$. By Theorem 8.26, we know that D has at most $1/12\,n^4/k^2 - 1/4\,n^3/k^2$ crossings. Randomly map the vertices of G to the vertices of K_n, and let D' be the induced k-page drawing of G. The probability that a crossing survives is the probability that both edges belong to G, which is at most $3\binom{m}{2}/\binom{n}{4} \le 4m^2/(n(n-1)(n-2)(n-3))$. Then the expected number of crossings is

$$\frac{4m^2}{n(n-1)(n-2)(n-3)}\left(\frac{n^4}{12k^2} - \frac{n^3}{4k^2}\right) \; \le \; \frac{m^2 n^2}{3k^2(n-1)(n-2)}$$

$$= \frac{m^2}{3k^2}\left(1 + O\left(\frac{1}{n}\right)\right).$$

$\square$

The bound is not optimal since Maltiz's theorem, Theorem 8.5, implies that $\mathrm{bkcr}_k(G) = 0$ when k is of order $m^{1/2}$.

Nevertheless, the corollary gives useful bounds, for example

$$\mathrm{bkcr}_k(K_{m,n}) \le \frac{1}{3}m^2 n^2/m^2,$$

which is not much worse than the best known upper bound (roughly $\frac{1}{4}\binom{m}{2}\binom{n}{2}$) using an explicit construction [104].

8.2   Monotonicity

8.2.1   Monotone Drawings

An edge in a drawing is *x-monotone*, or just *monotone*, if it intersects every vertical line at most once; a drawing of a graph is *x-monotone* if all its edges are *x*-monotone and no two vertices have the same *x*-coordinate. Rectilinear drawings are *x*-monotone, but the reverse is not necessarily true. A non-stretchable pseudoline arrangement cannot be realized using straight lines, but since every pseudoline arrangement is isomorphic to a wiring diagram (Lemma 5.10 in Chapter 6), it has an *x*-monotone realization. For embeddings, equivalence can be shown, but it comes at a price.

Theorem 8.28 (Pach and Tóth [266]). *Every x-monotone embedding of a graph is stretchable, that is, there is an isomorphic rectilinear embedding of the same graph in which the x-coordinates of the vertices are the same.*

Proof. We use induction on the number n of vertices of the graph. The cases of $n \leq 3$ are immediate, so we consider $n \geq 4$. We can assume that the graph is triangulated. Suppose there is an inner face which is not a triangle. Suppose the boundary of the face contains a vertex v both of whose neighbors on the boundary are on the same side of v, but v is not the leftmost or rightmost vertex of the face. In that case, move up (or down) starting at v until you hit an edge, then follow that edge to its endpoint u which lies on the other side of v than do its neighbors. In particular, u is not adjacent to v (any vertices adjacent to v must be on the other side), and we can add an x-monotone edge from v to u. We conclude that such a vertex v does not exist, so as we traverse the boundary of the face we only change direction twice, at its leftmost and at its rightmost vertex. Hence we can add an x-monotone edge between any two vertices on the boundary of the face, so we can triangulate it (without creating duplicate edges in the rest of the graph). Using a similar argument, we can also ensure that the outer face is a triangle.

If the drawing contains a separating triangle uvw, that is, a triangle which has vertices both inside and outside of it, we can split the graph into two subgraphs, the triangle with everything inside, and the triangle with everything outside of it. Inductively, we stretch both of these subgraphs. While this keeps the x-coordinates of all points unchanged, the y-coordinates of u, v and w may differ in the two embeddings. We can use an affine transformation of the form $f(x,y) = (x, ax + by + c)$ to match the two triangles, and combine the two embeddings into an embedding of the original graph. From this point on, we can assume that any triangle we encounter bounds an empty face (which may be the unbounded face).

Let u_1 and u_2 be the first and second vertex of the graph (as seen from the left). There must be an edge u_1u_2, since otherwise u_1u_2 would be part of a face of length larger than 3.

Contract u_1u_2 by moving u_1, and edges incident to it, along u_1u_2 towards u_2 and then identifying u_1 with u_2; we can perform this contraction while keeping edges incident to u_1 x-monotone. If this contraction leads to duplicate edges, the graph must contain a triangle u_1u_2v, which, as we argued before, bounds an empty face. So there can be at most two such triangles, one attaching above u_1u_2, and the other one below. Remove the duplicate edges, and apply induction to stretch the resulting drawing. Since the resulting embedding is straight-line, all edges leave u_2 towards the right, so we can separate u_1 and edges incident to it, by moving it slightly to the left, and we can add back any duplicate edges we removed, as well as u_1u_2. We may not be able to move u_1 all the way to its original x-coordinate, while doing so, we may collapse a triangle u_1vw, namely v, lying between u_1 and w, may come to rest on u_1w. Without loss of generality, let us suppose that v lies above u_1w, and both lie above the horizontal line through u_1u_2, which we can assume is the x-axis. Then the line through vw crosses the x-axis to the left of u_1. Using a linear transformation $f(x, y) = (x, cy)$ by a sufficiently small c we can move the crossing of vw with the x-axis arbitrarily far to the left, in particular, we can move it past the original x-coordinate of u_1. This removes v as an obstacle towards sliding u_1 to its original x-coordinate, and we can deal with all obstacles this way. □

The sliding and scaling during the proof may seem like a bad idea, leading to potentially small y-coordinates. It is known, however, that this is unavoidable; there are x-monotone graphs whose realization requires exponentially small y-coordinates [111, 229].

8.2.2 The Monotone Crossing Number

The *monotone crossing number*, mon-cr(G), of a graph G is the smallest number of crossings in an x-monotone drawing of G. By definition, cr(G) ≤ mon-cr(G), and mon-cr(G) ≤ $\widetilde{\text{cr}}(G)$ ≤ $\overline{\text{cr}}(G)$, since, as we mentioned earlier, every pseudoline arrangement is isomorphic to an x-monotone drawing (a wiring diagram). In particular, mon-cr(G) = 0 if and only if G is planar.

Since two edges in a page of a book cross if and only if their endpoints alternate along the spine, we can assume that each edge in a book drawing is x-monotone. So mon-cr(G) ≤ bkcr$_2$(G), and we can consider mon-cr as a relaxed version of bkcr$_2$(G).

For complete bipartite graphs, Zarankiewicz's conjecture implies that mon-cr($K_{m,n}$) = cr($K_{m,n}$), since the optimal drawings are monotone (even rectilinear). For complete graphs, the 2-page drawings are isomorphic to x-monotone drawings, so Hill's conjecture implies that mon-cr(K_n) = cr(K_n), indeed, they are equivalent, as we are about to see.

One way to relax monotonicity is to only require that edges lie between their endpoints, that is, an edge does not extend to the left or right of its endpoints. We called such drawings x-*bounded* earlier, and showed that x-bounded drawings of K_n require at least $Z(n)$ crossings (see Lemma 6.16), which implies the result for x-monotonicity.

Theorem 8.29 (Ábrego et al. [3, 4]; Balko, Fulek, and Kynčl [42]). *Any good x-monotone drawing of K_n has at least $Z(n)$ crossings.*

So if mon-cr(K_n) = cr(K_n), then Hill's conjecture is true. The proof of the theorem by Balko, Fulek, and Kynčl [42] shows that the result remains true for drawings which are not necessarily good if we only count pairs of edges which cross oddly, and require adjacent edges to cross an even number of times (the so-called monotone odd$_\pm$ crossing number of the graph).

For general graphs, the monotone crossing number can be polynomially bounded in the crossing number.

Theorem 8.30 (Pach and Tóth [270]). *We have* mon-cr(G) ≤ $\binom{2\,\text{cr}(G)}{2}$.

This result separates mon-cr from $\widetilde{cr}$, since, as we saw in Chapter 6, there are graphs G with $cr(G) = 4$, and $\widetilde{cr}$ arbitrarily large.

At the core of the proof is a result on redrawing plane graphs. Following Pach and Tóth we call an embedding of a planar graph v-*spinal* if the embedding is x-monotone, v is the leftmost vertex, any two vertices incident on the same face can be connected by an x-monotone curve lying inside the face, and any downwards vertical ray starting at a vertex of the outer face lies entirely in the outer face.

Theorem 8.31 (Pach and Tóth [270]). *Every plane graph has an isomorphic v-spinal embedding for every vertex v incident to the outer face.*

With this result we can easily complete the proof of Theorem 8.30.

Proof of Theorem 8.30. Fix a drawing of G with $k = cr(G)$ crossings, so at most $2k$ edges are involved in crossings. Let H be the subgraph of G consisting of the crossing-free edges. By Theorem 8.31, this H has a v-spinal embedding for an arbitrary vertex v on the outer face. Since the drawing is v-spinal, we can add all edges in $E(G) - E(H)$ as x-monotone curves, resulting in $\binom{2k}{2}$ pairs of crossing edges. Swapping arcs (as in Theorem 3.16, from Chapter 3), we can ensure that any two of these edges cross at most once, giving us a drawing with $\binom{2k}{2}$ crossings. (Note that there cannot be any self-crossings in x-monotone edges, and swapping two arcs formed by two edges keeps both edges x-monotone). □

The result on v-spinal embeddings is easily proved by induction over the blocks of the graph.

Proof of Theorem 8.31. Let D be the plane drawing of G, and v a vertex incident to the outer face. We can assume that G is connected; if not, we can add edges to G to make it connected, and without changing which vertices are incident to each face (there is no need to split a face). After obtaining a v-spinal drawing of the connected graph, we can remove the edges we added, and the resulting drawing is still v-spinal and isomorphic to the original drawing.

Suppose G has a cut-vertex v' so that $G = G_1 \cup G_2$ and $V(G_1) \cap V(G_2) = \{v'\}$. We can assume that $v \in V(G_1)$; $v = v'$ is allowed. By induction, G_1 has a v-spinal embedding, and G_2 has a v'-spinal embedding. We can combine these drawings by placing G_2 close to v' in G_1 and in the face of G_1 into which it belongs; by making the embedding of G_2 very thin, we can place it along one of the edges incident to v', so all vertices of G_1 can be connected to all vertices incident to the face using x-monotone curves, unless v' is the rightmost vertex of G_1, in which case we place G_2 to the right of G_1. See 8.4 for illustrations for this case.

So G is 2-connected. If G is a cycle, we can construct a v-spinal embedding directly: draw the edge preceding v as a long horizontal segment, and place the remaining vertices between, and below that segment. If G is not a cycle, it contains a path P, possibly of length 1, between two vertices xy so

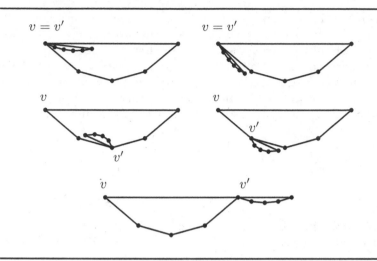

FIGURE 8.4: Combining G_1 and G_2. In the first row, $v = v'$, in the second and third row $v \neq v'$. In the first column, G_2 lies inside G_1, in the second column outside.

that all interior vertices of the path have degree 2, and removal of the path leaves a 2-connected graph G' (this is sometimes known as an open ear decomposition [114, Proposition 3.1.1]). If v belongs to G', we can recursively construct a v-spinal embedding of G'. Then there is an x-monotone curve connecting x and y in the face of G' that contained P. Add that curve to the drawing, and subdivide it, close to x, to reintroduce the vertices along P. If v does not belong to G', it must be one of the interior vertices of P. Suppose $P = xP_1vv'P_2y$. Take an x-spinal drawing of G', add the reverse of xP_1v to the left of it, connect it by a long edge vv' above all other vertices and edges to the far right of the drawing, and then add $v'P_2y$, which must be on the outer face. See Figure 8.5. □

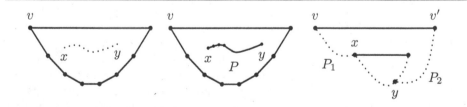

FIGURE 8.5: Recovering G from G'.

Theorem 8.30 begs the question whether cr can be bounded by a multiple of mon-cr. The following result shows that this is the best we can hope for.

Theorem 8.32 (Pach and Tóth [270]). *There is a graph G with $\mathrm{cr}(G) = 6 < 7 = \mathrm{mon\text{-}cr}(G)$.*

It immediately follows that there are graphs with arbitrarily large $\mathrm{cr}(G)$, and $\mathrm{mon\text{-}cr}(G) \geq 7/6\,\mathrm{cr}(G)$.

Proof. Take a wheel on 7 outer vertices $v_1, \ldots, v_7$, in that order, add edges $v_1 v_5$ and $v_3 v_7$ as well as a new vertex u connected to v_2, v_4, and v_6. Figure 8.6 shows two x-monotone drawings of this graph H with 3 crossings. Obtain H'

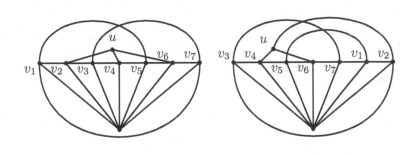

FIGURE 8.6: Two x-monotone drawings of H with three crossings.

from H by replacing every edge of the original wheel with 4 new paths of length 2. Note that in a cr-minimal, or mon-cr-minimal drawing of H', none of these paths are involved in a crossing. Since we know that H' has a mon-cr-minimal drawing with 3 crossings, at least one of the paths replacing an edge of the wheel must be free of crossings, but then we can reroute the other six paths close to the crossing-free paths. Because the (fattened) wheel is free of crossings, edges $v_1 v_5$ and $v_3 v_7$ must cross, and do so once in any drawing with 3 crossings. For u this leaves only the region bounded by $v_3 v_4 v_5$ and arcs from the two crossing edges, creating two crossings, call this the Δ-region. If u were placed in any of the other regions of the drawing, it would require at least 3 additional crossings to connect u to v_2, v_4, and v_6, resulting in at least 4 crossings. So a drawing of H' with 3 crossings is essentially as shown in Figure 8.6, and there are no drawings with fewer crossings.

For G, take two disjoint copies of H', call them H'_1 and H'_2, and connect their u vertices. The left drawing in Figure 8.7 shows that G has crossing number at most 6 (and, since each copy of H' requires 3 crossings, fewer crossings are not possible). We will next show that G satisfies $\mathrm{mon\text{-}cr}(G) = 7$. The upper bound follows from the drawing in Figure 8.7 on the right (the smaller copy of H is based on the right drawing in Figure 8.6, with u moved

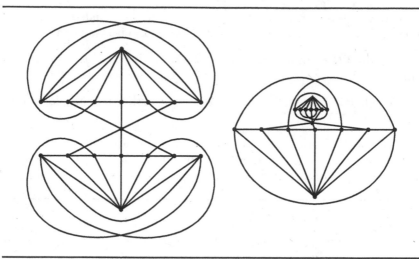

FIGURE 8.7: A cr-minimal drawing of G on the left, and a mon-cr-minimal drawing on the right.

to the outer region). Suppose for a contradiction that mon-cr$(G) = 6$, and fix such a drawing D. Then in this drawing mon-cr$_D(H_1') = $ mon-cr$_D(H_2') = 3$. As we argued earlier, u_1 must lie in the Δ-region of H_1'. For u_1 to be connected to u_2 without crossing (there are no crossings left), u_2 has to lie on the outer face of H_2'. But u_2 has to lie in the Δ-region of H_2', which forces that region to be the outer face of the drawing of H_2' in D. Let us argue that this cannot happen: for v_1v_5 and v_3v_7 to cross in the drawing of H_2', their endpoints must alternate; up to left-right mirroring, there are two possible orders: $v_1 < v_3 < v_5 < v_7$, and $v_3 < v_5 < v_7 < v_1$, where $s < t$ means that s lies to the left of t. Note that in an x-monotone drawing, edges lie entirely between their endpoints, so in the first case, the crossing has to occur between v_3 and v_5, and in the second case between v_5 and v_7. In both cases, the Δ-region is an inner region, which is a contradiction. $\qquad\square$

What is the largest k for which cr$(G) \leq k$ implies that mon-cr$(G) = $ cr(G)? Theorems 8.32 and 5.3 imply that $3 \leq k \leq 5$, but the exact value of k is unknown.

8.3   Notes

The book thickness of planar graphs was first shown to be finite by Buss and Shor [77], needing 9 pages. Heath improved the bound to 7, and Yannakakis to 4 [353]; an early version of Yannakakis paper claims that there are planar graphs requiring 4 pages, but this claim is no longer present in the published version. For a somewhat older survey on book thickness, see [66].

8.4 Exercises

(8.1) Show that every graph on n vertices has a topological embedding in a book with three pages, so that every edge crosses the spine at most $O(\log n)$ times. *Hint:* Binary tree.

(8.2) Show that every planar graph on n vertices has a topological embedding in a book with 2-pages in which at most n edges cross the spine, and each edge crosses the spine at most once.

(8.3) Show that $\operatorname{bt}(K_{m,m}) \leq m - 1$ for $3 \leq m \leq 4$.

(8.4) Show that if $n \geq 2k+1$, there is an n-vertex graph G with a good convex drawing, without $k + 1$ pairwise disjoint edges.

(8.5) Complete the proof of Theorem 8.18 by doing the calculations for even m.

(8.6) Show that $\operatorname{bkcr}_1(P_3 \square P_n) \leq \begin{cases} 2n - 3 & \text{if } n \text{ is odd} \\ 2n - 4 & \text{else.} \end{cases}$. *Note:* The lower bound is also true, but more painful to prove.

(8.7) Show that $\operatorname{bkcr}_1(P_4 \square P_n) \leq 4(n - 2)$. *Note:* Again, equality holds.

(8.8) Show that $\operatorname{bkcr}_k(Q_n) \leq 4^n/(2(2^k - 1))$. *Hint:* Group edges by dimension.

(8.9) Show that $\operatorname{bkcr}_2(C_m \square C_n) \leq (m - 2)n$ for $3 \leq m \leq n$. *Hint:* Find a drawing with crossing-free Hamiltonian cycles.

(8.10) Show that the convex crossing and the 2-page crossing numbers are **NP**-complete, even if the ordering of the vertices on the spine is known. *Hint:* In both cases, the crossing number also remains **NP**-complete to approximate within a constant factor.

(8.11) Show that there are planar graphs G with arbitrarily large $\operatorname{bkcr}_2(G)$. *Note 1:* This shows that $\operatorname{bkcr}_2(G)$ cannot be bounded in $\operatorname{mon-cr}(G)$, which is 0 for planar graphs. *Note 2:* As we mentioned earlier it is an open problem whether there are planar graphs with $\operatorname{bkcr}_3(G) > 0$.

(8.12) Show that crossing number of a random graph $G \sim G(n, 1/n)$ in a book with k pages is almost surely at least $\Omega((n/\log n)^{1/2}/k)$. *Note: G* is expected to have a linear number of edges, so this fits with Malitz's theorem if we let $k = 50n^{1/2}$.

(8.13) Show that mon-cr$(G) \leq \widetilde{\text{cr}}(G)$ for all graphs G. *Hint:* We did the hard work earlier.

(8.14) Show that there are graphs G for which mon-cr$(G) < \widetilde{\text{cr}}(G)$. *Hint:* Again, we have already done the hard work.

Chapter 9

The Pair Crossing Number

9.1 Counting Pairs

The crossing number is the smallest number of pairwise crossings of edges. Starting with this phrasing it is hardly surprising that early on the crossing number was sometimes defined as the smallest number of pairwise crossing edges, or to be more precise, the smallest number of pairs of edges crossing in a drawing. This may not sound too different from the usual definition of the crossing number—and sometimes has been used to define the crossing number—but we know remarkably little about this new notion, the pair crossing number, and how it relates to the standard crossing number.

Definition 9.1. The *pair crossing number*, pcr(D), of a drawing D of a graph is the number of pairs of edges which cross in the drawing. The *pair crossing number*, pcr(G), of a graph G is the smallest pcr(D) of a drawing D of G.

By definition, pcr $\leq$ cr, and it has been conjectured (and the author of this book is happy to follow suit) that pcr = cr, which would make irrelevant most of the results we are about to see. Intuitively, the conjecture seems plausible, as it seems difficult to imagine how multiple crossings between edges can be advantageous; however, attempts at proving equality, even for restricted models, via local redrawing very quickly run into obstacles (flipping two crossing arcs can increase the pair crossing number of the drawing), and current global redrawing techniques seem to have reached a limit, see Section 9.3.2.

In the absence of a proof that pcr(G) = cr(G), we may ask two other questions: how many crossings are required to realize a pcr-minimal drawing? And can cr(G) be bounded in pcr(G)? Let us start with the first question.

Lemma 9.2 (Schaefer and Štefankovič [301]). *If a drawing of a graph contains an edge e with at least 2^t crossings, where t is the number of edges crossing e, then the graph can be redrawn so that the number of crossings along e decreases, the number of crossings along no other edge increases, and no new pair of edges crosses.*

Proof. Let D be a drawing of a graph in which some edge $e = uv$ is involved in at least 2^t crossings with t edges. To every point x on uv we associate a binary vector $c_x \in \{0,1\}^t$ which records the parity of crossing between

ux and f for every edge f crossing e. The crossings split uv into at least $2^t + 1$ segments. Since there are only 2^t vectors, there must be points x and y belonging to different segments, for which $c_x = c_y$. This implies that every edge crosses the curve xy (as part of uv), an even number of times, and there is at least one crossing. Erase the drawing in a small neighborhood of the curve xy, keeping x and y on the boundary; to simplify the description, let us use a homeomorphism of the plane to ensure that the boundary of the neighborhood is convex (Theorem A.3). See Figure 9.1. Let f be an edge

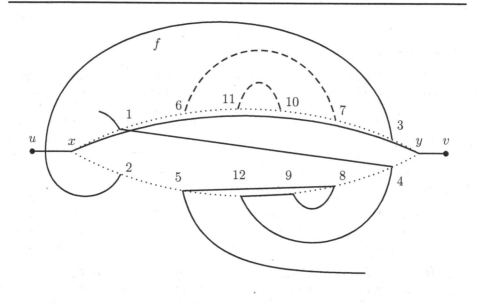

FIGURE 9.1: Reducing the number of crossings along xy. Crossings with one edge f are shown. The boundary of the erased neighborhood is dotted. The dashed parts of f are dropped when shortening f.

which used to intersect xy; then f now has half-edges bordering the small neighborhood of xy. Traverse f. Since f crossed xy an even number of times, we can pair up consecutive crossings of f with xy. Each such pair involves four ends. Connect the first and last end of each pair by a straight line (in the erased neighborhood), and remove the piece of f between the second and third end. We do this for all such pairs, and all edges which crossed uv. Suppose edges f and g cross inside the erased neighborhood after this redrawing. Then the pieces of f and g that were removed when the straight lines were inserted must have crossed outside the neighborhood, since their ends also alternated along the boundary. Since the outside crossing was removed, we have not increased the number of crossings along f or g. The boundary of the neighborhood has as many crossings as xy used to have, so one of the two ways of routing xy

along the boundary must have fewer crossings, and we choose that routing; this reduces the number of crossings along e. □

We conclude that if an edge in a drawing crosses at most t other edges, then it is involved in at most $2^t - 1$ crossings (without increasing the number of crossings along any edge, or adding new pairs of edges which cross). The following corollary answers the first question we asked. It is an open problem whether the exponential bound can be improved.

Corollary 9.3. *Every graph has a* pcr-*minimal drawing with local crossing number at most* $2^k - 1$*, where* $k = \mathrm{pcr}(G)$*. In particular, every graph G has a* pcr-*minimal drawing with at most* $k2^{k+1}$ *crossings.*

Proof. If $k = \mathrm{pcr}(G)$, then every edge crosses at most k other edges, and the bound on the local crossing number follows by applying Lemma 9.2 to each edge. Since there are at most $2k$ edges involved in drawings, the local bound implies the upper bound on the total number of crossings of the drawing. □

We can get a much better bound on pcr if we do not insist on a pcr-minimal drawing. The easiest such bound is $\mathrm{cr}(G) \leq \binom{2\,\mathrm{pcr}(G)}{2}$: in a cr-minimal drawing at most $2\,\mathrm{pcr}(G)$ edges are involved in crossings; Theorem 3.16 then immediately implies the upper bound. Valtr [343] and then Tóth [335] showed that this trivial bound can be improved asymptotically.

Theorem 9.4 (Tóth [335]). *If $k = \mathrm{pcr}(G)$, then $\mathrm{cr}(G) \leq 9k^2/\log^2 k$.*

In Section 9.3.2 we will see how to lower the bound even further using separator theorems. For the proof of Theorem 9.4 we use the following redrawing lemma, a version of Lemma 1.3.

Lemma 9.5. *Given a drawing D of a graph G with edges colored red and blue, we can find a new drawing D' of G so that the number of crossings along each red edge does not increase, and every two blue edges cross at most once.*

Proof. Suppose there are two blue edges e and e' which intersect more than once; as in Lemma 1.3, we can reduce the number of crossings between e and e' without introducing new crossings along any other edge. Remove self-crossings if necessary. □

Proof of Theorem 9.4. Fix a drawing of G with $k = \mathrm{pcr}(G)$ pairs of crossing edges. Color edges red if they are crossed by fewer than t edges, and blue otherwise; we will determine t later. By Lemma 9.5, there is a drawing in which every two blue edges cross at most once, and the number of crossings along no red edge has increased. In particular, crossing-free edges have remained free of crossings. There are at most $2k/t$ blue edges, since $k = \mathrm{pcr}(G)$. Red edges have fewer than 2^t crossings by Lemma 9.2, and there are at most $2k$ red edges involved in crossings. Hence, the total number of crossings is bounded by

$$\binom{2k/t}{2} + 2k2^t,$$

which is at most $9k^2/\log^2 k$ for $t = \log(k)/2$.   □

For x-monotone drawings, Valtr was able to give a stronger bound. In analogy with mon-cr, we define the *monotone pair crossing number*, mon-pcr(G), to be the smallest number of pairs of crossing edges in any x-monotone drawing of G.

Theorem 9.6 (Valtr [343]).

$$\text{mon-cr}(G) \leq \text{mon-pcr}^{\frac{4}{3}}(G).$$

Proof. We proceed as in the proof of Theorem 9.4. We start with an x-monotone drawing D of G in which at most $k = \text{mon-pcr}(G)$ pairs of edges cross and the total number of crossings is minimal. Color edges red if they are crossed by fewer than t edges, and blue otherwise. Let e be an arbitrary red edge and pick a smallest (in terms of enclosed region) bigon formed by $\gamma_e \subseteq e$ and $\gamma_f \subseteq f$ with some other edge f. If that bigon does not contain an endpoint of an edge crossing γ_e, we can reroute γ_f along γ_e, reducing the number of crossings (since any edge crossing γ_e must also cross γ_f), which is not possible. Hence, any bigon a red edge e is involved in must contain an endpoint of an edge crossing e. This implies that another edge f can cross a red edge e at most $2t-1$ times, since they can form at most $2t-2$ bigons (the two endpoints of f cannot lie inside such a bigon in an x-monotone drawing).

We conclude that in D there are at most $k(2t-1)$ crossing involving a red edge. We now apply Lemma 9.5, which gives us a drawing in which every two blue edges cross at most once, and the number of crossings along no red edge has increased. Note that swapping arcs of two monotone edges does not destroy monotonicity of the edges, so the drawing is still monotone.

Since the number of blue edges is at most $2k/t$, the total number of crossings is at most

$$\binom{2k/t}{2} + k(2t-1),$$

which is bounded by $4k^{4/3}$ for $t = k^{1/3}$.   □

The pair crossing number, like the crossing number, is related to the bisection width of the graph; this result, due to Kolman and Matoušek [213], is a consequence of Theorem 11.18, which we will prove in Chapter 11.

Theorem 9.7 (Kolman and Matoušek [213]). *There is a constant $c > 0$ so that*

$$\text{pcr}(G) + \text{ssqd}(G) \geq c \left(\frac{\text{bw}(G)}{\log n}\right)^2.$$

In Remark 1.4 we introduced Rule − according to which we can modify a crossing number so as to only count crossings between pairs of independent edges. Applying this rule to the pair crossing number gives us the *independent pair crossing number*, $\mathrm{pcr}_-(G)$, which is the smallest number of independent pairs of edges which cross in any drawing of G. Bounding cr in terms of pcr_- requires a different approach from the one we took for pcr. It is open, whether $\mathrm{cr}(G) = o(\mathrm{pcr}_-^2(G))$. The best known bound is $\mathrm{cr}(G) \leq \binom{2\,\mathrm{pcr}_-(G)}{2}$, which follows from Theorem 11.13.

9.2 A Crossing Lemma

The basic crossing lemma, Theorem 1.8, also holds for pcr.

Theorem 9.8 (Crossing Lemma. Pach and Tóth [265]). *If G is a graph with n vertices, and m edges, and so that $m \geq 9/2, n$, then*

$$\mathrm{pcr}(G) \geq \frac{1}{60.75}\frac{m^3}{n^2},$$

where $m = |E(G)|$ and $n = |V(G)|$.

The proof requires the Hanani-Tutte theorem, so we postpone it to Chapter 11, where Theorem 9.8 will be an easy corollary of Theorem 11.16. Since we do not know whether $\mathrm{pcr}(G) = \mathrm{cr}(G)$ even for complete or complete bipartite graphs, Theorem 9.8 can be used for lower bounds.

Corollary 9.9. *We have $\mathrm{pcr}(K_n) = \Omega(n^4)$, $\mathrm{pcr}(K_{m,n}) \geq \Omega(m^2 n^2)$.*

The coefficient of $1/60.75$ in the crossing lemma for pcr is weaker than the bounds we have seen for cr. It appears that the improvements which have been achieved for cr do not seem to carry over easily to pcr. In this section, we establish an improved crossing lemma constant for pcr_+, which is the variant of pcr in which we only allow drawings in which no two adjacent edges are allowed to cross (Rule + from Remark 1.4).

The approach mimics the successful route taken for improving the crossing lemma constant, by proving bounds on the local crossing number. The *local pair crossing number* of a drawing is the largest number of edges that any edge in the drawing is being crossed by. The *local pair crossing number*, $\mathrm{lpcr}(G)$, of a graph G is the smallest $\mathrm{lpcr}(D)$ of any drawing D of G.

Lemma 9.10 (Ackerman and Schaefer [15]). *If $\mathrm{lpcr}(G) \leq 2$, then $\mathrm{lpcr}(G) = \mathrm{lcr}(G)$.*

Proof. Fix a cr-minimal drawing of G which realizes $\mathrm{lpcr}(G)$. We will show that if an edge e is crossed by at most two other edges, it crosses each such edge at most once. If e crosses at most one edge, this immediately follows from Lemma 9.2. If e crosses two edges, say f and g, Lemma 9.2 implies that there are at most three crossings. Hence, if e is crossed by some edge, say f, twice, the pattern of crossings along e must be fgf. (We can exclude orders such as ffg in which two crossings with the same edge are consecutive; we can always simplify such a drawing locally; alternatively, applying Lemma 9.2 to that part of e will work). In that case, we can swap the arcs between the two crossings of e and f, see Figure 9.2 (the other case in which f crosses e in the same direction twice is similar). Since f may have had a crossing with

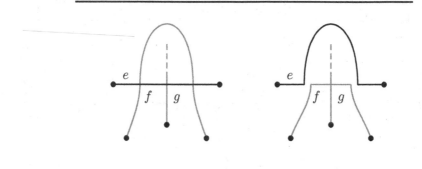

FIGURE 9.2: Removing crossings along an edge.

one other edge, e now crosses at most one edge; f itself now crosses g (which it may not have crossed before), but since f no longer has any crossing with e, f does not cross more edges than it did before. Finally, while we added a crossing of g with f, we also removed the only crossing between g and e, so the number of edges g crosses did not increase. Hence, the resulting drawing still realizes $\mathrm{lpcr}(G)$, but since we reduced the number of crossings, we have a contradiction to the original drawing having been cr-minimal. Hence no edge crosses e more than once. □

If G can be drawn so that every edge crosses at most k other edges, where $0 \leq k \leq 2$, then G has at most $(k+3)(n-2)$ crossings (combine Lemma 9.10 with Theorem 7.3). These bounds allow us to improve the lower bound on pcr, just as we did for cr in Lemma 7.4, with small modifications to the argument only.

Corollary 9.11. *If G is a graph on $n \geq 3$ vertices and m edges, then $\mathrm{pcr}(G) \geq 3m - 12(n-2)$.*

Proof. Fix a pcr-minimal drawing of G. We can assume that $m > 3(n-1)$, since otherwise the right-hand side of the bound is negative. Since for a planar graph $m \leq 3(n-2)$, the drawing must contain at least $m - 3(n-2)$ pairs of crossing edges (if there were fewer pairs, we could remove one edge from each pair, making the drawing crossing-free, but leaving us with too many edges for a planar graph). Hence $\mathrm{pcr}(G) \geq m - 3(n-2)$. If $m > 4(n-2)$, then there must be $m - 4(n-2)$ edges which cross at least two other edges. If not, we could remove fewer than $m - 4(n-2)$ edges to obtain a drawing in which every edge crosses at most one other edge. By Lemma 9.10, the resulting graph is 1-planar, so, by Theorem 7.3, contains at most $4(n-2)$ edges, which is a contradiction. So $\mathrm{pcr}(G) \geq 2(m-4(n-2))+(4(n-2)-3(n-2)) = 2m-7(n-2)$, which is larger than $3m - 12(n-2)$ for $m > 4(n-2)$. Finally, if $m > 5(n-2)$, Theorem 7.3 and Lemma 9.10 combined imply that we have to remove at least $m-5(n-2)$ each crossing at least three other edges, before we get a 2-planar graph, so, we conclude that $\mathrm{pcr}(G) \geq 3(m-5(n-2))+(2(5(n-2))-7(n-2) = 3m-12(n-2)$ in this case. $\qquad\square$

We now have all the tools required to prove a crossing lemma for pcr_+.

Theorem 9.12 (Crossing lemma. Ackerman and Schaefer [15]). *If G is a graph with n vertices, and m edges, and so that $m \geq 6n$, then*

$$\mathrm{pcr}_+(G) \geq \frac{1}{36}\frac{m^3}{n^2}.$$

The coefficient can be further improved to $1/32.4$ using a discharging argument [15], which implies that a graph with $\mathrm{lpcr}(G) = 3$ has at most $6(n-2)$ edges. This would also follow, if Lemma 9.10 could be extended to $\mathrm{lpcr}(G) = 3$, which seems quite likely.

Proof. We adapt the proof of the traditional crossing lemma, Theorem 1.8. Fix a drawing of G which satisfies Rule $+$, that is, there are no crossings between adjacent edges, and which realizes $\mathrm{pcr}_+(G)$. Choosing each vertex of G with probability p we obtain a random subgraph H of G with an expected number pn of vertices of pn, p^2m of edges, and a $p^4\mathrm{pcr}_+(G)$ crossings (any crossing is between pairs of independent edges, because the drawing satisfies Rule $+$).

Then $p^4\mathrm{pcr}_+(G) \geq 3mp^2 - 12(pn-2) \geq 3mp^2 - 12pn$ by Lemma 9.11. Choosing $p = 6n/m$ gives us the lower bound we claimed. $\qquad\square$

Why do we have to work with pcr_+ in place of pcr? The relevant step occurs in the proof of Theorem 9.12 when we invoke Rule $+$ to ensure that a crossing occurs between two independent edges, which therefore have four endpoints. A crossing between two adjacent edges would not give us the same asymptotic bound. For cr this was not an issue, since we could remove adjacent crossings; for pcr we do now know how to do that. Even the question whether $\mathrm{lpcr}_-(G) = 1$ implies $\mathrm{lpcr}(G) = 1$ is open. The case of $\mathrm{lpcr}_-(G) = 0$ is settled by Theorem 11.8.

9.3 String Graphs

If for some set of objects we have a notion of intersection, we can define a corresponding intersection graph with vertices representing objects and edges representing object intersection. This notion has mostly been studied for geometric objects such as intervals on the line (interval graphs), arcs on a circle (circular arc graphs), disks (disk intersection graphs), and so on. We are interested in *strings*, simple curves in the plane.

Definition 9.13. A graph is a *string graph* if it is the intersection graph of a set of simple curves (strings) in the plane.

A set of simple curves represented by an intersection graph is sometimes called its *(string intersection) model*. Every planar graph is a string graph: take a planar rectilinear drawing of the graph, and for each vertex draw a string starting close to the vertex, and traversing each edge incident to the vertex to just beyond the midpoint before turning back and continuing with the next edge in the rotation at the vertex. An edge in this drawing results in the two strings corresponding to its end-vertices crossing twice. Two strings belonging to non-adjacent vertices do not cross, so we have constructed a model of the planar graph, see Figure 9.3 for an example.

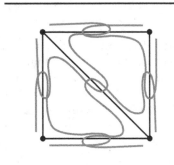

FIGURE 9.3: Constructing a string model of K_4 (strings as gray curves).

In his 1984 thesis, Scheinerman [303] conjectured that every planar graph is the intersection graph of straight-line segments in the plane. Settling this conjecture has turned out to be a hard problem; if the conjecture is strengthened to require the straight-line segments to only use four possible slopes, then its truth implies the four-color theorem. The first, major, break into the problem is due to Chalopin, Gonçalves and Ochem [88] who were able to show that every planar graph can be realized as a string graph in which every pair of strings crosses just once (for a simplified proof, see [62]). Chalopin and Gonçalves [87] have also announced a proof of Scheinerman's conjecture, but the full version of their paper has not appeared yet.

Many natural graph families are string graphs, K_n, $K_{m,n}$, so one may wonder whether there are any graphs which are not. This question was answered by Sinden who found a simple construction to turn a non-planar graph into a non-string graph.

Lemma 9.14 (Sinden [315]). *Let $\hat{G}$ be the graph obtained from G by subdividing each edge once. If $\hat{G}$ is a string graph, then G is planar.*

Proof. Take a string model of $\hat{G}$, let $c(v)$ be the string corresponding to $v \in V(\hat{G})$. The string $c(v)$, where $v \in V(\hat{G}) - V(G)$, may have multiple crossings, but there must be one subarc of $c(v)$ which intersects two strings belonging to vertices of G. Shorten $c(v)$ to that subarc for all $v \in V(\hat{G}) - V(G)$. Contracting each string $c(v)$ for $v \in V(G)$ to a single point $p(v) \in c(v)$ results in a planar drawing of G. $\qquad\square$

What makes string graphs so difficult? The following example due to Kratochvíl and Matoušek [218] gives us a first hint.

Theorem 9.15 (Kratochvíl and Matoušek [218]). *There are string graphs on n vertices for which any model requires at least $2^{n/8-2}$ crossings.*

For the proof we exploit the close relationship between string graphs and the realizability of weak isomorphism types.

Lemma 9.16 (Kratochvíl [216]). *Given a weak isomorphism type on a graph G with m edges, we can construct a graph $\hat{G}$ on at most $2m$ vertices so that the weak isomorphism type on G is realizable if and only if $\hat{G}$ is a string graph. The number of crossings required to realize the weak isomorphism type is at most the number of crossings in the model of $\hat{G}$.*

Proof. Let (G, R), $R \subset E(G)^2$ be the weak isomorphism type. We construct a new graph $\hat{G}$ on vertex set $V(G) \cup E(G)$ and edges $R \cup \{\{v, e\} : v \in e \in E(G)\}$. Then $\hat{G}$ has at most $n + m \leq 2m$ vertices.

We claim that (G, R) is realizable if and only if $\hat{G}$ is a string graph. If (G, R) is realizable, fix a realization of it. For each $v \in V(G)$ erase a small neighborhood of v (turning all edges into strings), and replace v itself with a string that crosses all edges it was incident to, see Figure 9.4. This shows that

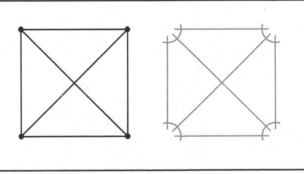

FIGURE 9.4: K_4 and $\hat{K}_4$.

$\hat{G}$ is a string graph. For the other direction, suppose that $\hat{G}$ is a string graph.

Contract all strings corresponding to vertices in $V(G)$ to a point, this gives us a realization of (G, R). □

Our proof of Theorem 9.15 simplifies the original construction, building on an idea from [238].

Proof of Theorem 9.15. We recursively build a weak isomorphism type (G_k, R_k) containing an edge $e_k = x_k y_k$ so that in any realization of (G_k, R_k), edge e_k is involved in at least 2^k crossings.

For $k = 0$ we start with a K_4 with a pair of crossing edges, and let e_0 be one of the two crossing edges. Suppose we have already constructed (G_k, R_k, e_k) as required. To build G_{k+1} we take G_k and add three vertices, x_{k+1}, y_{k+1}, and z_{k+1}, as well as edges $x_k x_{k+1}$, $x_k y_{k+1}$, $y_k z_k$ and $e_{k+1} = x_{k+1} y_{k+1}$. We require that $x_k x_{k+1}$ and $x_k y_{k+1}$ are free of crossings, e_{k+1} crosses the same edges as e_k as well as $y_k z_{k+1}$, and $y_k z_{k+1}$ crosses e_{k+1} only. Figure 9.5 shows that we can extend a realization of (G_k, R_k) to a realization of (G_{k+1}, R_{k+1}) in which e_{k+1} has at least twice as many crossings as e_k. We need to show that in any

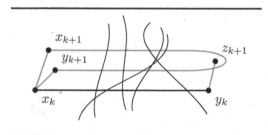

FIGURE 9.5: Edge $x_{k+1} y_{k+1}$ picking up twice the number of crossings of $x_k y_k$.

realization of (G_{k+1}, R_{k+1}), edge e_{k+1} is involved in at least 2^{k+1} crossings, but this is now easy to see: $y_k z_{k+1}$ may only cross e_{k+1}, so, we can shorten it (by moving z_{k+1}), so the two edges cross exactly once. This crossing splits e_{k+1} into two arcs; we can reroute e_k along either arc using the crossing-free part of $y_k z_{k+1}$ at one end, and either $x_{k+1} x_k$ or $y_{k+1} x_k$ at the other end. Since both reroutings give us a realization of (G_k, R_k), both arcs must contain at least 2^k crossings, and e_{k+1} has at least 2^{k+1} crossings.

Graph G_k has $6+4k$ edges, so for every m we can build a weak isomorphism type on a graph with m edges, which requires at least $2^{\lfloor (m-6)/4 \rfloor} \geq 2^{m/8-2}$ crossings. By Lemma 9.16 this gives us a string graph on m vertices, which requires at least $2^{m/8-2}$ crossings. □

The exponential lower bound on string graphs opened up the disturbing possibility that even more crossings may be necessary; and in the absence of any type of upper bound on the number of crossings, it was not even known whether recognizing string graphs was decidable or not. Lemma 9.2 holds the key to a positive answer.

Theorem 9.17 (Schaefer and Štefankovič [301]). *If a string graph with m edges is realizable, there is a string model with at most $m^2 \cdot 2^m$ crossings.*

Proof. If G is a string graph, there is a string model $c(v)$, $v \in V(G)$. For each edge $uv \in E(G)$, fix a crossing $x_{uv} \in c(u) \cap c(v)$. Let H be the graph on vertices x_{uv} with edges corresponding to subarcs of strings connecting the x_{uv}. So H has $m = |E(G)|$ vertices, and at most $2m$ edges (each vertex has degree at most 4). We also have a drawing of H in the plane. Suppose any edge of H is involved in at least 2^m crossings. We then apply Lemma 9.2 to reduce the number of crossings, without introducing any new crossings between edges. Repeating this process for all edges gives us a drawing of H in which every edge has fewer than 2^m crossings. Reinterpreting the edges as subarcs of strings now gives us a string model of G in which every string has at most $m2^m$ crossings. (Fixing the points x_{uv} guarantees that two strings $c(u)$ and $c(v)$ still intersect after the redrawing.) □

An alternative approach to an exponential upper bound was suggested by Kratochvíl [217]; he conjectured that in a cr-minimal drawing of a graph in a given weak isomorphism type, any edge involved in crossings crosses some other edge exactly once. If true, this would imply an exponential upper bound on models of string graphs, see Exercise (9.8). Kratochvíl's problem remains open.

Since Theorem 9.17 gives us an upper bound on the size of a model of a string graph, we can exhaustively test all possible models, implying the following corollary.

Corollary 9.18 (Schaefer and Štefankovič [301]). *The string graph recognition problem is decidable.*

The exponential upper bound of Theorem 9.17 implies that string graphs can be recognized in non-deterministic exponential time, the exponential-time analogue of **NP**. Using some computational topology on compressed representations of curves, it can be shown that the recognition problem lies in **NP**, see [300]. This is somewhat surprising, since, as we saw, models may require exponential size, so there is no time in **NP** to build models explicitly. Together with an earlier result that recognizing string graphs is **NP**-hard [216], this settles the complexity of the string graph recognition problem as **NP**-complete. The string graph problem may appear to be a very specialized visualization problem, but it turns out to be equivalent to the *topological inference problem,* one of the most expressive visualization problems, which is therefore **NP**-complete as well [300, 301].

9.3.1   A Separator Theorem

According to the planar separator theorem, Theorem A.11, the vertices of any planar graph G can be partitioned into three sets A, B, and S so that there are no edges between A and B in $G - S$, and $|S| = O(|E(G)|^{1/2})$. S, or sometimes (S, A, B), is called a *separator* of G, and the separator in this case is *balanced* in the sense that $|A|, |B| \leq 2/3\,|V(G)|$. Our goal in this section is to establish a separator theorem for string graphs, which, as we saw earlier, generalize planar graphs. Let $\sigma(m)$ be the smallest size $|S|$ of a balanced separator (S, A, B) of a string graph with m edges.

Lemma 9.19 (Fox and Pach [137]). *If G is a string graph, then $\mathrm{pcr}(G)$ is bounded above by the number of paths of lengths 2 and 3 in G.*

Proof. Take a string model of G, with string $c(v)$ corresponding to vertex $v \in V(G)$. For every string pick a point $p(v) \in c(v)$. For every edge $uv \in G$ construct a curve γ_{uv} as follows: start at $p(u)$ and follow $c(u)$ to its first crossing with $c(v)$, and then continue along $c(v)$ to $p(v)$. Perturb each curve slightly, so two curves do not share common arcs. This gives us a drawing of G, with vertices at $p(u)$ and edges drawn as γ_{uv}. Suppose two curves γ_{uv} and γ_{wx} cross. Then $c(u) \cup c(v)$ must cross $c(w) \cup c(x)$. If $|\{u, v, w, x\}|$ is three, then there is a path of length 2, with the middle vertex being the common shared endpoint. If $|\{u, v, w, x\}|$ is four, then there is a path of length 3 including edges uv, wx and the edge corresponding to the crossing between $c(u) \cup c(v)$ and $c(w) \cup c(x)$. $\qquad\square$

Theorem 9.20 (Fox and Pach [137]). *Let G be a string graph with $n = |V(G)|$ and $m = |E(G)|$.*

  (i) *If G has p paths of length at most 3, then $\mathrm{bw}(G) = O(p^{1/2}\log n)$.*

 (ii) *If G has maximum degree Δ, then $\sigma(m) = O(\Delta m^{1/2}\log m)$.*

(iii) $\sigma(m) = O(m^{3/4}\log m)$.

Proof. (i) Theorem 9.7 implies that $\mathrm{bw}(g) \leq c\log n (\mathrm{pcr}(G) + \mathrm{ssqd}(G))^{1/2}$ for some constant c. Every vertex v contributes $\deg^2(v)$ to $\mathrm{ssqd}(G)$ which is the same as twice the number of paths of length two with v as middle vertex, plus the number of paths of length one (an edge) incident to v. So $\mathrm{ssqd}(G)$ is twice the number of paths of length one or two, and $\mathrm{ssqd}(G) \leq 2p$. Together with $\mathrm{pcr}(G) \leq p$, by Lemma 9.19, this gives us that $\mathrm{bw}(G) \leq c'p^{1/2}\log n$.

(ii) Every edge is involved in at most $2(\Delta - 1)$ paths of length 2, so there are at most $m(\Delta - 1)$ paths of length 2 in G. Every edge is the middle edge of at most $(\Delta - 1)^2$ paths of length 3, so there are at most $m(\Delta - 1)^2$ such paths. Overall, we have at most $m + m(\Delta - 1) + m(\Delta - 1)^2 \leq m\Delta^2$ paths of length at most 3. Together with (i) we get $\mathrm{bw}(G) \leq c\Delta m^{1/2}\log n$, and the result follows, since $\sigma(m) \leq \mathrm{bw}(G)$. Suppose there is a partition $V(G) = V_1 \cup V_2$ so that $|V_1|, |V_2| \leq 1/3\,|V(G)|$—which implies $|V_1|, |V_2| \geq 2/3\,|V(G)|$—and

$|E(V_1, V_2)| \leq \mathrm{bw}(G)$. Let S be the set of endpoints of edges $E(V_1, V_2)$ in V_1. Then $|S| \leq \mathrm{bw}(G)$, and (S, V_1, V_2) is a separator for G.

(*iii*) Set $d = m^{1/4}/\log^{1/2} m$, and with that define $V_{\leq d} := \{v \in V(G) : \deg(v) \leq d\}$, and $V_{>d} := \{v \in V(G) : \deg(v) > d\}$. Then $|V_{>d}| < m/(2d) < m^{3/4}\log^{1/2} m$, and $G[V_{\leq d}]$ has maximum degree d, so, by (*i*), it contains a balanced separator (S, A, B) with $|S| \leq cdm^{1/2}\log m = cm^{3/4}\log^{1/2} m$ for some constant c. Then $(S \cup V_{>d}, A, B)$ is a balanced separator for G with $|S| \leq (c+1)m^{3/4}\log^{1/2} m$. □

Matoušek [238] was able to reduce the size of a string graph separator to $\sigma(m) = O(m^{1/2}\log m)$. More recently, Lee [224] announced an upper bound of $m^{1/2}$, which is asymptotically optimal.

9.3.2 Improving the pcr-Bound

Theorem 9.4 showed that $\mathrm{cr}(G) = O(k^2/\log^2 k$, where $k = \mathrm{pcr}(G))$; in this section we will lower the upper bound using separators for string graphs. We use the following redrawing lemma [238, 336], a more sophisticated version of Lemma 9.5.

Lemma 9.21 (Tóth [336]; Matoušek [238]). *Given a drawing D of a graph G with edges colored red and blue, we can find a new drawing D' of G so that all edges are drawn arbitrarily close to edges in D, D' is good, and the number of crossings between blue edges has not increased.*

Proof. Start with D. If there are two edges e and e' which intersect more than once, let p_1 and p_2 be two such intersections (allowing a common endpoint) consecutive on e, and let $\gamma \subseteq e$, $\gamma' \subseteq e'$ be two arcs connecting p_1 and p_2. Note that γ and γ' intersect in p_1 and p_2 only, but e may cross γ'. Let (b, r) and (b', r') be the number of crossings with blue and red edges on γ and γ'. If (b, r) is at most (b', r') in the lexicographic order, then we reroute γ' along γ, otherwise, we reroute γ along γ'. In either case we remove at least one crossing between e and e', and do not increase the number of crossings between pairs of blue edges. To see that we are making progress, we look at (c_{bb}, c_{br}, c_{rr}), the number of blue/blue, blue/red and red/red crossings. If both e and e' are blue, then we either reduce c_{bb}, or c_{bb} stays the same and c_{br} is reduced. If both are red, c_{bb} stays the same, and either c_{br} is reduced, or, if it remains the same, c_{rr} is reduced (since we remove at least one of p_1 and p_2, only one of them can be an endpoint). If one of e, e' is blue and the other red, we reduce c_{bb}, unless it stays the same, in which case we reduce c_{br} by removing a crossing between e and e'. □

For the following theorem, recall that $\sigma(m)$ is the smallest size of a balanced separator for a string graph with m edges. Also, we let $\log 0 = 0$ for the purposes of this theorem.

Theorem 9.22 (Tóth [336]). *Given a drawing D of a graph G with $\mathrm{pcr}(D) = k$ and ℓ edges involved in crossings, we can find a drawing of D' of D, in which edges are redrawn close to edges in D, and $\mathrm{cr}(D') \leq ck\sigma(2k)\log\ell$, where $c = 2/\log(3/2)$ is a constant.*

For the proof note that σ is sublinear, so $\sigma(x) + \sigma(y) \leq \sigma(x+y)$, since the union of a string graph with x strings and a string graph with y strings is a string graph with $x + y$ strings.

Proof. Let F be the set of ℓ edges involved in crossings. If $\ell = 0$, then $k = 0$, and there is nothing to prove. Since $\ell = 1$ is not possible (unless we have self-crossings, which we can remove), we can assume that $\ell \geq 2$. Consider the ℓ edges involved in crossings as strings (simple curves); by erasing a small neighborhood of each vertex, we avoid including a common endpoint as a string crossing. We now know that there is a (balanced) separator for the resulting string graph of size $s(\ell)$, so $F = F_0 \cup F_1 \cup F_2$, with $\ell_0 := |F_0| \leq s(\ell)$, and no edge in F_1 crossing an edge in F_2. Let ℓ_i, $i = 1, 2$ be the number of edges in F_i which cross another edge in F_i. We apply induction to D_i, the drawing of the graph induced by F_i in D. With $k_i = \mathrm{pcr}(D_i)$, the redrawing of D_i has $ck_i\sigma(2k_i)\log(\ell_i)$ many crossings, $i = 1, 2$. We add back the edges in F_0 as they were drawn originally. Each edge in F_0 may cross every edge in F, possibly multiple times. At this point, we apply Lemma 9.21 with the blue edges being the edges in $F_1 \cup F_2$, and the red edges being the edges in F_0. After that redrawing, the number of crossings between edges in $F_1 \cup F_2$ has not increased, and the drawing is good, so every edge in F_0 crosses every edge in $F_1 \cup F_2$ at most once, for at most $\ell_0\ell$ additional crossings.

Since all edges are redrawn close to the edges in the original drawing, edges not in F are not affected, and we can add them back without causing new crossings. So the total number of crossings is at most

$$\sum_{i=1}^{2}(ck_i\sigma(2k_i)\log\ell_i) + \ell_0\ell \leq ck\sigma(2k)\log(2/3\,\ell) + \sigma(2k)2k,$$

using (i) the fact that σ is sublinear, (ii) we have $\ell_i \leq 2/3\ell$, since F_0 is a balanced separator, and (iii) $\ell \leq 2k$. Simplifying the expression, we get

$$ck\sigma2k(\log(2/3\,\ell) + 2/c].$$

For this to be at most $ck\sigma(2k)\log\ell$, which is what we have to prove, we need $\log(2/3\,\ell) + 2/c \leq \log\ell$, or $c \geq 2/\log(3/2)$. $\square$

Depending on which separator we use, we get increasingly better results; the last result in the corollary seems to be the limit of this approach, since $k^{1/2}$ is asymptotically optimal for $\sigma(k)$.

Corollary 9.23. *If $k = \mathrm{pcr}(G)$, then for some constant c,*

$$\mathrm{cr}(G) \leq \begin{cases} ck^{7/4} \log^{3/2} k & \textit{using the Fox-Pach separator } [137] \\ ck^{3/2} \log^2 k & \textit{using Matoušek's separator } [238] \\ ck^{3/2} \log k & \textit{using Lee's separator } [224] \end{cases}$$

9.4 Notes

String graphs trace their history to a paper on genetics by Seymour Benzer [58]. The pair crossing number was introduced by Mohar, at an AMS meeting in 1995 [213], but did not appear in print until Pach and Tóth [265] pointed out that researchers in the past had not always carefully distinguished pcr from cr. And we still do not know whether that makes a difference, or not. For a nicely written survey of string graphs and the pair crossing number, see Matoušek's paper on "String Graphs and Separators" [238].

9.5 Exercises

(9.1) Show that $\mathrm{pcr}_- = \mathrm{pcr}$ implies that $\mathrm{pcr} = \mathrm{pcr}_+$.

(9.2) Show that every complete k-partite graph is a string graph.

(9.3) Show that every planar graph has a string model in which every two strings intersect at most once (allowing touching points).

(9.4) Show that string graphs are closed under the induced minor relation (vertex deletions, and edge contractions), but not the minor relation.

(9.5) Show that string graphs are exactly the intersection graphs of paths in a grid graph.

(9.6) Show that if a graph has a k-quasi-planc drawing, then it has at most $O(n \log^{2k-2} n)$ edges. *Hint:* Adapt the proof of Theorem 7.20. Theorem 9.7 will be useful.

(9.7) Given a partial order $\preceq$ on a set V, its incomparability graph has vertices V and edges uv for every u and v which are *incompatible* in $\preceq$, that is, neither $u \preceq v$, nor $v \preceq u$. Show that every incompatibility graph is a string graph. *Hint:* Use the fact that every finite partial order can be written as the intersection of a finite set of total orders.

(9.8) Suppose that in any cr-minimal drawing of a graph in a given weak isomorphism type, any edge which is involved in crossings, is crossed by some edge exactly once. Under this assumption, show that any edge crossing t other edges, has at most $2^t - 1$ crossings. *Note:* Prove this directly, without using Lemma 9.2.

(9.9) Given a weak isomorphism type (G, R) on a graph G, we saw a drawing D *weakly realizes* (G, R) if any pair of edges crossing in D belongs to R (not all pairs of edges in R have to cross in D). Show that a weak realization of a weak isomorphism type may require an exponential number of crossings.

(9.10) Is every hypercube Q_n a string graph? Prove or find the smallest counterexample.

(9.11) Show that there is a string graph which is not the intersection graph of straight-line segments.

Chapter 10

The k-planar Crossing Number

10.1 Drawing in k Planes

What if instead of one plane we have k planes available to us, and we are willing to partition the graph into separate pieces each drawn in its own plane?

Definition 10.1 (k-planar Crossing Number). The *k-planar crossing number*, $\mathrm{cr}_k(G)$, of G, is the smallest value of $\sum_{i=1}^{k} \mathrm{cr}(G_i)$ for any partition G_i of a graph G, that is, $G = \bigcup_{i=1}^{n} G_i$. The function cr_2 is called the *biplanar crossing number*.

We do not require vertices to lie in the same positions on each plane, since we can isomorphically redraw each G_i in its plane to place the vertices anywhere we want; in particular, the number of crossings does not change. The k-planar crossing number is closely related to the book crossing number: if we have a drawing of G on $2k$ pages, then we can combine pairs of pages into a single sheet to obtain a drawing of G on k planes.

Lemma 10.2 (Winterbach [350]). *For $k \geq 1$,*

$$\mathrm{cr}_k(G) \leq \mathrm{bkcr}_{2k}(G).$$

The inequality can be strict: for $k = 1$ we saw in Theorem 8.3 that not every planar graph can be embedded in a book with two pages. For arbitrary k, consider $G = K_{m,m^2/4 + 25m^{7/4}}$ where $m = 2k + 1$. Theorem 8.9(*ii*) implies that $\mathrm{bkcr}_{2k}(G) = 2k + 1$. On the other hand, Theorem 3.18(*ii*) implies that the thickness of G is at most $2k$, so $\mathrm{cr}_k(G) = 0$ [104].[1]

Lemma 10.2 combined with Corollary 8.27 allows us to bound cr_k in terms of the number of edges of the graph.

Corollary 10.3 (Shahrokhi, Sýkora, Székely, and Vrt'o [308]). *If G is a graph on n vertices and m edges, and $0 < k \leq n$, then*

$$\mathrm{cr}_k(G) \leq \frac{1}{12}\frac{m^2}{k^2}\left(1 + O\left(\frac{1}{n}\right)\right).$$

---

[1] We are not aware of any elementary construction separating cr_k and bkcr_{2k} for arbitrary k.

The dependence on k in the corollary is not optimal. Theorem 3.19 showed that for $k = \theta(G) \leq \lfloor (m/3)^{1/2} + 1/6 \rfloor$ every graph with m edges can be embedded in a book with k pages.

We can get a first lower bound on cr_k from Euler's formula.

Lemma 10.4. *If G is a graph drawn in $k \geq 1$ planes, then*

$$\mathrm{cr}_k(G) \geq m - k\frac{g}{g-2}(n-2),$$

where g is the girth of G, unless G is a forest in which case we let $g = 3$, and $n = |V(G)| \geq 3$, $m = |E(G)|$.

Proof. If G has a crossing-free drawing in k planes, then each plane can contain at most $g/(g-2)\,(n-2)$ edges (Corollary A.5), so the total number of edges is at most $kg/(g-2)\,(n-2)$. The result then follows as usual. $\square$

For sufficiently large m we have a crossing lemma for cr_k.

Theorem 10.5 (Crossing Lemma. Shahrokhi, Sýkora, Székely, and Vrt'o [312]). *If $m \geq 4.5kn$, then*

$$\mathrm{cr}_k(G) \geq \frac{1}{60.75}\frac{m^3}{k^2n^2} - kn.$$

Proof. Suppose $G = \bigcup_{i=1}^{k} G_i$, so that

$$\mathrm{cr}_k(G) = \sum_{i=1}^{k} \mathrm{cr}(G_i),$$

and let $m_i := |E(G_i)|$, and $m = |E(G)|$. The function

$$f(x) := \begin{cases} \frac{1}{60.75}\frac{x^3}{n^2} - n & \text{if } x \geq 4.5n \\ 0 & \text{otherwise.} \end{cases}$$

satisfies $\mathrm{cr}(H) \geq f(|H|)$ for all graphs H, applying the crossing lemma, Theorem 1.8, with $\lambda = 4.5$. Since f is convex, we can bound $\sum_{i=1}^{k} \mathrm{cr}(G_i)$ from below using Jensen's inequality, as we did for the book crossing number in Theorem 8.23:

$$\mathrm{cr}_k(G) = \sum_{i=1}^{k} \mathrm{cr}(G_i) \geq \sum_{i=1}^{k} f(m_i) \geq k\sum_{i=1}^{k} f\left(\frac{m}{k}\right) \geq \frac{1}{60.75}\frac{m^3}{k^2n^2} - kn.$$

$\square$

Adding k planes reduces the crossing number, by a factor asymptotic to $1/k^2$ (an improvement by a factor of k over the bound we had for books, Theorem 8.24).

Theorem 10.6 (Pach, Székely, Tóth, and Tóth [261]). *If $k \geq 1$, then*

$$\mathrm{cr}_k(G) \leq \left(\frac{2}{k^2} - \frac{1}{k^3} \right) \mathrm{cr}(G).$$

For $k = 2$ this bound matches an earlier bound on the biplanar crossing number [98].

Proof. The bound is based on a clever partitioning argument. Fix a cr-minimal drawing D of G in the plane. To each vertex $v \in V(G)$ assign a random number $c(v)$ in the range 0 to $k - 1$. Let G_i be the graph consisting of edges uv so that $c(u) + c(v) \equiv i \bmod k$, where $0 \leq i \leq k - 1$. We draw G_i in the i-th plane as follows: draw each connected component of G_i separately (so it does not intersect with any other component of G_i) as it is drawn in D. All edges in a component of G_i are of the same type, where the *type* of an edge uv is the (unordered) pair of labels $\{c(u), c(v)\}$.

The probability that a pair of crossing edges in D survives in our drawing of G in k planes is the same as the probability that the two edges have the same type. This probability is $k(1/k^2)^2 = 1/k^3$ for types where $c(u) = c(v)$, and $\binom{k}{2}(2/k^2)^2 = 2(1/k^2 - 1/k^3)$ for types with $c(u) \neq c(v)$, so $2/k^2 - 1/k^3$ in total. Then the expected number of crossings in the k-plane drawing is $(2/k^2 - 1/k^3)\,\mathrm{cr}(G)$, so there has to be such a drawing. $\qquad\square$

On the other hand, we cannot bound $\mathrm{cr}(G)$ in $\mathrm{cr}_k(G)$ in general: take a K_5 and replace each edge with $m/20$ paths of length two. The resulting graph has m edges, and crossing number at least $m^2/400$, but its biplanar crossing number is 0. Czabarka, Sýkora, Székely, and Vrt'o [98] show that for sufficiently large n and m there are graphs G with $\mathrm{cr}(G) = \Omega(m^2)$ and $\mathrm{cr}_2(G) = O(m^3/n^2)$.

The bound in Theorem 10.6 is asymptotically sharp, that is, there are graphs for which the decay of cr_k is proportional to $1/k^2$.

Theorem 10.7 (Pach, Székely, Tóth, and Tóth [261]). *There are graphs G for which*

$$\mathrm{cr}_k(G) \geq \left(\frac{1}{k^2} - o(1) \right) \mathrm{cr}(G).$$

Proof. We will use $G = K_n$ to show that $\mathrm{cr}_k(G) \geq (c - o(1))/k^2\,\mathrm{cr}(G)$ for some $c > 0$. The stronger result with $c = 1$ uses graphs with density between n and n^2 and requires the fact that the midrange crossing constant exists (see Theorem 1.9).

Fix an $\varepsilon > 0$, and let $n \geq 8k/\varepsilon + 1$. Suppose $\mathrm{cr}_k(K_n) = \sum_{i=1}^{k} \mathrm{cr}(G_i)$ for some partitioning G_i of K_n: $K_n = \bigcup_{i=1}^{k} G_i$. If $|E(G_i)| \geq \frac{\varepsilon}{k}\binom{n}{2}$, then by the crossing lemma, $\mathrm{cr}(G_i) \geq c|E(G_i)|^3/n^2$ (the lemma applies since we chose $n \geq 8k/\varepsilon + 1$). Let I be the set of indices i for which $|E(G_i)| \geq \frac{\varepsilon}{k}\binom{n}{2}$.

The total number of edges in those G_i for which $i \notin I$ is at most $\varepsilon\binom{n}{2}$, so $(1 - \varepsilon)\binom{n}{2}$ edges lie in those G_i with $i \in I$.

Then

$$\mathrm{cr}_k(K_n) \;\geq\; \sum_{i \in I} c \frac{|E(G_i)|^3}{n^2}$$

$$\geq\; \frac{c}{n^2} \sum_{i \in I} |E(G_i)|^3$$

$$\geq\; \frac{c}{n^2} |I| \left(\sum_{i \in I} \frac{|E(G_i)|}{|I|} \right)^3$$

$$\geq\; c \frac{(1-\varepsilon)^3 \binom{n}{2}^3}{k^2 \; n^2},$$

where we used Jensen's inequality for the convex function x^3 in the penultimate inequality. As we let ε go to 0 we get a lower bound of c'/k^2 for $\mathrm{cr}_k(K_n)/\mathrm{cr}(K_n)$. $\qquad\square$

Theorem 10.6 allows us to connect cr_k to the thickness, θ, of the graph.

Corollary 10.8. *For every graph G,*

$$\theta(G) \leq (k\,\mathrm{cr}_k(G))^{1/2} + \frac{7}{6}k.$$

Proof. Let $(G_i)_{i=1}^k$ so that $\mathrm{cr}_k(G) = \sum_{i=1}^k \mathrm{cr}(G_i)$, and $G = \bigcup_{i=1}^k G_i$. Fix a cr-minimal drawing D_i of G_i, so $\mathrm{cr}(D_i) = \mathrm{cr}(G_i)$. For every D_i remove any crossing-free edges, and draw them on a new page; this requires k pages. let G_i' denote the remaining graph and let D_i' be its drawing. Each D_i' contains only edges involved in crossings now, so $m_i' := |E(G_i')| \leq 2\,\mathrm{cr}(G_i') \leq 2\,\mathrm{cr}(G_i)$. By Theorem 3.19, each G_i' can be embedded in a book with $k_i = (m_i'/3)^{1/2} + 1/6$ pages, for a total of

$$\sum_{i=1}^k k_i \;\leq\; k \left(\sum_{i=1}^k \frac{m_i}{3k} \right)^{1/2} + \frac{k}{6}$$

$$\leq\; k \left(\frac{2\,\mathrm{cr}_k(G)}{3k} \right)^{1/2} + \frac{k}{6}$$

$$\leq\; (k\,\mathrm{cr}_k(G))^{1/2} + \frac{k}{6}.$$

Together with the k pages we needed to draw the crossing-free edges, this yields the bound we claimed. $\qquad\square$

The exponent in the corollary is not optimal; Czabarka, Sýkora, Székely, and Vrt'o [98] show that $\theta(G) = 2 + O(\mathrm{cr}_2(G)^{0.406} \log(n))$. They conjecture that the right exponent should be 0.25.

10.2 Bounds and Values

10.2.1 Complete Bipartite Graphs

Theorem 10.9 (Czabarka, Sýkora, Székely, and Vrt'o [97]). *If $m, n \geq ck - 1$, where $c \geq 5$, then*

$$\mathrm{cr}_k(K_{m,n}) \geq \frac{4(c-4)}{c(ck-2)^2} \binom{m}{2}\binom{n}{2}.$$

Asymptotically, we get the best coefficient for $c = (9 - 4/k)^{1/2} + 3 \to 6$ as $k \to \infty$; using $c = 6$ we get a lower bound of $1/(3(3k-1)^2)\binom{m}{2}\binom{n}{2}$ for $\mathrm{cr}_k(K_{m,n})$, as long as $n \geq 6k - 1$.

Proof. By Lemma 10.4, $\mathrm{cr}_k(K_{m,n}) \geq mn - 2k(m+n-2)$, so $\mathrm{cr}_k(K_{ck-1,ck-1}) \geq (ck-1)^2 - 4k(ck-2)$. Using the standard counting argument with $K_{ck-1,ck-1}$ as the subgraph,

$$
\begin{aligned}
\mathrm{cr}_k(K_{m,n}) &\geq \frac{\binom{m}{ck-1}\binom{n}{ck-1}}{\binom{m-2}{ck-3}\binom{n-2}{ck-3}} \mathrm{cr}_k(K_{ck-1,ck-1}) \\
&\geq 4\binom{m}{2}\binom{n}{2}\frac{(ck-1)^2 - 4k(ck-2)}{((ck-1)(ck-2))^2} \\
&\geq \frac{4(c-4)}{c(ck-2)^2}\binom{m}{2}\binom{n}{2},
\end{aligned}
$$

where we used that $(ck-1)^2 - 4k(ck-2)/((ck-1)(ck-2))^2 \geq (c-4)/(c(ck-2)^2)$; this is equivalent to $c((ck-1)^2 - 4k(ck-2)) - (c-4)(ck-1)^2 > 0$, which is true, since $c((ck-1)^2 - 4k(ck-2)) - (c-4)(ck-1)^2 = 4$. □

Extending a thickness construction for complete bipartite graphs, Shahrokhi, Sýkora, Székely, and Vrt'o [312] managed to determine the k-planar crossing number of $K_{2k+1,m}$ for all m, and for $K_{2k+2,m}$ for $1 \leq m \leq 4k^2$. These are rare exact results.

Theorem 10.10 (Shahrokhi, Sýkora, Székely, and Vrt'o [312]). *If $k \geq 2$, then*

(i) $\mathrm{cr}_k(K_{2k+1,n}) = \left\lfloor \frac{n}{2k(2k-1)} \right\rfloor (n - k(2k-1)(\left\lfloor \frac{n}{2k(2k-1)} \right\rfloor - 1))$, *and*

(ii) $\mathrm{cr}_k(K_{2k+2,n}) = 2\left\lfloor \frac{n}{2k^2} \right\rfloor (n - k^2(\left\lfloor \frac{n}{2k^2} \right\rfloor - k^2))$, *with equality for $1 \leq 4k^2$.*

To get a flavor of the construction, we start with the case $k = 2$, and will then show how to prove (ii) in the general case.

Theorem 10.11 (Czabarka, Sýkora, Székely, and Vrt'o [97]). *If $n \geq 1$, then*

$$\mathrm{cr}_2(K_{5,n}) = \left\lfloor \frac{n}{12} \right\rfloor (n - 6\left\lfloor \frac{n}{12} \right\rfloor - 6).$$

Let us call a 2-page embedding *uniform* (also known as as *self-complementary*) if the two subgraphs are isomorphic. The theorem is based on the fact that there is a uniform embedding of $K_{5,12}$ in which even the two embeddings are isomorphic, see Figure 10.1. The rectangular boxes represent the five vertices of the left partition, and the ovals the twelve vertices of the right partition. The notation $a : b$ means that node represents vertex a in the first page, and vertex b in the second page.

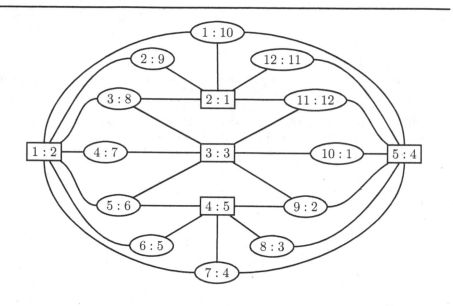

FIGURE 10.1: A uniform 2-page embedding of $K_{5,12}$. Left-partition vertices are in boxes; right-partition vertices in ovals. The pair $a : b$ represents vertex a on the first page, and vertex b on the second page.

Proof. Let $n = 12q + r$ where $0 \leq r < 12$, and split the n vertices into 12 blocks B_i of roughly even size, that is r blocks of size $q+1$ and $12-r$ blocks of size q. To get a 2-page drawing of $K_{5,n}$ replace vertex i (of the right partition) with the vertices in block B_i, $1 \leq i \leq 12$. If we place the vertices of the block along a circular arc, the only intersections are caused by nodes like $3 : 8$, where two vertices of the left partition have to be connected to all vertices in the block from one side. Since each block occurs exactly once in a node like this (the nodes are $1 : 10$, $3 : 8$, $5 : 6$, $7 : 4$, $9 : 2$, $11 : 12$), we obtain a 2-page drawing of $K_{5,12}$ with

$$\sum_{i=1}^{12} \binom{|B_i|}{2} = r\binom{q+1}{2} + (12-r)\binom{q}{2}$$

crossings; substituting $q = \lfloor n/12 \rfloor$ and observing that $n - 6\lfloor n/12 \rfloor = 6\lfloor n/12 \rfloor + r$ yields the upper bound of the theorem.

For the lower bound we use induction on n. For $12 \leq n < 24$, we need to show that $\mathrm{cr}_2(K_{5,n}) \geq n-12$, but this follows from Lemma 10.4 with $k = 2$ and $g = 4$. So we can assume that $n \geq 24$. By the standard counting argument, $\mathrm{cr}_2(K_{5,n+1})/\mathrm{cr}_2(K_{5,n}) \geq \binom{n+1}{n}/\binom{n-1}{n-2} \geq (n+1)/(n-1)$. Define $f(n) :=$ $\lfloor n/12 \rfloor(n - 6\lfloor n/12 \rfloor - 6)$. If we can show that $f(n+1)/f(n) \geq (n+1)/(n-1)$, then $\mathrm{cr}_2(K_{5,n}) \geq f(n)$ follows. If $n \not\equiv 11 \bmod 12$, then $\lfloor (n+1)/12 \rfloor = \lfloor n/12 \rfloor$, and $f(n+1)/f(n) = (n - 6\lfloor n/12 \rfloor - 5)/(n - 6\lfloor n/12 \rfloor - 6) = 1 + 1/(n - 6\lfloor n/12 \rfloor - 6) \geq 1 + 1/(n/2 - 6)$ which is at least $(n+1)/(n-1)$ for $n > 12$. If $n \equiv 11 \bmod 12$, we have $\lfloor (n+1)/12 \rfloor = (n+1)/12$, and $\lfloor n/12 \rfloor = (n-11)/12$. With that $f(n+1)/f(n) = (n+1)/(n-11)(n/2 - 11/2)/(n/2 - 1/2) = (n+1)/(n-1)$. □

Theorem 10.12 (Czabarka, Sýkora, Székely, and Vrt'o [97]). *If $n \geq 1$, then*

$$\mathrm{cr}_2(K_{6,n}) = 2 \left\lfloor \frac{n}{8} \right\rfloor \left(n - 4 \left\lfloor \frac{n}{8} \right\rfloor - 4 \right),$$

with equality for $n \leq 16$.

We establish the upper bound using a uniform 2-page embedding of $K_{6,8}$, see Figure 10.2. We abbreviate $i : i$ to i (for the vertices on the x-axis).

Proof. Lemma 10.4, with $k = 2$ and $g = 4$, shows that $\mathrm{cr}_2(K_{6,n}) \geq 6n - 4(4 + n) = 2n - 16$ which agrees with the formula in the theorem up to $n = 16$.

For the upper bound, let $n = 8q + r$ with $0 \leq r < 8$. Let $b_i := q + 1$ for $1 \leq i \leq r$ and $b_i := q$ for $r + 1 \leq i \leq 8$. The 2-page drawing of $K_{6,n}$ consists of replacing vertex i with b_i vertices placed on the y-axis close to i (on both pages). This results in crossings at nodes $2 : 4$, $3 : 1$, $6 : 8$, and $7 : 5$. For example, connecting node 2 on the y-axis to 1, 2, and 5 and 6 on the x-axis, causes $2\binom{b_2}{2}$ crossings, for a total of

$$2 \sum_{i=1}^{8} \binom{b_i}{2} = 2r \binom{q+1}{2} + 2(8 - r) \binom{q}{2}$$

crossings. With $q = \lfloor n/8 \rfloor$, this formula evaluates to $2\lfloor n/8 \rfloor(4\lfloor n/8 \rfloor + r - 4)$ which equals the formula in the theorem since $n - 4\lfloor n/8 \rfloor = 4\lfloor n/8 \rfloor + r$. □

To generalize this result to arbitrary k, we work with a uniform k-page embedding of $K_{2(k+1),k^2}$. Figure 10.3 shows the upper left quadrant of the embedding; mirroring it with respect to both axes gives the full embedding we are looking for. A node containing $a_1 : \cdots : a_k$ means that in the embedding vertex a_i is placed in the location of this node in page i (and a single number, as before, means that the node has the same location on all pages).

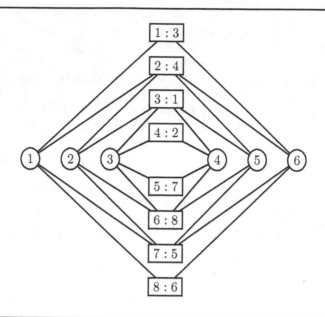

FIGURE 10.2: A uniform 2-page embedding of $K_{6,8}$. Left-partition vertices are in ovals, right-partition vertices in boxes. The pair $a : b$ represents vertex a on the first page, and vertex b on the second page; a is short for $a : a$.

Proof of Theorem 10.10(ii). Lemma 10.4, with $g = 4$, shows that $\mathrm{cr}_k(K_{2k+2,n}) \geq (2k + 2)n - 2k(n + 2k + 2 - 2) = 2n - 4k^2$ which equals the formula in part (*ii*) up to $n = 4k^2$.

To see the upper bound, let $n = 2k^2q + r$ with $0 \leq r < 2k^2$, and $b_i := q+1$ for $1 \leq i \leq r$ and $b_i := q$ for $r + 1 \leq i \leq 2k^2$. Take the k-page embedding of $K_{2k+2,2k^2}$ described by Figure 10.3, with the vertices on the x-axis labeled $1, \ldots, 2k+2$, and the y-axis $1, \ldots, 2k^2$ (the figure shows only half the vertices). On each page, replace vertex i on the y-axis with b_i vertices also on the y-axis, and close to i. There are $2k$ nodes in the embedding of the K_{2k+2,k^2} which are incident to two vertices on each side of the y-axis. Since each of these nodes contains each of the blocks of vertices in its half exactly once, the drawing has

$$2\sum_{i=1}^{2k^2} \binom{b_i}{2} = 2r\binom{q+1}{2} + 2(2k^2 - r)\binom{q}{2}$$

crossings. Substituting $q = \lfloor n/(2k^2) \rfloor$ yields the bound in (*ii*).   □

The proof of (*i*), giving the exact result for $\mathrm{cr}_k(K_{2k+1,n})$ is not dissimilar, but requires some more effort since we do not know any uniform k-page embedding of $K_{2k+1,n}$.

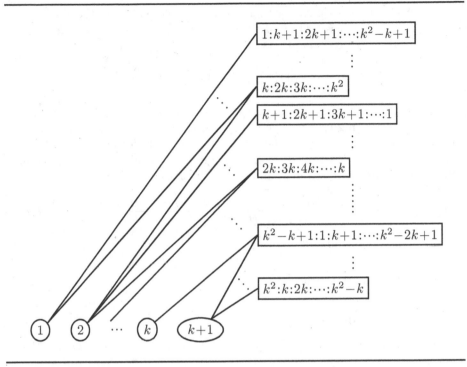

FIGURE 10.3: The upper left quadrant of a uniform k-page embedding of $K_{2k+2,2k^2}$.

We can exploit the uniform k-page embedding for a general upper bound on $\mathrm{cr}_k(K_{m,n})$.

Theorem 10.13 (Shahrokhi, Sýkora, Székely, and Vrt'o [312]). *If* $m \geq 2k+2$ *and* $n \geq 2k^2$, *then*

$$\mathrm{cr}_k(K_{m,n}) \leq \frac{k^2+k+2}{16(k(k+1))^2} M^2 N^2,$$

where $M = (2k+2)\lceil m/(2k+2) \rceil$ *and* $N = 2k^2 \lceil n/(2k^2) \rceil$.

Asymptotically, this implies that $\mathrm{cr}_k(K_{m,n}) \leq 1/(16k^2)(mn)^2 + O((m+n)\max(m,n))$. Theorem 10.9 gives a lower bound of $\mathrm{cr}_k(K_{m,n}) \geq 1/(3(3k-1)^2)\binom{m}{2}\binom{n}{2} = 1/(12(3k-1)^2)((mn)^2 - O((m+n)\max(m,n))$ for $m,n \geq 6k-1$. So the *bipartite k-page crossing constant* satisfies $1/(16k^2) \leq \lim_{m\to\infty} \mathrm{cr}_k(K_{m,m})/m^4 \leq 1/(12(3k-1)^2)$. For $k=2$, for example, the bounds are $1/300 \leq \lim_{m\to\infty} \mathrm{cr}_2(K_{m,m})/(mn)^2 \leq 1/72$.

Proof. Take the k-page embedding of $K_{2k+2,2k^2}$ from Figure 10.3 and replace each vertex on the x-axis with $m' := M/(2k+2)$ vertices on the x-axis, close to the original vertices; similarly, we replace each vertex on the y-axis with $n' := N/(2k^2)$ vertices on the y-axis, close to the original vertex. The number of crossings in this drawing is

$$4k\left((k+1)\binom{m'}{2}\binom{kn'}{2} + km'^2\binom{n'}{2}\right),$$

where the first term in the sum counts crossings incident to two endpoints on the x-axis belonging to the same block, and the second term counts crossings whose endpoints on the x-axis belong to two different blocks, all for one quadrant on one page. Multiplied by $4k$, this gives the total number of crossings. Substituting $m' = M/(2k+2)$ and $n' = N/(2k^2)$ yields

$$\frac{(k^2+k+2)MN - 2k(k(k(-2k+m+n-4)+3m+2n-2)+n)}{16(k(k+1))^2}MN$$

$$\leq \frac{(k^2+k+2)}{16(k(k+1))^2}M^2N^2,$$

the upper bound of the theorem. $\qquad\square$

10.2.2 Complete Graphs

Theorem 10.14 (Shahrokhi, Sýkora, Székely, and Vrt'o [312]). *If $n \geq 12k-1$, then*

$$\mathrm{cr}_k(K_n) \geq \frac{1}{2(3k-1)^2}\binom{n}{4}.$$

By Theorem 10.6 we have

$$\mathrm{cr}_k(K_n) \leq \left(\frac{2}{k^2} - \frac{1}{k^3}\right)Z(n) = \frac{1}{128}\left(\frac{2}{k^2} - \frac{1}{k^3}\right)n^4 + O(n^3),$$

so we understand $\mathrm{cr}_k(K_n)$ asymptotically, and we can bound the *k-page crossing constant* by $\frac{1}{2}(3k-1)^2 \leq \lim_{n\to\infty}\mathrm{cr}_k(K_n)/\binom{n}{4} \leq 3/(8k^2)$. Using finite geometries, the upper bound can be improved by a small factor for some k [312].

Proof. Let us write $a = \lfloor n/2 \rfloor$; by assumption, both a and $n-a$ are at least $6k-1$, so $K_{a,n-a}$ has crossing number at least $1/(3(3k-1)^2)\binom{a}{2}\binom{n-a}{2}$. Since $K_{a,n-a}$ occurs $\binom{n}{a}$ times in K_n, and each pair of crossing edges is counted in $4\binom{n-4}{a-2}$ subgraphs $K_{a,n-a}$, we conclude that K_n has at least

$$\frac{1}{3(3k-1)^2}\binom{a}{2}\binom{n-a}{2}\frac{\binom{n}{a}}{4\binom{n-4}{a-2}}$$

$$= \frac{1}{48(3k-1)^2}n(n-1)(n-2)(n-3)$$

$$= \frac{1}{2(3k-1)^2}\binom{n}{4},$$

which is what we claimed. □

The current best upper bound on $\mathrm{cr}_k(K_n)$ in general is achieved by the $(k/2)$-page book drawing we saw in Theorem 8.26. In the biplanar case we know more. Durocher, Gethner, and Mondal [119] showed that $\mathrm{cr}_2(K_{10}) = 10$ and $4 \leq \mathrm{cr}_2(K_{11}) \leq 6$. They are also able to show that $\mathrm{cr}_2(K_n) < \mathrm{bkcr}_4(K_n)$; the difference is of order n^3, so it does not lead to an improvement in the biplanar crossing constant.

10.2.3 Random Graphs

We saw in Theorem 3.7 that a random graph almost surely has crossing number at least $\Omega(p^2 n^4)$ if $p = c/n$ and c is sufficiently large. Spencer showed that this result can be extended to cr_2.

Theorem 10.15 (Spencer [318]). *If $G \sim G(n,p)$ with $p = c/n$ and $c \geq 400$, then almost surely*

$$\mathrm{cr}_2(G) = \Omega(p^2 n^4).$$

Proof. Let G_1 and G_2 be so that $G = G_1 \cup G_2$ and $\mathrm{cr}_2(G) = \mathrm{cr}(G_1) + \mathrm{cr}(G_2)$. We obtain a lower bound on $\mathrm{cr}_2(G)$ by finding a lower bound on $\mathrm{bw}(G_1) + \mathrm{bw}(G_2)$. Let (X_1, Y_1) and (X_2, Y_2) be partitions of $V = V(G)$ so that all sets in the partitions have size at least $|V|/3$, and $\mathrm{bw}(G_i) = E(X_i, Y_i)$, $i = 1, 2$. If $|X_1 \cap X_2| < |V|/6$ or $|Y_1 \cap Y_2| < |V|/6$, then both $X_1 \cap Y_2$ and $X_2 \cap Y_1$ must have size at least $|V|/6$. Since $c \geq 400 \geq 8\ln(4)36$, Lemma 3.6 implies that $E(X_1 \cap Y_2, X_2 \cap Y_1)$ almost surely has size at least $p/2 \, (n/6)^2$. Every edge in this set lies in both $E(X_1, Y_1)$ and $E(X_2, Y_2)$, so if it belongs to G_1, such an edge contributes to $\mathrm{bw}(G_1)$, and if it belongs to G_2, it contributes to $\mathrm{bw}(G_2)$. We conclude that $\mathrm{bw}(G_1) + \mathrm{bw}(G_2) \geq p/2 \, (n/6)^2$. The same argument as in Theorem 3.7 then concludes the proof.

So both $|X_1 \cap X_2|$ and $|Y_1 \cap Y_2|$ have size at least $|V|/6$, and we can then use the same argument with $X_1 \cap X_2$ and $Y_1 \cap Y_2$ in place of $X_1 \cap Y_2$ and $X_2 \cap Y_1$. □

10.3 Rectilinear and Geometric k-Planar Crossing Numbers

If we require the drawings on each page to be rectilinear, we get the *rectilinear k-planar crossing number*, $\overline{\mathrm{cr}}_k$. Some results carry over, for example Theorem 10.6, in which the drawings on each page are subdrawings of the initial drawing. If the initial drawing is rectilinear, then so are all the subdrawings.

Corollary 10.16 (Pach, Székely, Tóth, and Tóth [261]). *If $k \geq 1$, then*

$$\overline{\mathrm{cr}}_k(G) \leq \left(\frac{2}{k^2} - \frac{1}{k^3}\right) \overline{\mathrm{cr}}(G).$$

Similarly, Theorem 10.7 can be extended to the rectilinear case, showing that there are graphs for which $\overline{\mathrm{cr}}_k(G) \geq (1/k^2 - o(1))\,\overline{\mathrm{cr}}(G)$, settling the asymptotics of $\overline{\mathrm{cr}}_k$.

The rectilinear k-page crossing number can be bounded in the k-page crossing number for sufficiently dense bounded degree graphs, as the following result shows.

Theorem 10.17 (Shahrokhi, Sýkora, Székely, and Vrt'o [312]). *If G is a graph on n vertices, and $k \geq 1$, then*

$$\overline{\mathrm{cr}}_k(G) = O(\log(n)(\mathrm{cr}_k(G) + \mathrm{ssqd}(G))).$$

Proof. By Theorem 8.17,

$$\overline{\mathrm{cr}}_1(H) \leq \mathrm{bkcr}_1(H) = O(\log(n)(\mathrm{cr}(H) + \mathrm{ssqd}(H)))$$

for any graph H. If $G = \bigcup_{i=1}^{k}$ so that $\overline{\mathrm{cr}}_k(G) = \sum_{i=1}^{k} \overline{\mathrm{cr}}_1(G_i)$, we can apply Theorem 8.17 to each G_i and add up the results to obtain the desired bound. $\square$

We turn to our final variant of the k-planar crossing number. What happens if we not only require that drawings on every page are rectilinear, but that vertices have the same location on each page (imagine the book closed, all copies of a vertex must lie on top of each other)? We call this the *geometric k-planar crossing number*, $\overline{\overline{\mathrm{cr}}}_k(G)$, of a graph G. In a way, this is the most natural variant if we think of applications, say to circuit design, and *geometric thickness*, the smallest number k for which $\overline{\overline{\mathrm{cr}}}_k(G) = 0$, has been studied, but little is known about the associated crossing number. The upper bound in Theorem 10.6 is not known to carry over (we can no longer move disconnected subdrawings on the same page apart), but a weaker bound can be salvaged.

Theorem 10.18 (Pach, Székely, Tóth, and Tóth [261]). *If $k \geq 1$, then*

$$\overline{\overline{\mathrm{cr}}}_k(G) \leq \frac{1}{k}\,\overline{\mathrm{cr}}(G).$$

On the other hand, the lower bound carries over from $\overline{\mathrm{cr}}_k$. For complete graphs, we can prove better asymptotics.

Theorem 10.19. *If $n \geq k^2 + k + 1$, we have*

$$\overline{\overline{\mathrm{cr}}}_k(K_n) \leq \frac{512}{k^2} \binom{n}{4}.$$

Proof. Let p be a prime between $k/2$ and $k - 1$, and let us assume that n is a multiple of $p^2 + p + 1$ (this can be achieved by at most doubling n). Arbitrarily position the vertices of K_n in a plane (in general position), leading to a rectilinear drawing with at most $\binom{n}{4}$ crossings. To distribute the edges of the K_n to k planes, we make use of a design. Since p is a prime, there is a finite projective plane with $p^2 + p + 1$ nodes and lines so that each line consists of $p + 1$ nodes and each node belongs to $p + 1$ lines. Let $n' = n/(p^2 + p + 1)$, and split the n vertices of K_n into $p^2 + p + 1$ groups of size n'. Assign each group to a node of the finite projective plane. Draw (rectilinear) edges between vertices of groups belonging to the same line on the same plane. This gives us a drawing of K_n on $p + 1 \le k$ planes. Each plane contains edges between $p + 1$ nodes, so a complete graph on at most $(p + 1)n'$ vertices, requiring at most $\binom{(p+1)n'}{4}$ crossings on that plane. Since each node occurs on two planes (since it belongs to two lines), the total number of crossings is at most $2(p^2 + p + 1)\binom{(p+1)n'}{4} \le 2(p + 1)/p^3 \binom{n}{4}$. Using that p lies between $k/2$ and $k - 1$, this gives us an upper bound of $16/k^2 \binom{n}{4}$. Since we may have doubled n, this is $512/k^2 \binom{n}{4}$ in the original n. $\qquad\square$

10.4 Notes

For some results on hypercubes and grids, see [97]. The survey papers by Czabarka, Shahrokhi, Sýkora, Székely, and Vrt'o [97, 98, 312] still make for good reading on the subject of k-planar crossing numbers.

10.5 Exercises

(10.1) Show the following variant of the crossing lemma, Theorem 10.5: $\text{cr}_k(G) \ge \frac{1}{64} m^3/n^2$ as long as $m \ge 6kn$, where $m = |E(G)|$ and $n = |V(G)|$.

(10.2) We saw that $\text{cr}_2(K_{m,n}) \ge \frac{1}{300}(mn)^2 - O((m + n)\max(m, n))$. Improve the factor $1/300$. *Hint:* Pick your subgraph $K_{a,b}$ wisely.

(10.3) Show that $\text{cr}_2(K_{7,12}) = 16$. *Hint:* There is a uniform drawing of $K_{7,12}$ with 8 crossings.

(10.4) Show that $\text{cr}_2(K_{8,12}) = 24$. *Hint:* There is a uniform drawing of $K_{8,12}$ with 12 crossings.

(10.5) Show that $\mathrm{cr}_2(K_{8,n}) \leq 4\lfloor n/6 \rfloor (n - 3\lfloor n/6 \rfloor - 3)$ with equality for $n \leq 12$. *Hint:* Solve the previous exercise first.

(10.6) Show that $\mathrm{cr}_2(K_{10}) = 2$.

(10.7) State and prove a crossing lemma for $\overline{\mathrm{cr}}_k$. *Note:* Use the lower bound on the rectilinear crossing constant.

(10.8) Prove Theorem 10.18.

Chapter 11

The Independent Odd Crossing Number

11.1 Removing Even Crossings

When two edges cross twice in a drawing, we can swap the arcs involved, and reduce the number of crossings along both edges by two (we first encountered this move in Lemma 1.3). It seems as if we can think of crossings along an edge modulo 2. This thought inspires a different way to count crossings: the *odd crossing number*, $\mathrm{ocr}(D)$, of a drawing D is the number of pairs of edges that cross an odd number of times (*oddly*, from now on). When distinguishing between two edges crossing evenly or oddly, we call this the *crossing parity* of the two edges. The *odd crossing number* of a graph G, $\mathrm{ocr}(G)$, is the smallest $\mathrm{ocr}(D)$ of any drawing D of G. By the argument we sketched we should have $\mathrm{ocr}(G) = \mathrm{cr}(G)$. We will see presently that this is not true (what is wrong with our intuitive argument that we can remove two crossings from a pair of edges?), but some of the intuition can be salvaged.

We write ocr_Σ for the odd crossing number on surface Σ. An edge in a drawing *even* if it crosses every other edge an even number of times (including not at all). Otherwise, the edge is *odd*. A drawing is *even* if all edges in the drawing are even, or, equivalently, the odd crossing number of the drawing is zero.

Theorem 11.1 ((Weak) Hanani-Tutte Theorem. Cairns and Nikolayevsky [82], Pelsmajer, Schaefer, and Štefankovič [277]). *If* $\mathrm{ocr}_\Sigma(G) = 0$, *then* $\mathrm{cr}_\Sigma(G) = 0$. *Moreover, if G has an even drawing in Σ, then the embedding scheme of that drawing describes an embedding of G in Σ.*

At the heart of the proof is a simple redrawing idea (based on [277]).

Lemma 11.2. *Suppose G is a multigraph drawn in surface Σ.*

(i) *If $e \in E(G)$ is an even edge so that $\Sigma - e$ is connected, then we can redraw $G - e$ so that e is free of crossings, and the crossing parity of no pairs of edges changes.*

(ii) *If $F \subseteq G$ is a forest of even edges, then we can redraw G so that F is free of crossings, and the crossing parity of no pairs of edges changes.*

Proof. (*i*) Cut all edges crossing e. Since e was even, every edge f crossing it has an even number of ends on each side of e. Pair up these ends on each side, and connect them (along e). This does not change the parity of crossing between any two edges, but some edges may now consist of multiple components (all but one of them closed). Since $\Sigma - e$ is connected, we can reconnect all components of an edge to each other without changing the parity of crossing between any pair of edges while avoiding e: let γ be a curve in $\Sigma - e$ connecting the two disconnected components (and not passing through any vertices). We can then reconnect the components along that curve. This does not change the crossing parity between any pair of edges, since crossings introduced along γ come in pairs. If we introduced any self-crossings of edges, these can be removed locally just as in the case of cr.

(*ii*) Let $E(F) = \{e_1, \ldots, e_k\}$, and let $F_i := \{e_1, \ldots, e_i\}$, $0 \leq i \leq k$. We show by induction that G can be redrawn so that F_i is free of crossings. Since $F_0 = \emptyset$, the base case is true trivially. So suppose we have already redrawn G, without changing crossing parity of any pair of edges, so that F_{i-1} is free of crossings. We can now use the redrawing procedure from part (*i*) to clear e_i of all crossings; when reconnecting closed components we can avoid not just e_i but also the crossing-free edges in F_{i-1}, since F_i is a forest, so $\Sigma - F_i$ is connected, since edges of F_i are crossing-free.   □

A simple observation: a closed curve which is separating (removing it separates the surface into more than one piece) crosses every other closed curve an even number of times.

Proof of Theorem 11.1. We will show the theorem by induction on the number of vertices and edges, and the Euler genus of Σ. To make the induction work we extend the claim to multigraphs:

> If a multigraph G has a drawing on Σ in which every pair of edges crosses evenly, then the embedding scheme of the drawing describes an embedding of G on Σ.

Suppose we have a counterexample G and Σ, where Σ has minimal Euler genus, and G minimal size. Fix a drawing D of G on Σ with $\mathrm{ocr}(D) = 0$. Suppose G contains an edge uv which is not a loop. We can assume that uv has signature 1; if not, we flip the rotation at v as well as the signature of all edges incident to v, resulting in an equivalent embedding scheme. We now contract uv by moving v towards u along uv, identifying u and v, and joining their rotations. Since uv was even, contracting it does not change the parity of crossing between any pair of edges. Embed the resulting graph inductively, and uncontract uv slightly (giving it signature 1) to get an embedding of G with an embedding scheme which is equivalent to the original embedding scheme, contradicting the assumption that G is a counterexample.

Hence, we know that all edges in G are loops. Suppose there are more than two vertices. Connect any two of them, say u and v by a curve γ; as above, we can assume that γ has signature 1. Move v towards u along γ, identify the vertices, and join the rotation system of u and v. In the resulting drawing every pair of edges still crosses evenly: while γ may have odd crossings, all edges incident to v are loops, so the parity of crossing between no pair of edges changes as v is moved through a crossing with an edge. At this point we know that we have a single vertex u with loops, a bouquet.

If Σ is the sphere, we are done: pick a loop whose ends are closest in the rotation at u. Then the two ends must be consecutive (any edge starting between the two ends of u has to end between them too, since any two loops cross evenly, and those two ends would be closer). We can then remove the edge, apply induction, and reintroduce the edge in its original position in the rotation, since its two ends are consecutive. So we know that Σ is not the sphere.

If u is incident to a non-separating loop e, we clear e of crossings using Lemma 11.2(i). We can then cut the surface along e. This splits u into two vertices, and we obtain a drawing of a modified graph G' (with all pairs of edges crossing evenly) on a surface with hole(s). If e is two-sided, there are two holes, each with a copy of u on the boundary. Filling in both holes with disks gives us a surface with smaller Euler genus than Σ, so we can apply induction to embed G' with its induced embedding scheme. We then drill two holes close to the two copies of u, in the same place in the rotation where the original holes were located, and identify the boundaries of the two holes and the two copies of u to get an embedding of G with the original embedding scheme. If e is one-sided, there is one hole containing both copies of u on its boundary. Every edge between the two copies has flipped signature. Fill in the hole with a disk, and merge the two copies of u on the disk. Since the resulting surface has smaller Euler genus than Σ, we can apply induction to get an embedding of the resulting graph. Separate u into two copies (recovering the rotations of the two copies), cut a hole with both copies of u on its boundary (not interfering with the rest of the embedding), and identify the boundary with itself so it becomes a one-sided curve and the two copies of u become one. We can then reinsert e along the one-sided curve. This gives an embedding of G with an embedding scheme which is equivalent to the original embedding scheme.

We are left with the case that u is incident to separating curves only. Since Σ is not the sphere, it contains a non-separating curve γ (avoiding u). This curve must cross every loop incident to u an even number of times, otherwise that loop would be non-separating by the observation before the proof. Use Lemma 11.2(i) to free γ of crossings. We can then cut Σ along γ, and fill in the hole using a disk. But this is impossible, since we now have an even drawing of G on a surface of smaller Euler genus, contradicting the choice of Σ. □

The proof of Theorem 11.1 can be pushed farther to deal with odd crossing number up to two; we leave the proof to the exercises.

Theorem 11.3 (Pelsmajer, Schaefer, and Štefankovič [277]). *If* $\mathrm{ocr}_\Sigma(G) \leq 2$, *then* $\mathrm{ocr}_\Sigma(G) = \mathrm{cr}_\Sigma(G)$.

The theorem may suggest that ocr and cr are the same, but this turns out to be false, even on the sphere as we will see in Section 11.5.3. The two cannot be arbitrarily far apart, as the following theorem shows. In the sphere, this was first shown by Pach and Tóth [265].

Theorem 11.4 (Pelsmajer, Schaefer, and Štefankovič [277]). *For any surface* Σ,

$$\mathrm{cr}_\Sigma(G) \leq \binom{2\,\mathrm{ocr}_\Sigma(G)}{2}.$$

Lemma 11.2 is not enough to prove this theorem; we need to clean all even edges of crossings.

Lemma 11.5. *Given a drawing of a graph on surface Σ with even edges E_0, we can remove all crossings with edges in E_0.*

Proof of Theorem 11.4. Fix an ocr-minimal drawing of G on Σ. Let E_0 be the set of even edges in the drawing. By Lemma 11.5, we can assume that all edges in E_0 are free of crossings. Now Theorem 3.16, which works for surface as well as for the plane, implies the result. $\square$

Lemma 11.5 does not claim that the crossing parities of all pairs of edges remains the same, as we did in Lemma 11.2. This, provably, cannot be done even on a torus [277]. It is an open question whether Lemma 11.5 remains true under the weaker condition that ocr of the drawing does not increase.

Proof. Fix a drawing D of G on surface Σ, and let E_0 be the set of even edges. In this proof, we allow G to be a multigraph (allowing both loops and multiple edges). We will show by induction on G and the Euler genus of Σ that G has a drawing on Σ with the same embedding scheme in which all edges in E_0 are free of crossings.

We start by contracting all even edges (merging vertices and rotations, and adjusting signatures as in the proof of Theorem 11.1). At this point any edge which is not a loop must be odd. We also contract the odd edges (again merging vertices and rotations, and adjusting signatures). This does not create new odd edges, since all even edges are loops, so contraction does not change their parity of crossing with an odd edge. The drawing now consists of a set of isolated vertices with loops. Connect the vertices by edges (which may be odd), and contract the new edges as well. Again, no edge in E_0 becomes odd. The graph now consists of a single vertex v with a bouquet of loops. If there is an even loop e so that $\Sigma - e$ is connected, we use Lemma 11.2 to remove all crossings with e. As we did in Theorem 11.1 we cut the surface along e splitting v into two vertices, and, depending on whether e is one-sided or two-sided, adding one or two boundary disks to fill the holes. As before, if e is one-sided, we identify the two copies of v using the disk. In either case, we have a graph G' on a surface of Euler genus smaller than Σ, and we can use induction to find a drawing of G' on that surface in which all even edges are free of crossings. From this drawing, we can construct a drawing of G on Σ, as in Theorem 11.1: if e was two-sided, we introduce two holes next to the two copies of v, where the ends of e were located, and identify the boundaries of the two holes (so that the two copies of v become one again); we can then add e without crossings along the identified boundaries; if e was one-sided, we separate v into its two copies, and cut a hole with both copies of v on its boundary; we then turn the boundary into a one-sided curve so that the two copies of v overlap, and the original embedding scheme is recovered. We can add e without crossings along the one-sided curve that was the boundary.

For any remaining even loop e at v, $\Sigma - e$ is not connected. Pick an even loop e so that its ends do not enclose another even loop (we orient e so there is a first and second end, and we can distinguish between the two sides of the loop). No loop f starting between the ends of e can be even. Remove e and all loops enclosed by it. Apply induction, then add e close to v bounding a disk. We can now add back all loops f which were enclosed by e within that disk, leaving e free of crossings. Since these loops were all odd, it does not matter how they cross each other.

We now have a drawing in which all loops that were edges in E_0 are free of crossings. We uncontract all contracted edges. This does not introduce any new crossings, so all edges of E_0 are free of crossings, and the embedding scheme is equivalent to the original one. $\square$

We do not have a better bound of cr in terms of ocr than Theorem 11.4, even in the sphere; this is surprising, given that we do have such bounds for the pair crossing number, which is much harder to handle.

11.2 The Independent Odd Crossing Number

11.2.1 An Algebraic Invariance of Good Drawings

The *independent odd crossing number*, $\mathrm{iocr}(D)$, of a drawing D is the number of pairs of *independent* edges which cross an odd number of times in D. The *independent odd crossing number*, $\mathrm{iocr}(G)$, of a graph G is the smallest $\mathrm{iocr}(D)$ of a drawing D of G.[1] By definition, $\mathrm{iocr}(G) \le \mathrm{ocr}(G) \le \mathrm{pcr}(G)$.

It is easy to see that $\mathrm{iocr}(K_5) = \mathrm{iocr}(K_{3,3}) = 1$, and more is true. (Compare this result to Theorem 1.12.)

Theorem 11.6. *We have* $\mathrm{iocr}(K_n) \equiv \binom{n}{4} \bmod 2$, *and* $\mathrm{iocr}(K_{m,n}) \equiv \binom{n}{2}\binom{m}{2} \bmod 2$ *if both n and m are odd.*

The theorem implies $\mathrm{iocr}(K_5) = \mathrm{iocr}(K_{3,3}) = 1$ since both numbers are

---

[1] In other words, we are applying Rule $-$ from Remark 1.4 to ocr.

odd, and both graphs can be drawn with a single crossing. The proof consists of two parts: showing that there are drawings of K_n and $K_{m,n}$ for which the claim holds, and, then, showing that the parity of good drawings of K_n and $K_{m,n}$ is invariant.

An (e, v)-*move* in a drawing of a graph is performed as follows: connect e and v via a curve γ, avoiding all other vertices. Then reroute e along γ, and very close to it, so it can be made to cycle around v, see Figure 11.1.

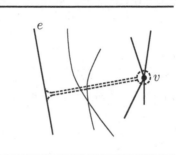

FIGURE 11.1: An (e, v)-move.

The effect of such an (e, v)-move is that the parity of crossing between e and every edge with one end at v changes, and no other crossing parity is affected: any edge that crosses γ will be crossed by e an even number of times. The *independent crossing parities* of a drawing D of a graph G is the vector consisting of $\mathrm{iocr}_D(e, f)$ for every pair (e, f) of independent edges in G.

Lemma 11.7 (Tutte [340]). *The independent crossing parities for any two drawings of a graph differ by a set of (e, v)-moves.*

In other words, any two vectors of independent crossing parities of a graph are equivalent under (e, v)-moves.

Proof. Fix two drawings D and D' of G. By Lemma A.22 we can assume that both drawings share the same vertex locations and all edges are poly-lines. Define D_i to be the drawing in which the first i edges are drawn as in D' and the remaining edges as in D. So $D_0 = D$ and $D_{|E(G)|} = D'$. In each step the drawing of a single edge changes; suppose we can prove the lemma in that case, so there are sets S_i of (e, v)-moves so that D_i and D_{i+1} differ by S_i. Then the independent crossing parities of D_0 and D' differ by the symmetric difference of all S_i, that is, by the set of (e, v)-moves which occur on an odd number of the S_i. Hence we can assume that D and D' differ only on edge e. We can also assume that the two drawings of e intersect at most finitely often (perturbing bend vertices slightly if necessary, this does not affect the crossing parities). The two curves representing e form a closed curve γ, so $\mathbb{R}^2 - \gamma$ has a 2-coloring by Lemma A.2. The coloring partitions $V(G)$ into two color classes: V_0 and V_1. Any edge whose endpoints lie in the same color classes crosses C an even number of times, so it has the same parity of crossing with both representations of e. On the other hand, any edge f whose endpoints lie in different color classes, crosses γ oddly, so the parity of crossing with e changes as we go from D to D'. In other words, the change in independent crossing parities is equivalent to performing (e, v)-moves for all $v \in V_0$ (or all $v \in V_1$, this has the same result). $\qquad \Box$

Proof of Theorem 11.6. As we mentioned earlier, there are two parts to the proof: first we need to find drawings of K_n and $K_{m,n}$ which satisfy the claim of the theorem, and, second, we need to show that the parity of iocr for two drawings of K_n and $K_{m,n}$ is invariant. For the first part, consider a convex, rectilinear drawing of K_n: every K_4-subgraph contributes a unique crossing, so the crossing number (and independent odd crossing number) of that drawing is $\binom{n}{4}$. For $K_{m,n}$ consider a bipartite, rectilinear drawing; again every K_4-subgraph contributes exactly one crossing, and every such graph is determined by 2 vertices in each partition, so there are $\binom{n}{2}\binom{m}{2}$ crossings in the drawing.

For the second part let $G = K_n$ or $G = K_{m,n}$. By Lemma 11.7, the independent crossing parities in two drawings of G differ by a set of (e, v)-moves. Since each vertex of G is incident to an even number of edges independent of an edge e, namely, $n - 3$ for K_n, and $m - 1$ or $n - 1$ for $G = K_{m,n}$, none of these moves changes the parity of iocr of the drawing, and so $\mathrm{iocr}(D) \bmod 2$ is invariant. $\qquad \Box$

The proof showed that the parity of $\mathrm{iocr}(G)$ is the same for all drawings of $G = K_n$ or $G = K_{m,n}$ if n and m are odd. The requirement that n and m be odd is necessary: K_4 and $K_{2,2}$ can both be drawn with one or no crossings. Indeed, K_n and $K_{m,n}$ with both m and n odd are the only graphs for which the parity of iocr is the same for all drawings as shown by Archdeacon and Richter [35].

11.2.2 The Strong Hanani-Tutte Theorem

Theorem 11.8 ((Strong) Hanani-Tutte Theorem [92, 340]). *If* $\mathrm{iocr}(G) = 0$, *then* $\mathrm{cr}(G) = 0$.

In other words, if a graph can be drawn so that any two independent edges cross evenly, then the graph is planar. The following short and elegant argument is due to Kleitman [208].

Proof. Since $\mathrm{iocr}(G) \leq \mathrm{cr}(G)$ one direction is immediate. For the other direction assume $\mathrm{cr}(G) > 0$. Then G is non-planar, and by Kuratowski's Theorem (Theorem A.6) contains a subdivision of a K_5 or a $K_{3,3}$. Note that if G contains a subdivision of H, then $\mathrm{iocr}(G) \geq \mathrm{iocr}(H)$: we obtain a drawing of H from G by taking the drawing of the subdivision and suppressing the subdividing vertices. This can only decrease iocr of the drawing. Since we already saw that $\mathrm{iocr}(K_5) = \mathrm{iocr}(K_{3,3}) = 1$, this completes the proof. $\qquad \Box$

The strong Hanani-Tutte theorem reduces a topological problem, planarity, to an algebraic one.

Corollary 11.9. *Testing planarity is equivalent to solving a linear system of equations over* $\mathrm{GF}(2)$, *the field of two elements.*

Proof. Fix a good drawing D of a given graph G (for example, a convex, rectilinear drawing). Theorem 11.8 implies that a graph G is planar if and only if it has a drawing D' in which every pair of independent edges crosses evenly. By Lemma 11.7, the vectors of independent crossing parities of D and D' differs by a set of (e, v)-moves, so we can set up a system of linear equations as follows: create a variable $x(e, v)$ for every $e \in E(G)$, $v \in V(G)$. Define $c(e, f) := \mathrm{iocr}_D(e, f)$.

Then the system consisting of equations

$$c(ab, cd) + x(ab, c) + x(ab, d) + x(cd, a) + x(cd, b) \equiv 0 \bmod 2$$

for every pair of independent edges ab, cd in G is solvable if and only if G is planar. $\square$

Unfortunately, though, iocr in general is intractable. This follows easily from Theorem 4.1.

Theorem 11.10 (Pelsmajer, Schaefer, and Štefankovič [279]). *Deciding* $\mathrm{iocr}(G) \le k$ *is* **NP**-*complete, even for cubic graphs.*

Even in spite of this, iocr opens up an alternative, algebraic approach to crossing numbers, so we naturally ask how relevant this is for cr itself, or, in other words, how close iocr and cr are.

Theorem 11.11 (Pelsmajer, Schaefer, and Štefankovič [278]). *We have* $\mathrm{cr}(G) = \mathrm{iocr}(G)$ *if* $\mathrm{iocr}(G) \le 2$.

We will only prove the case $\mathrm{iocr}(G) \le 1$, since the full proof is lengthy.

Lemma 11.12. *Let C be a cycle in a drawing of a graph so that no edge of C is involved in an independent odd crossing. Then the drawing can be modified so that C is free of crossings and no new pairs of independent edges crossing oddly are introduced.*

Proof. Since no edge of C is involved in an independent odd crossing, we can ensure that every edge of C is crossed evenly by every other edge; this can be done by modifying the rotation at the endpoints of each edge (note that this does not create new pairs of independent edges crossing oddly). For every edge e on C we cut all edges crossing e; since all other edges cross e an even number of times, this leaves us with an even number of half-edges for each edge on either side of e. We can reconnect them along e (without crossing e). Note that this does not change the parity of crossing between any pair of edges (because any new crossings are paired up, one on each side of e). Perform this operation for all edges $e \in E(C)$. During the process we may separate edges into multiple curves: one simple curve which connects the endpoints together with several closed curves. Let $D(C)$ be the drawing of the simple curves representing edges in C. Then $D(C)$ is crossing-free, so dropping all closed components belonging to edges in C does not lead to new pairs of independent edges crossing oddly. $D(C)$ separates $\mathbb{R}^2$ into two regions. If a closed component of an edge lies in the same region as its simple curve, we can reconnect the closed component and the simple curve: connect them by a curve which does not cross $D(C)$ and avoids vertices. In a small neighborhood of this curve we can reconnect the closed component and the simple curve. Note that this does not change the parity of crossing between any pair of edges. Any closed components which lie in the other region of $\mathbb{R}^2 - D(C)$ we remove. This may change the parity of crossing between two edges, but it will not create a new pair of independent edges crossing oddly: suppose e and f cross oddly in the final drawing. Then they must belong to the same region of $\mathbb{R}^2 - D$. Any closed components of e and f lying in different regions are closed curves and thus crossed an even number of times. Removing them did not change the parity of crossing between e and f, so e and f crossed oddly in the initial drawing. $\square$

Theorem 11.13 (Pelsmajer, Schaefer, and Štefankovič [278]). *We have* $\mathrm{cr}(G) \leq \binom{2\,\mathrm{iocr}(G)}{2}$.

The theorem implies Theorem 11.11 for $\mathrm{iocr}(G) \leq 1$, in particular it gives an alternative proof of the strong Hanani-Tutte theorem, Theorem 11.8 without invoking Kuratowski's theorem. It also implies $\mathrm{cr}(G) \leq \binom{2\gamma(G)}{2}$ for any crossing number between $\mathrm{iocr}(G)$ and $\mathrm{cr}(G)$, notably cr_- and pcr_-, for which a direct proof is not known.

Proof. We show by induction that a drawing of G can be redrawn so that only edges involved in independent odd crossings in the original drawing cross, and cross at most once. If we start with an iocr-minimal drawing of G this immediately implies the result since there are at most $2\,\mathrm{iocr}(G)$ involved in independent odd crossings in this drawing, so the redrawing contains at most $\binom{2\,\mathrm{iocr}(G)}{2}$ crossings.

We prove this result by induction on the value $w(G) := \sum_{v \in V(G)} \deg(v)^3$. The explanation for this expression is that $w(G)$ decreases if we split a vertex of degree at least 4 into two vertices of degree at least 3 connected by an edge, since $64 = 4^3 > 3^3 + 3^3 = 54$.

Fix a drawing of G. Suppose the drawing contains a cycle C which does not contain edges involved in independent odd crossings, but C is not free of crossings. Apply Lemma 11.12 to make C crossing-free. If C contains a vertex of degree at least 4 we proceed as follows: if v is incident to edges on both sides of C we split v into two vertices along C, one incident to edges in the interior of C, the other incident to edges in the exterior of C. These two groups of edges cannot cross since C is crossing-free, so this change does not increase iocr of the drawing. However, it does decrease $w(G)$, so by induction we can find the desired drawing for the new graph. Contracting the edge we added yields the desired drawing of G (this edge was free of crossings when introduced, so it remains free of crossings in the redrawing). If v is incident to edges on one side only, we again split v, but away from C, so the new vertex on C has degree 3, and the second vertex is incident to all the non-C edges v was incident to. Again, this new graph has lower weight than G so we can find the claimed drawing. The new edge we added was again free of crossings, so remains so in the redrawing, so again we can contract it to find the desired drawing of G.

We conclude that all vertices on C have degree at most 3. Note that in this case the redrawing did not introduce crossings with edges incident to C: there is at most one such edge at each vertex of C and any new crossing would have to be with an adjacent edge sharing the same vertex at C, but the only two such edges lie on C and are free of crossings. Hence we can repeat this process until all cycles which do not contain edges involved in independent odd crossings are free of crossings and have vertices of degree at most 3 only.

Let E_0 be the set of edges lying on crossing-free cycles in the current drawing. Let E_1 be the set of edges involved in independent odd crossings. This leaves edges $F = E(G) - (E_0 \cup E_1)$ which are cut-edges of $G - E_1$. Take the plane drawing of $G[E_0]$ and add crossing-free drawings of edges in F. Note that this can be done since the two endpoints of F lie in the same face, and, since the edges in F are cut-edges of $G - E_1$, adding them cannot separate a face. But this means we can now add edges in E_1: any edge in E_1 has its endpoints in a face of $G[E_0 \cup F]$, so we can add each edge in E_1 into its face. The only resulting crossings are between edges of E_1. If two edges cross more than once, we can reduce the number of crossings by swapping arcs as we did in Lemma 1.3.   □

The strong Hanani-Tutte theorem is also known to be true on the projective plane [106, 275], but for no other surfaces. Theorem 11.13 is not known to hold even on the projective plane.

Question 11.14. Is $\mathrm{iocr}(G) = \mathrm{ocr}(G)$ for cubic graphs?

11.2.3   A Crossing Lemma

There is a crossing lemma for iocr, the smallest of the main crossing number variants. To prove the crossing lemma, we adapt Lemma 1.7.

Lemma 11.15. *If G is a graph with $n \geq 3$ vertices, and m edges, then* $\mathrm{iocr}(G) \geq m - 3n + 6$.

Proof. Let D be an iocr-minimal drawing. For every pair of independent edges which cross oddly, pick one of the edges of the pair. Removing those edges from D leaves us with a drawing in which the remaining pairs of edges are either adjacent, or cross evenly. By Theorem 11.8 we are looking at a drawing of a planar graph. Hence, $m - \mathrm{iocr}(G) \leq 3n - 6$, by Euler's formula, and the conclusion follows. □

A careful review of the proof of Theorem 1.8 shows that it can be used without change for iocr. There is one important detail: a crossing is always caused by two edges with four endpoints. This is true for iocr, but not, as far as we know, pcr or ocr, so we need the strong Hanani-Tutte theorem to complete the proof, which unfortunately means that, different from the case of cr, we do not have crossing lemmas for iocr, ocr, pcr and pcr_ for any surface for which the strong Hanani-Tutte theorem has not been established (the plane and projective plane are the only exceptions at this point [275]).

Theorem 11.16 (Crossing Lemma. Pach and Tóth [265]). *If G is a graph with n vertices, and m edges, and so that $m \geq \lambda n$, then*

$$\mathrm{iocr}(G) \geq c\frac{m^3}{n^2},$$

with $c = (\lambda^{-2} - 3\lambda^{-3})$.

As before, we get the best constant $c = 1/60.75$ for $\lambda = 9/2$, and $c = 1/64$ for $\lambda = 4$. The theorem implies the same crossing lemma for any crossing number which is at least iocr, in particular, pcr, pcr_, and ocr.

Corollary 11.17. *We have $\mathrm{ocr}(K_n) \geq \mathrm{iocr}(K_n) \geq \frac{1}{486}n(n-1)^3$.*

Spencer and Tóth [319] show that $\mathrm{iocr}(G) = \Omega(p^2 n^2)$ for a random graph $G \sim G(n,p)$ with $p = n^{\varepsilon - 1}$ by counting certain subdivisions of K_5 (they prove the bound for pcr, but the same argument works for iocr).

11.3   Lower Bounds on Odd Crossings

The following theorem is the analogue of Theorem 3.1 for the odd crossing number; the bound is weaker, by a $\log^2 n$ factor, but it is still good enough to obtain some interesting consequences.

Theorem 11.18 (Pach and Tóth [269]). *If G is a graph on n vertices, then*

$$\mathrm{ocr}(G) + \mathrm{ssqd}(G) \geq c\left(\frac{\mathrm{bw}(G)}{\log n}\right)^2,$$

for some constant $c > 0$.

The theorem immediately implies the same result for pcr, which we stated as Theorem 9.7, since $\mathrm{pcr}(G) \geq \mathrm{iocr}(G)$. It is not clear, on the other hand, whether the result extends to any of the independent crossings numbers: iocr, pcr_, or even cr_.

Recall that the edge expansion, $\beta(G)$, of a graph G is the minimum $|E(V_1, V_2)|/\min(|V_1|, |V_2|)$ over all partitions $V_1 \cup V_2 = V(G)$.

Lemma 11.19 (Kolman and Matoušek [213]). *Every graph G has a subgraph with at least $2n/3$ vertices and expansion at least $\mathrm{bw}(G)/n$, where $n = |V(G)|$.*

Proof. Let $G_1 := G$. If G_1 has expansion $\mathrm{bw}(G)/n$, we are done. If not, then there is a partition $U_1 \cup V_1 = V(G_1)$ so that $|E(U_1, V_1)|/\min(|U_1|, |V_1|) < \mathrm{bw}(G)/n$; we assume $|U_1| \leq |V_1|$, so $|U_1| \leq n/2$, and $|E(U_1, V_1)| < \mathrm{bw}(G)|U_1|/n$. In other words, U_1 can be separated from the rest of the graph by removing fewer than $\mathrm{bw}(G)|U_1|/n$ edges. This implies that $|U_1| < n/3$, since otherwise (U_1, V_1) would be a separator with a bisection width smaller than $\mathrm{bw}(G)$.

We continue this process with $G_{i+1} := G_i - U_i$. If at any point we find a G_i with expansion $\mathrm{bw}(G)/n$, we are done, G_i is the subgraph we are looking for. Since the U_i are non-empty, there must be a first stage i at which $|V(G_i)| \leq 2n/3$; since before that $|V(G_{i-1})| > 2n/3$, and U_i contains at most half the vertices in $V(G_{i-1})$, we also know that $|V(G_i)| \geq n/3$. Now $U := U_1 \cup \ldots \cup U_{i-1}$ is separated from $V(G_i)$ by removing fewer than $\mathrm{bw}(G)|V(G_i)|/n \leq 2\,\mathrm{bw}(g)/3$ edges, which would be a bisection with width less than $\mathrm{bw}(G)$, which is a contradiction. □

We again work with an embedding of the complete graph, for that we need to adapt Lemma 3.8 to ocr.

Lemma 11.20. *For any two graphs G and H we have*

$$\mathrm{ocr}(H) \leq \phi(H \hookrightarrow G)^2(\mathrm{ssqd}(G) + \mathrm{ocr}(G)).$$

Proof. We use the same construction as in the proof of Lemma 3.8 to turn a drawing of G realizing $\mathrm{ocr}(G)$, into a drawing of H. If the drawing of H has two edges which cross oddly, then they either cross close to a vertex of G or they cross close to a crossing of two edges in G which cross oddly. This observation, as before, implies the bound. □

Proof of Theorem 11.18. We use Lemma 11.19 to replace G with a subgraph F on at least $2n/3$ vertices, and expansion $\beta >= \mathrm{bw}(G)/n$, where $n = |V(G)|$. To obtain the bound of the theorem, it is then sufficient to show that

$$\mathrm{ssqd}(F) + \mathrm{ocr}(F) \geq c\left(\frac{\beta n}{\log n}\right)^2,$$

for $n = |V(F)|$ and some constant $c > 0$, since ocr and ssqd are monotone for subgraphs.

We know, from Theorem 3.10, that $\phi(K_n \hookrightarrow F) = O(n \log(n)/\beta)$, so by Lemma 11.20,

$$\mathrm{ocr}(K_n) = O\left(\left(\frac{n \log n}{\beta}\right)^2 (\mathrm{ssqd}(F) + \mathrm{ocr}(F))\right).$$

Since $\mathrm{ocr}(K_n) = \Omega(n^4)$, as we saw in Corollary 11.17, it follows that

$$\mathrm{ssqd}(F) + \mathrm{ocr}(F) \geq c \left(\frac{\beta n}{\log n}\right)^2,$$

for some constant $c > 0$, which is what we needed. $\qquad\square$

We saw in Theorem 3.2 that we can find a rectilinear, convex drawing of any graph with $O(\log(n)(\mathrm{cr}(G) + \mathrm{ssqd}(G)))$ crossings; using Theorem 11.18 a similar result can be proved with ocr in place of cr. We have to replace the application of the edge separator theorem in Theorem 3.2 with an application of Theorem 11.18. We follow the proof of Kolman and Matoušek [213]; the bound can be improved by a $\log(n)$-factor.

Theorem 11.21 (Kolman and Matoušek [213]). *Every graph has a convex, rectilinear drawing with at most $c \log^4(n)(\mathrm{ocr}(G) + \mathrm{ssqd}(G))$ crossings.*

In the proof we use the notion of the *hereditary bisection width*, $\mathrm{hbw}(G)$, of a graph G which is $\max_{H \subseteq G} \mathrm{bw}(H)$. Theorem 11.18 implies that $\mathrm{bw}^2(G) \leq c \log^2(n)(\mathrm{ocr}(G) + \mathrm{ssqd}(G))$, for some constant c. Since $\log^2(n)$, $\mathrm{ocr}(G)$ and $\mathrm{ssqd}(G)$ are monotone for subgraphs, we also have $\mathrm{hbw}^2(G) \leq c \log^2(n)(\mathrm{ocr}(G) + \mathrm{ssqd}(G))$.

Proof. As before, we draw G recursively, placing its vertices on a convex arc, and drawing edges as straight-line segments. For such a drawing D, we let $\ell(D)$ be the largest number of edges whose endpoints are separated on the arc by some vertex.

Let (V_1, V_2) be a bisection of G realizing $\mathrm{bw}(G)$, and let D_i be a recursively drawn G_i, where G_i is the graph induced by G on V_i, $i = 1, 2$. When we combine the two drawings, and add the $\mathrm{bw}(G)$ edges in the separator, we obtain a drawing D of G with $\ell(D) \leq \mathrm{bw}(G) + \max(\ell(D_1), \ell(D_2))$. So, by induction, $\ell(D) \leq c' \, \mathrm{hbw}(G) \log n$, for some constant c'.

The number of crossings is bounded as follows:

$$\mathrm{cr}(D) \leq \binom{\mathrm{bw}(G)}{2} + \mathrm{cr}(D_1) + \mathrm{cr}(D_2) + \mathrm{bw}(G)(\ell(D_1) + \ell(D_2)),$$

since each edge in the separator can cross at most $\ell(D_i)$ edges in D_i. Let us bound the terms in this expression. By induction we have

$$\mathrm{cr}(D_1) + \mathrm{cr}(D_2) \leq c \log^4\left(\frac{2n}{3}\right)(\mathrm{ocr}(G_1) + \mathrm{ocr}(G_2) + \mathrm{ssqd}(G_1) + \mathrm{ssqd}(G_2))$$

$$\leq c \log^4\left(\frac{2n}{3}\right)(\mathrm{ocr}(G) + \mathrm{ssqd}(G)).$$

Using the estimate for ℓ, we get

$$\mathrm{bw}(G)(\ell(D_1) + \ell(D_2) \le 2c'\,\mathrm{hbw}^2(G)\log n,$$

which implies that

$$
\begin{aligned}
\mathrm{cr}(D) \quad &\le \quad \mathrm{hbw}^2(G)\left(\frac{1}{2} + 2c'\log n\right) + c\log^4\left(\frac{2n}{3}\right)(\mathrm{ocr}(G) + \mathrm{ssqd}(G)) \\
&\le \quad c''(\mathrm{ocr}(G) + \mathrm{ssqd}(G))\log^3(n) + c\log^4\left(\frac{2n}{3}\right)(\mathrm{ocr}(G) + \mathrm{ssqd}(G)) \\
&\le \quad (\mathrm{ocr}(G) + \mathrm{ssqd}(G))\left(c\log^4(n) + c''\log^3(n) - c\log\left(\frac{3}{2}\right)\log^3(n)\right).
\end{aligned}
$$

which is at most $c\log^4(n)(\mathrm{cr}(G) + \mathrm{ssqd}(G))$ for sufficiently large c. $\qquad\square$

11.4 Algebraic Crossing Numbers

There is another, related way of counting crossings. Orient all edges of the graph. This allows us to distinguish between a left-to-right and a right-to-left crossing along an edge (recall Figure 4.5). Along an edge e label all left-to-right crossings with an edge f by $+1$, all right-to-left crossings by -1, and add up the total. We call the absolute value of that total $\mathrm{acr}(e, f)$, the algebraic crossing number of e and f. This number is independent of the orientation and whether we count crossings along e or f. The *algebraic crossing number*, $\mathrm{acr}(D)$, of a drawing D is the sum of all $\mathrm{acr}(e, f)$ for all ordered pairs of edges (e, f). The *algebraic crossing number*, $\mathrm{acr}(G)$, of a graph is the smallest value of $\mathrm{acr}(D)$ for any drawing D of G. If we sum over ordered pairs of *independent* edges only, we get the *independent algebraic crossing number*, iacr.

The algebraic crossing number variants were introduced by Tutte [340] in an attempt to lay an algebraic foundation for the study of the crossing number. We now know that there are graphs for which $\mathrm{acr}(G) < \mathrm{cr}(G)$, as we will see in the next section, so we are not capturing cr with this new notion. That does not preclude the possibility though that for some interesting classes of graphs we do, and Tutte's algebraic theory will apply. We include a simple example.

Theorem 11.22 (Pelsmajer, Schaefer, and Štefankovič [279]). *For 2-vertex graphs with rotation, algebraic crossing number and crossing number are the same.*

This result is the limit of our current knowledge in several directions: we do not know whether crossing number and pair crossing number agree for 2-vertex graphs with rotation, and we do not know whether the theorem remains true for 3-vertex graphs.

Proof. Fix a drawing of G with the given rotation that minimizes first acr, and then cr. This drawing cannot contain an empty bigon formed by two crossings, that is, a bigon not containing a vertex in its interior or on its boundary. If it did, pick a smallest such bigon (in terms of region enclosed). Then every edge that crosses a side of the bigon has to cross the other side next, since otherwise it would form a smaller bigon. We say that all edges cross the bigon transversally. Then swapping the two arcs of the bigon does not affect acr of the drawing (or the rotation, since the bigon is between two crossings), but reduces cr, contradicting our assumption.

If $\mathrm{acr}(e) = \mathrm{cr}(e)$ for every edge, then the drawing is also cr-minimal, and we are done. So there is some edge e for which $\mathrm{acr}(e) < \mathrm{cr}(e)$, and then there must be another edge f for which $\mathrm{acr}(e, f) < \mathrm{cr}(e, f)$. We claim that e and f form an empty bigon between two crossings, completing the proof, since we already showed that no such bigon exists. To see the truth of the claim note that $\mathrm{acr}(e, f) < \mathrm{cr}(e, f)$ implies that at least one of the crossings between e and f counts as $+1$ one, and another one counts as -1. Pick two crossings along e that have opposite signs and have minimal distance. There cannot be a crossing between them on e, since that would give us two crossings of opposite sign of smaller distance. So the e-arc does not cross f at all, so we have two arcs between two crossings which do not cross each other. We claim that this bigon is empty: it cannot contain either of the endpoints, since the ends of f cannot lie inside this bigon (since the e-arc is free of crossings). $\square$

An important advantage of the algebraic crossing number over the crossing number is that it often leads to simpler algebraic and computational models. The following theorem is an example of this.

Theorem 11.23 (Pelsmajer, Schaefer, and Štefankovič [279]). *The algebraic crossing number of a 2-vertex graph with rotation can be computed in polynomial time.*

By Theorem 11.22 then, the crossing number of a 2-vertex graph with rotation can be computed in polynomial time. In combination with Theorem 11.23, this completes the proof of Theorem 4.10.

Proof. Let u and v be the two vertices of the graph and E its set of edges. By renaming the edges, we can assume that the rotation at u is $(123 \cdots n)$. Let γ be a curve connecting u and v; we will use γ as a reference line for the other edges, so we may as well imagine it drawn as a straight-line segment. For any two edges e and f between u and v, set $a_{e,f} = 0$ if the ends of γ, e and f have inverse orders at u and v, and 1 otherwise. Then

$$\mathrm{acr}(e, f) = x_f - x_e + a_{e,f},$$

where x_e and x_f denote the number of times e, respectively f, wind around u. Hence, the algebraic crossing number of the drawing can be computed as

Rule +	ocr$_+$	acr$_+$	cr
	ocr	acr	
Rule −	iocr	acr$_-$	cr$_-$

TABLE 11.1: Odd, algebraic, and standard crossing numbers.

the minimun of

$$\sum_{e,f \in E} |x_f - x + e + a_{e,f}|,$$

over $x_e \in \mathbb{Z}$. Solving a linear integer program is an **NP**-complete problem, but in this case, it turns out that the linear relaxation of the program, only requiring $x_e \in \mathbb{R}$, yields the same value, so we can solve the problem using linear programming, which can be done in polynomial time. To see that for every real solution of the program, there is an integer solution with the same value, fix a minimal real solution $x*$ which also satisfies the largest number of equalities $x_f - x_e + a_{e,f} = 0$; without loss of generality, we can assume that one of the x_e^* is 0. Let G be the graph with vertex-set E and edges ef if $x_f^* - x_e^* + a_{e,f} = 0$. If G is connected, then all values in x^* are integers, so we can assume that there is a subset $E' \subseteq E$ for which there are no edges between E' and $E - E'$ in G. Define $c(\lambda)$ to be the value of the linear program on $x^* + \lambda 1_{E'}$, where $1_{E'}$ is the vector which contains 1 for $e \in E'$ and 0 otherwise. Let I be the interval of $\mathbb{R}$, for which the sign of no $x_f - x_e + a_{e,f}$ changes compared to $x^*f - x_e^* + a_{e,f}$. Then I is an open interval, containing 0. Since $c(\lambda)$ is a linear function taking on its minimal value at $\lambda = 0$, it must be constant on I. Since G is not connected, I cannot be all of $\mathbb{R}$, so it must have a boundary point λ'. Then $x^* + \lambda' 1_{E'}$ satisfies at least one additional constraint, which is a contradiction. $\qquad\square$

11.5   Separations

Table 11.1 lists the crossing number variants we are mostly interested in in this chapter: all combinations of Rule + and − with counting crossings by parity, algebraically, or the standard way; there are only eight, rather than nine variants, since cr = cr$_+$ as we saw in Remark 1.4. The crossing numbers increase from left to right and bottom to top.

Our goal in this section is to show that some of these counting methods differ: there is a graph G for which iocr$(G) <$ ocr$(G) <$ acr$(G) <$ cr(G). We obtain this result by working with restricted drawing models, in particular drawings with rotation system, anchored drawings, and x-monotone drawings. To prove a separation is then a two-step procedure: we prove the separation for the restricted drawing model, and show that such a separation can be transferred to the unrestricted setting. In this section, γ will denote an arbitrary crossing number.

We will work with weighted graphs to prove separations; the construction we saw in Lemma 3.34 for turning a weighted graph into an unweighted graph not only leaves cr unchanged, but also pcr, ocr, and acr. The proof follows along the same lines.

Lemma 11.24. *For every integer-weighted graph (G, w) there is an unweighted graph G' so that $\gamma(G, w) = \gamma(G')$, for $\gamma \in \{\mathrm{cr}, \mathrm{pcr}, \mathrm{ocr}, \mathrm{acr}\}$. The size of G' is bounded by a polynomial in the maximum weight (represented in unary) and the size of G.*

Note that iocr and iacr are missing from the lemma. Subdividing an edge may change their value, so the standard construction for unweighting these graphs does not work.

11.5.1 Translating Separations

We start with anchored drawings, which will give us a separation of algebraic and pair crossing number. We write $\gamma(G, A)$ for the smallest $\gamma(D)$ for any A-anchored drawing of G (vertices in A are incident to the outer region of the drawing in the same order as in A).

Lemma 11.25. *Given a graph G and a set $A \subseteq V(G)$, we can construct a graph G' so that $\gamma(G, A) = \gamma(G')$ for $\gamma \in \{\mathrm{cr}, \mathrm{ocr}, \mathrm{acr}, \mathrm{pcr}\}$. If G is weighted, then so is G'.*

Proof. Suppose $A = (a_0, \ldots, a_k)$. To G add a new vertex x and connect it by $|V(G)|^4$ paths of length 2 to each of the vertices in A; also, add $|V(G)|^4$ paths of length 2 between a_i and a_{i+1} for $0 \leq i < k$ as well as between a_k and a_0. Any cr-minimal A-anchored drawing of G can be extended to a drawing of G' without introducing any crossings, so γ remains unchanged.

For the other direction, fix a γ-minimal drawing D of G'. Since $\gamma(G) < |V(G)|^4$, there must be paths from x to each vertex of A so that $\gamma_D(e) = 0$ for every edge e in those paths. Let S be the union of these paths. Similarly, there must be paths between a_i and a_{i+1}, for $0 \leq i < k$, as well as a_k and a_0, so that $\gamma_D(e) = 0$ for any edge on these paths. Let C be the union of these paths (a cycle). If $\gamma \in \{\mathrm{cr}, \mathrm{pcr}\}$, then $S \cup C$ is free of crossings. If $\gamma \in \{\mathrm{ocr}, \mathrm{acr}\}$, then, since ocr $\leq$ acr, we know that ocr$(e) = 0$ for any edge in $S \cup C$. Using Lemma 11.2(ii) applied to the tree S in $G \cup S \cup C$, we can redraw $G \cup S \cup C$ so that S is free of crossings, and we can do so without changing the parity of crossing between any pair of edges. (While we did not state this in Lemma 11.2, the number of algebraic crossings between two edges also does not change, since it is not affected by the tunneling operation.) Lemma 11.12 then tells us that we could remove all crossings with C without redrawing the edges in S, which did not cross C. We do not actually perform this redrawing, but it allows us to conclude that the vertices of A appear along C in the same order as they appear in the rotation around x (since $C \cup S$ is a subdivision of a 3-connected graph, so it has a unique embedding). Hence, we can draw the edges of C along the spokes of S to remove all crossings with edges in C (and this works for acr as well as ocr). We can now remove $S \cup C$ to obtain an A-anchored drawing of G with γ as we did before. $\qquad\square$

Next are drawings of graphs with a given rotation system. We write $\gamma(G, \rho)$ for the smallest $\gamma(D)$ for any drawing D with rotation system ρ. We will use drawings of 2-vertex graphs with rotation system to separate odd and algebraic crossing number.

Lemma 11.26. *Given a 2-vertex integer-weighted multigraph (G, w) with rotation system ρ for which $\gamma(G, \rho, w) \leq \sum_{e \in E(G)} \binom{w(e)}{2}$ and $\sum_{e \in E(G)} w(e) \leq 2 \max_{e \in E(G)} w(e)$, we can construct a cubic, 3-connected graph G' so that $\gamma(G, \rho, w) = \gamma(G')$ for $\gamma \in \{\mathrm{cr}, \mathrm{ocr}, \mathrm{acr}, \mathrm{pcr}\}$.*

Proof. Let $V(G) = \{u, v\}$. Create hexagonal grids H_u and H_v of length and width $W := \max\{2 \sum_{e \in E(G)} w(e), 1 + \gamma(G, \rho, w)\}$. On one side of the boundary of H_x, $x \in \{u, v\}$, mark of disjoint paths $P_{x,e}$, each containing $w(e)$ degree-2 vertices, for every edge e between u and v, in the order determined by the rotation at x.

For each edge e connect the $w(e)$ degree-2 vertices of $P_{u,e}$ and $P_{v,e}$ by edges, in opposite order around the boundaries of H_u and H_v (so no crossings between edges replacing the same edge e are necessary). Let G' be the resulting graph.

It is easy to see that a drawing of G with the given rotation system and weights can be turned into a drawing of G' with the same value of γ by replacing each edge e in G with $w(e)$ parallel edges and replacing u and v with H_u and H_v, and attaching the parallel edges to their groups. A crossing between edges e and f in the drawing of G turns into $w(e)w(f)$ crossings of G', for any way γ of measuring it. So $\gamma(G') \leq \gamma(G, \rho, w)$.

To show that $\gamma(G') \geq \gamma(G, \rho, w)$, assume that there is a drawing D of G' with fewer than $\gamma(G, \rho, w)$ crossings. Consider one of the H_x, $x \in \{u, v\}$. Since there are W paths of length W parallel to the side used for attachments, and $\gamma(G, \rho, w) < W$, there must be a path R_x in H_x parallel to the side of attachments, so that $\gamma_D(R_x) = 0$. As in Lemma 11.25 we argue that we can assume that R_x is free of crossings for any of the $\gamma \in \{\mathrm{cr}, \mathrm{ocr}, \mathrm{acr}, \mathrm{pcr}\}$. For each edge $e \in E(G)$ we can pick $w(e)$ paths through H_x connecting vertices of attachment belonging to e to R_x so that all these paths are pairwise vertex-disjoint.

For both $x = u$ and $x = v$ erase R_x and bundle the ends of the paths we choose to connect to a new vertex x; we can do so without adding any crossings, since R_x was crossing-free. We obtain a drawing of G in which each edge e has been replaced by $w(e)$ paths with disjoint interiors. Moreover, every vertex either has a rotation given by ρ or its reverse since the path

R_x in each H_x was free of crossings. Suppose that the two rotations are as in ρ or both are flipped; in the latter case, flip both, so they are both as required by ρ. For any e, there is a path P_e from u to v representing it that minimizes $\gamma_D(P_e)$. We can remove all other paths representing e, suppress all interior vertices of P_e, and assign the resulting edge a weight of $w(e)$ without increasing the value of γ of the drawing. Performing this replacement for all edges gives us a drawing of G with rotation ρ and weights as specified by w with crossing number less than $\gamma(G, \rho, w)$, which is a contradiction. We conclude that exactly one of the two rotations is flipped. This means that all paths from R_u to R_v belonging to the same edge e are twisted, forcing $\sum_{e \in E(G)} \binom{w(e)}{2}$ crossings, using Corollary 4.13 (here we need $\sum_{e \in E(G)} w(e) \leq 2 \max_{e \in E(G)} w(e)$). By assumption, this number is at least $\gamma(G, \rho, w)$, so again we have a contradiction. We conclude that $\gamma(G') \geq \gamma(G, \rho, w)$.

Suppressing all degree-2 vertices turns the graph into a cubic, 3-connected graph without changing the value of γ. $\qquad\square$

Finally, we let $\text{mon-}\gamma(G)$ be the smallest $\gamma(D)$ of any x-monotone drawing D of G; x-monotone drawings will allow us to separate iocr from ocr.

Lemma 11.27 (Fulek, Pelsmajer, Schaefer, and Števankovič [142]). *For every integer-weighted graph G we can construct a simple graph G' so that $\gamma(G') = \text{mon-}\gamma(G) + c$ for some constant c, and for every $\gamma \in \{\text{iocr}, \text{ocr}, \text{pcr}, \text{acr}\}$.*

The proof, which we omit, constructs a framework enforcing the x-monotone structure of the drawing; cleaning the framework of even crossings requires an extension of Lemma 11.2.

11.5.2 Algebraic Crossings Matter

Theorem 11.28 (Tóth [335]).

(i) *There is a graph G for which $9 = \text{ocr}(G) = \text{acr}(G) < \text{pcr}(G) = \text{cr}(G) = 10$.*

(ii) *There are graphs for which $\text{ocr}(G) = \text{acr}(G) \leq \frac{3}{4}\text{pcr}(G) = \text{cr}(G)$.*

We use a simplified version of Tóth's construction [335]; see Exercise (11.4) to see how to get a better bound for part (ii).

Proof. Start with a C_8 on vertices $A = \{v_1, \ldots, v_8\}$. Add two new vertices: x connected to v_1, v_3, and v_6, and y connected to v_2, v_5, and v_7. Assign the following weights: $w(xv_1) = a = w(yv_7)$, $w(xv_3) = b = w(yv_5)$, and $w(xv_6) = c = w(yv_2)$, for parameters a, b, c we will choose later. We will satisfy conditions $a \leq b$, $c \leq a + b$, and $b \leq a + c$, $a \leq b + c$, and $a^2 + b^2 \leq c(a + b)$. Finally, add an edge v_4v_8 of weight one. Let this graph be G. See the left illustration in Figure 11.2

We calculate $\text{pcr}(G, A)$ as follows: in the A-anchored drawing induced by

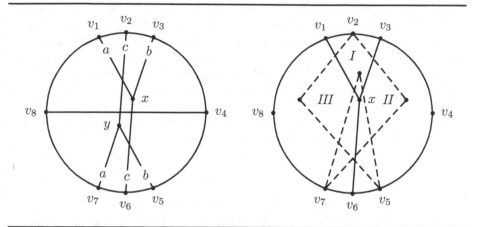

FIGURE 11.2: *Left:* Anchored graph G separating acr and pcr. *Right:* Crossings caused by placing y in region I, II or III.

$A \cup \{x\}$ there are three regions for placing y, two of them incurring $c(a+b)$, and the third incurring $a^2 + b^2$ crossings (see the right illustration in Figure 11.2). Edge $v_4 v_8$ always incurs at least $2c$ crossings, so in the first two cases, we have at least $c(a+b+2)$ crossings, and that number can be realized. In the third case, edge $v_4 v_8$ incurs at least $a+b+c$ crossings, leading to a total of $a^2+b^2+a+b+c$ crossings. So $\mathrm{pcr}(G, A) = \mathrm{cr}(G, A) = \min(c(a + b + 2), a^2 + b^2 + a + b + c)$.

For $\mathrm{acr}(G, A)$ consider the drawing D in Figure 11.3. One can think of it as starting with the third, suboptimal, drawing discussed above, and then performing a $(v_4 v_8, x)$-move. Analyzing it shows that $\mathrm{ocr}(D) = \mathrm{acr}(D) = a^2 + b^2 + 2c$, so $\mathrm{ocr}(G, A)$ and $\mathrm{acr}(G, A)$ are at most $a^2 + b^2 + 2c$, in fact, they are equal to it: as the analysis of pcr and cr showed, in any of the three ways of placing y, there are at least $\min\{c(a + b), a^2 + b^2\} = a^2 + b^2$ crossings between the two stars centered at x and y, and this remains true for ocr and acr. Also, edge $v_4 v_8$ crosses either $x v_6$ or both $x v_1$ and $x v_3$ oddly, causing at least c odd crossings in either case, and the same argument can be made for the star centered at y accounting for $2c$ odd crossings overall, so $\mathrm{acr}(G, A) \geq \mathrm{ocr}(G, A) \geq a^2 + b^2 + 2c$.

For (i) we want to minimize cr, so we instantiate $a, b, c = 1, 2, 2$, satisfying all the conditions, we get an anchored graph G with $\mathrm{pcr}(G, A) = \mathrm{cr}(G, A) = 10$, and $\mathrm{ocr}(G, A) = \mathrm{acr}(G, A) \leq 9$. We then use Lemma 11.25 to remove the anchoring, and Lemma 11.24 to remove the weights, obtaining a simple graph G' with $\mathrm{pcr}(G) = \mathrm{cr}(G) = 10$, and $\mathrm{acr}(G), \mathrm{ocr}(G) \leq 9$.

For (ii) we want to maximize cr / acr, so we use $a, b, c = 1, 1, 4/3$, again satisfying all conditions. We again get an anchored graph, which we can unanchor. To unweight it, we first multiply the weight of all edges by 3, to get rid of the $c = 4/3$, we can then apply Lemma 11.24. The resulting simple graph G' sat-

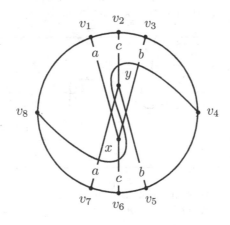

FIGURE 11.3: An acr-minimal drawing of G.

isfies $\mathrm{pcr}(G) = \mathrm{cr}(G) = 9 \cdot 8 = 72$ crossings, and $\mathrm{ocr}(G) = \mathrm{acr}(G) \leq 9 \cdot 6 = 54$, for a ratio of $\lambda = 3/4$. $\qquad\square$

We have seen that $\mathrm{acr}(G) < \mathrm{pcr}(G)$ is possible. Is it also possible that $\mathrm{pcr}(G) < \mathrm{acr}(G)$? That is a hard question, of course, since it would require separating pcr from cr, a problem which has resisted efforts for many years. Maybe, though, it is always true that $\mathrm{acr}(G) \leq \mathrm{pcr}(G)$?

11.5.3 Odd Crossings Matter

Theorem 11.28 separates ocr from cr, so we already know that the odd crossing number differs from cr, but the example does not separate ocr from acr. We will see next that it does matter whether we count crossing by parity or algebraically.

Theorem 11.29 (Pelsmajer, Schaefer, and Števankovič [276]).

(i) *There is a graph G for which $13 = \mathrm{ocr}(G) < \mathrm{acr}(G) = \mathrm{pcr}(G) = \mathrm{cr}(G) = 15$.*

(ii) *There are graphs for which $\mathrm{ocr}(G) < (\frac{3^{1/2}}{2} - o(1))\,\mathrm{acr}(G)$, and $\mathrm{acr}(G) = \mathrm{pcr}(G) = \mathrm{cr}(G)$.*
All graphs can be assumed to be cubic and 3-connected.

Proof. Let $G = (\{u,v\}, \{a,b,c,d\})$ with $\rho(u) = \rho(v) = (abcd)$. If $w(c) \leq w(d) \leq w(b) \leq w(a)$ and $w(b) + w(c) \geq w(a)$, then

$$\mathrm{cr}(G, \rho, w) = \mathrm{acr}(G, \rho, w) = w(a)w(d) + w(b)w(c).$$

By Theorem 11.22, it is sufficient to show this claim for cr. The left drawing in Figure 11.4 shows that $\mathrm{cr}(G, \rho, w) \leq w(a)w(d) + w(b)w(c)$. So we have to

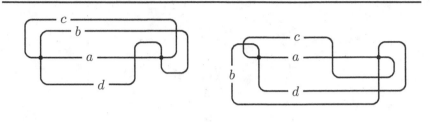

FIGURE 11.4: G separating ocr and acr. *Left:* cr-minimal drawing of G. *Right:* ocr-minimal drawing of G.

show that $\mathrm{cr}(G, \rho, w) \geq w(a)w(d) + w(b)w(c)$. Fix a drawing D of the graph. If a and d do not cross, then the ends of b and c are on opposite sides of $a \cup d$. If b and c do not cross each other either, this means that b crosses a and c crosses d, resulting in $w(a)w(b) + w(c)w(d) \geq w(a)w(d) + w(b)w(c)$. (Evaluate $(w(a) - w(c))(w(b) - w(d)) \geq 0$.) So b and c do cross, and both cross $a \cup d$ forcing at least $w(b)w(c) + ((w(b) + w(c))w(d) \geq w(a)w(d) + w(b)w(c)$ crossings (using $w(b) + w(c) \geq w(a)$). We conclude that a and d do cross, so the ends of b and c are on the same side of $a \cup d$. So one of them must cross both a and d, or they must cross each other. In either case, this adds at least $w(b)w(c)$ crossings to $w(a)w(d)$.

The right drawing in Figure 11.4 shows that $\mathrm{ocr}(G, \rho, w) \leq w(a)w(c) + w(b)w(d)$. If there are two disjoint pairs of edges crossing oddly, then of the three choices: $\{(a, b), (c, d)\}$, $\{(a, c), (b, d)\}$, and $\{(a, d), (b, c)\}$, the middle one leads to the smallest value $w(a)w(c) + w(b)w(d)$ (we already argued that $w(a)w(d) + w(b)w(c)$ is at least as large, for the last choice consider $(w(a) - w(d))(w(b) - w(c)) \geq 0$). Since two or fewer crossing pairs involving the same edge cannot get us from $(abcd)$ to its reverse, there must be at least three pairs of edges crossing oddly. In that case the smallest obtainable value is $w(c)w(d) + w(c)w(b) + w(d)w(b) = w(d)(w(b) + w(c)) + w(b)w(c) \geq w(a)w(d) + w(b)w(c)$ which is at least as large as $w(a)w(c) + w(b)w(d)$, as we already saw. So $w(a)w(c) + w(b)w(d)$ is a lower bound in all cases on $\mathrm{ocr}(G, \rho, w)$.

Choosing $w(c) = 1$, $w(d) = w(b) = 3$, and $w(a) = 4$, and applying Lemma 11.26 yields part (i). For part (ii), we let $w(c) = \lfloor \frac{3^{1/2} - 1}{2} k \rfloor$, $w(d) = w(b) = k$, and $w(a) = \lfloor \frac{\sqrt{3} + 1}{2} k \rfloor$. Then $\mathrm{cr}(G, \rho, w) = w(a)w(d) + w(b)w(c) \sim \sqrt{3}k^2$, and $\mathrm{ocr}(G, \rho, w) = w(a)w(c) + w(b)w(d) \sim 3/2k^2$. So as $k \to \infty$, $\mathrm{ocr}(g)/\mathrm{cr}(G) \to \frac{3^{1/2}}{2}$. Using Lemma 11.26 gives us a cubic, 3-connected graph achieving the same separation. $\square$

11.5.4 Adjacent Crossings Matter

Tutte [340] famously wrote that "crossings of adjacent edges are trivial, and easily got rid of". This may have been overly optimistic. While for cr versus cr_ the question remains open, for iocr we can show that counting adjacent crossings does make a difference.[2]

Theorem 11.30 (Fulek, Pelsmajer, Schaefer, and Števankovič [142]). *There is a graph G with* iocr$(G < ocr(G)$.

The separating graph is based on the conference version of [142].

Proof. Consider the weighted graph G drawn in Figure 11.5. The drawing

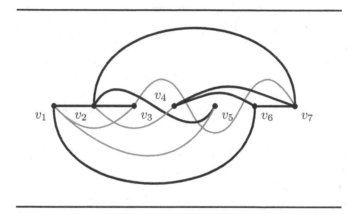

FIGURE 11.5: Separating mon-iocr from mon-ocr. Heavy black edges have weight 3, gray edges have weight 1.

shows that G has mon-iocr at most 2; the two independent odd crossings occur between (v_1v_3, v_2v_4), and (v_1v_5, v_3v_7). On the other hand, we can show that mon-ocr$(G) \geq 3$. Suppose, to the contrary, that mon-ocr$(G) \leq 2$ and fix a drawing witnessing this bound. Since the heavy black edges have weight 3, they cannot cross any other edge oddly. We can assume that the outer cycle $(v_1v_2, v_2v_7, v_7v_6, v_6v_1)$ is drawn essentially as shown in Figure 11.5: we use Lemma 11.2 to remove all crossings with $(v_1v_2, v_2v_7, v_7v_6, v_6v_1)$, while keeping the drawing x-monotone (after cutting, all vertices are on the same side of the resulting cycle, so we do not have to reconnect closed components, that is, we can drop them without changing the crossing parity between any pair of edges). Then v_1v_3 must pass below v_2, as shown in Figure 11.6, since otherwise it would cross v_2v_7 oddly. Then v_2v_5 must pass above v_3, since it cannot cross v_1v_3 oddly. Now v_3v_7 must cross both v_2v_5 and v_4v_6 evenly, which implies that it passes above v_4 and below v_5, implying that the heavy

---

[2]We should mention that Tutte's comment refers to iacr; the separating example below also separates iacr from acr, a fact we do not prove here.

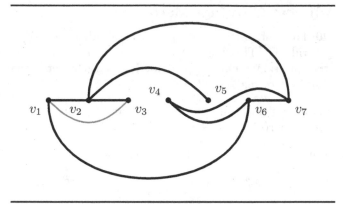

FIGURE 11.6: Drawing of the heavy edges.

edges are drawn as shown in Figure 11.6. Adding v_1v_5, v_2v_4, and v_3v_7, to this drawing causes at least three crossings. Lemma 11.27 now gives us a simple, unweighted graph with independent odd crossing number strictly smaller than the crossing number.   □

The x-monotone counterexample we constructed satisfies mon-iocr$(G) = 2$, and mon-ocr$(G) = 3$, begging the question whether there are examples with smaller mon-iocr. For mon-iocr$(G) = 0$ it can be shown that this is not the case [143, 266]. The question whether mon-iocr$(G) = 1$ implies that mon-ocr$(G) = 1$ is open, as far as we know.

Theorems 11.28, 11.29, and 11.30 tell us that there are graphs G_1, G_2, G_3 so that acr$(G_1) < $ cr(G_1), ocr$(G_2) < $ acr(G_2), and iocr$(G_3) < $ ocr(G_3). Then the union $G = G_1 \cup G_2 \cup G_3$ satisfies iocr$(G) < $ ocr$(G) < $ acr$(G) < $ pcr$(G) = $ cr(G), since each pair of crossing number notions is separated by one of the G_i, and iocr $\leq$ ocr $\leq$ acr $\leq$ cr.

We do not know whether either of the Rule $+$ variants can be separated from another crossing number. In the x-monotone drawing model, it is known that there are graphs for which mon-ocr$(G) < $ mon-ocr$_+(G)$, but the translation lemma, Lemma 11.27 does not work for mon-ocr$_+$.

11.6   Disjoint Edges in Topological Graphs

We saw earlier that a graph with sufficiently many edges must contain k pairwise crossing edges. Using the odd crossing number, we can show that the conclusion remains true if we require the crossing edges to be independent.

Theorem 11.31 (Pach and Tóth [269])**.** *If a graph has a drawing which contains no $k \geq 2$ independent edges so that each pair crosses oddly, then $|E(G)| \leq c_k n (\log(n))^{4k-2}$.*

The proof is an adaptation of the proof of Theorem 7.20.

Proof. We show inductively, on k and n, that if a graph can be drawn so that no k edges cross pairwise, then $|E(G)| \leq c_k n \log^{2k-4}(n)$, where c_k will be determined later. For $k = 2$, the Hanani-Tutte theorem, Theorem 11.8, implies that the graph is planar, so the result is true for all n as long as $c_2 \geq 3$, using Euler's formula.

Suppose that G has a drawing D in which no $k \geq 3$ edges cross pairwise, and that the result has been established inductively for $k' \leq k$ and $n' \leq n = |V(G)|$, where at least one of the inequalities is strict.

If we let G_e be the graph induced by all edges independent of e and which cross e oddly, then G_e cannot contain $k - 1$ pairs of independent edges which cross oddly. So by inductive assumption, $|E(G_e)| \leq c_{k-1} n \log^{4k-6}(n)$. We have

$$
\begin{aligned}
\mathrm{ocr}(D) &\leq \frac{1}{2}\left(\mathrm{ssqd}(G) + \sum_{e \in E(G)} |E(G_e)|\right) \\
&\leq \mathrm{ssqd}(G) + \frac{c_{k-1}}{2} |E(G)| n \log^{4k-6}(n),
\end{aligned}
$$

since every pair of oddly crossing edges is counted twice in $\sum_{e \in E(G)} |E(G_e)|$, if the pair is independent, and (at least) twice in $\mathrm{ssqd}(G)$, if the pair is adjacent.

By Theorem 11.18, $\mathrm{cr}(G) + \mathrm{ssqd}(G) \geq (c\,\mathrm{bw}(G)/\log(n))^2$ for some constant $c > 0$, so using $\mathrm{cr}(G) \leq \mathrm{cr}(D)$ and $\mathrm{ssqd}(G) \leq |E(G)| n$, we get

$$
\begin{aligned}
\mathrm{bw}(G) &\leq c \log(n)(\mathrm{cr}(G) + \mathrm{ssqd}(G))^{1/2} \\
&\leq c(|E(G)| n)^{1/2} \log(n)(c_{k-1}/2 \log^{4k-6}(n) + 2)^{1/2} \\
&\leq c(|E(G)| n)^{1/2} \log(n)(c_{k-1} \log^{4k-6}(n))^{1/2} \\
&\leq c\, c_{k-1}^{1/2}(|E(G)| n)^{1/2} \log^{2k-2}(n).
\end{aligned}
$$

as long as $c_{k-1}/2 \geq 2$. A bisection (V_1, V_2) of G realizing $\mathrm{bw}(G)$ results in two graphs $G_i = G[V_i]$, $i = 1, 2$, which are smaller than G, so induction applies, and we have $\mathrm{bw}(G) = |E(G)| - (|E(G_1)| + |E(G_2)|)$. We bound

$$
\begin{aligned}
|E(G_1)| + |E(G_2)| &\leq c_k \left(\frac{n}{3} \log^{4k-2}\left(\frac{n}{3}\right) + \frac{2n}{3} \log^{4k-2}\left(\frac{2n}{3}\right)\right) \\
&\leq c_k n \log^{4k-2}(n) \left(\frac{1}{3}\left(\frac{\log\left(\frac{n}{3}\right)}{\log(n)}\right)^{4k-2} + \frac{2}{3}\left(\frac{\log\left(\frac{2n}{3}\right)}{\log(n)}\right)^{4k-2}\right) \\
&\leq c_k n \log^{4k-2}(n) \left(\frac{1}{3}\left(1 - \frac{\log(3)}{\log(n)}\right)^{4k-2} + \frac{2}{3}\left(1 - \frac{\log\left(\frac{3}{2}\right)}{\log(n)}\right)^{4k-2}\right) \\
&\leq c_k n \log^{4k-2}(n)\left(1 - \frac{k}{\log n}\right)),
\end{aligned}
$$

using $(1 - a)^x \leq 1 - ax$ and $(4k - 2)\log(3) \geq k$ for $k \geq 3$. So

$$|E(G)| - c\, c_{k-1}^{1/2}(|E(G)|n)^{1/2}\log^{2k-2}(n) \leq c_k n \log^{4k-2}(n)\left(1 - \frac{k}{\log n}\right).$$

If we consider this inequality as an inequality in $x = |E(G)|$, it follows that the inequality holds for $x = 0$, since we can assume that $k < \log(n)$—otherwise the bound of the theorem is trivially true—so there is a point $x_0 \geq 0$ so that the inequality holds between 0 and x_0 and fails for $x > x_0$. If we try $x = c_k n \log^{4k-2}(n)$, we get

$$\frac{k}{\log n} c_k n \log^{4k-2}(n) \leq c(c_{k-1}c_k)^{1/2} n \log^{4k-3}(n)$$

which is equivalent to $k \leq c(c_{k-1}/c_k)^{1/2}$, so fails if $c_k \geq c^2 c_{k-1}$. Hence, $x_0 < c_k n \log^{4k-2}(n)$ which implies that $|E(G)| < c_k n \log^{4k-2}(n)$.

It remains to show that we can choose c_k to satisfy all requirements, namely $c_2 \geq 3$, and $c_k \geq c^2 c_{k-1}$. Setting $c_k = 3c^{2k-4}$ satisfies all requirements.   □

Pach, Radoičić, and Tóth [257] show how to improve the upper bound to $c_k n \log^{4(k-3)}(n)$, which implies a linear bound for $k = 3$.

We say two edges in a drawing are *disjoint* if they have no point in common (including a common endpoint). An elegant argument using odd crossings allows us to turn Theorem 11.31 on pairwise crossing edges, into a result on pairwise disjoint edges.

Corollary 11.32 (Pach and Tóth [269]). *If a graph has a good drawing which contains no $k \geq 2$ pairwise disjoint edges, then $|E(G)| \leq c_k n(\log(n))^{4k-2}$.*

We cannot drop the requirement that the graph is simple: Exercise (2.3) shows that every complete graph has a drawing in which every two edges cross at most twice.

Proof. Suppose G has a good drawing which contains no $k \geq 2$ pairwise disjoint edges. Let H be a largest bipartite subgraph of G. Then $|E(H)| \geq |E(G)|/2$, since every vertex of G must be incident to at least as many edges in $E(H)$ as in $E(G) - E(H)$, or it could be moved to the other partition of H. The induced drawing of H still does not contain k pairwise disjoint edges. We can move the vertices of H, keeping the drawing isomorphic, so that all vertices of one partition lie above the x-axis, and the remaining vertices below it (and all intersections with the x-axis are crossings). Sever all edges crossing the x-axis, mirror the drawing above the x-axis, and reconnect severed ends by straight-line segments. This changes the parity of crossing between any pair of edges (since every edge has to cross the x-axis an odd number of times). We claim that the resulting drawing does not contain k independent pairs of edges crossing oddly. If it did, then the drawing of H would contain k independent pairs of edges crossing evenly, and, since the drawing was simple, this implies the edges are disjoint. Therefore, $|E(H)| \leq c_k n(\log(n))^{4k-2}$, by Theorem 11.31, so $|E(G)| \leq 2c_k n(\log(n))^{4k-2}$.   □

Corollary 11.33 (Pach and Tóth [269]). *Every good drawing of a complete graph contains* $\Omega(\log(n)/\log\log(n))$ *pairwise disjoint edges.*

The current best bound for complete graphs is $\Omega(n^{1/3})$ as we saw in Theorem 2.9. For general graphs, and sparse graphs in particular, Corollary 11.32 has only been slightly improved to $\log^{1+\varepsilon}(n)$, by Fox and Sudakov [139].

11.7 Notes

In his 1970 paper, Tutte [340] suggested a "Theory of Crossing Numbers", which amounted to an algebraic study of iacr(D). In this paper he reproved what is now known as the (strong) Hanani-Tutte theorem, see [299] for a survey. In spite of Tutte's preference for counting crossings algebraically, it appears that the parity variants, iocr and ocr have received more attention over the years. We still have no deep understanding of the relationship between counting crossings by parity or algebraically, let alone the relationship of either variant to the standard crossing number.

11.8 Exercises

(11.1) Give a proof of Theorem 11.3. *Hint:* Extend the proof of Theorem 11.1.

(11.2) Show the following: if G has a drawing in which every Eulerian subgraph has an even number of crossings (with itself), then G is planar. *Hint:* The rotation system does not change.

(11.3) Does the statement in Exercise (11.2) remain true if we require that the rotation system not change?

(11.4) Improve the asymptotic separation in Theorem 11.28 by doubling the two-star device.

(11.5) Show that every graph has an ocr-minimal drawing with at most $2^{O(k)}$ crossings, where $k = \text{ocr}(G)$. *Note:* We proved an analogous result for the pair crossing number in Corollary 9.3. It is an open question whether a similar bound can be achieved for iocr-minimal drawings.

(11.6) Give a proof of Theorem 11.3. *Hint:* Extend the proof of Theorem 11.1.

(11.7) Prove a crossing lemma for iocr on the projective plane. *Hint:* You can use the fact that the strong Hanani-Tutte theorem is true on the projective plane [275].

(11.8) Show that mon-cr$(G) \leq \binom{2\,\mathrm{iocr}(G)}{2}$. This improves Theorem 8.30 from Chapter 8. *Hint:* Use the proof of Theorem 11.13.

Chapter 12

Maximum Crossing Numbers

12.1 Polygons and Cycles

We begin with one of the oldest known crossing number problems; its roots go back to a note by Richard Baltzer from 1885. In that note, Baltzer asked, and partially answered, the following question: what is the largest number of crossings in an n-gon? We generalize this problem by introducing the *maximum rectilinear crossing number*, max-$\overline{\text{cr}}(G)$ of a graph G as the largest number of crossings in a rectilinear drawing of a graph. Baltzer's problem then asks for max-$\overline{\text{cr}}(C_n)$.

Theorem 12.1 (Steinitz [321]). *We have*

$$\text{max-}\overline{\text{cr}}(C_n) = \begin{cases} n(n-3)/2 & \textit{if } n \textit{ is odd,} \\ n(n-4)/2+1 & \textit{if } n \textit{ is even.} \end{cases}$$

Proof. First note that max-$\overline{\text{cr}}(C_n) \leq n(n-3)/2$ for all $n \geq 3$, since only independent edges are allowed to cross, and for every edge there are $n-3$ independent edges in C_n for $n \geq 5$.

Next we will see that max-$\overline{\text{cr}}(C_n) \geq n(n-3)/2$ for odd n. Place the n vertices at the corners of a convex n-gon, and add edges between any two vertices of distance $\lceil n/2 \rceil$ along the boundary (these drawings are known as regular star polygons, with Schläffli symbol $\{n/\lceil n/2 \rceil\}$). Note that the resulting drawing is a drawing of a C_n, since every vertex has degree 2, and the graph is connected, since n and $\lceil n/2 \rceil$ are coprime (for odd n). Every edge crosses every edge it is not incident to, so the total number of crossings is $n(n-3)/2$.

For even n we use a construction due to Furry and Kleitman [144]: arrange vertices on two parallel lines in order $1\ (n-1)\ 3\ (n-3)\ \ldots$ on the first, and $\ldots 4\ (n-2)\ 2\ n$ on the second line. We claim that the resulting rectilinear drawing has $n(n-4)/2+1$ crossings: flip the order of the vertices on the second line, so it becomes $n\ 2\ (n-2)\ 4\ldots$. It is easy to see that *that* drawing has $n/2-1$ crossings. But then flipping the order of vertices on one side changes every pair of independent edges from crossing to non-crossing, and vice versa. Since there are $n(n-3)/2$ such pairs, as we saw earlier, this implies that we have $n(n-3)/2 - (n/2-1) = n(n-4)/2+1$ crossings in the original

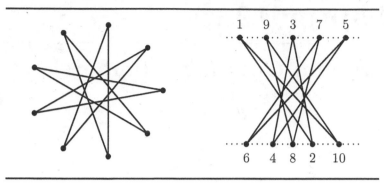

FIGURE 12.1: Crossing-maximal rectilinear drawings of C_9 *(left)* and C_{10} *(right)*.

drawing. To see the upper bound for even n we follow [28]. We already know that every edge has at most $n-3$ crossings. Suppose there are three edges each with exactly $n-3$ crossings. These edges cannot be adjacent, since they have to cross each other to achieve the $n-3$ bound. So the picture looks as in Figure 12.2. If an edge has exactly $n-3$ crossings, it must be in the middle

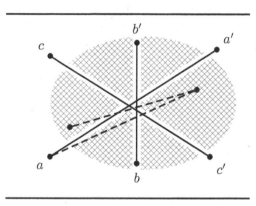

FIGURE 12.2: Three pairwise crossing edges aa', bb', and cc'. The lines through these edges create six outer regions, crosshatched.

of an N-shape, that is, the vertices preceding and succeeding it in the cycle must lie on opposite sides of it since n is even, and it must be crossed by all $n-3$ remaining edges. Follow the cycle starting at vertex a, and directed away from a'. Since the next edge has to cross both bb' and cc', the next vertex must lie in the region between a' and b' or in the region between a' and c'. Let us assume, without loss of generality, that the latter is the case (as pictured in the figure above). The cycle may now go back and forth between the region $a'c'$ and ac, but eventually it enters c from the $a'c'$ region, or c'

from the ac region. It then traverses cc' and has to leave on the opposite side from which it entered, meaning towards region $b'c$ if we entered c', and region bc' if we entered c. So we are stuck in these regions, and cannot reach b or b'. This shows that there are most two edges with $n - 3$ crossings, which readily implies the upper bound. $\qquad\square$

If we do not insist on rectilinear drawings, a maximum crossing number only makes sense if we restrict the drawings, in particular, we will use good drawings. Define the *maximum crossing number* of a graph G to be the largest number of crossings in a good drawing of G.

Theorem 12.2 (Harborth [177]). *If $n = 3$ or $n \geq 5$, we have*

$$\text{max-cr}(C_n) = n(n-3)/2,$$

and $\text{max-cr}(C_4) = 1$.

For this result we will use a clever replacement idea.

Lemma 12.3. *Suppose we are given a good drawing of a graph. Any edge uv can be replaced with a path $uxyv$ of length 3 so that the new drawing is good, the new edges cross all edges which crossed uv, and ux crosses all edges incident to v and vy all edges incident to u (in particular, they cross each other).*

Proof. Replace uv with a path $uxyv$, closely to uv so that each of the three new edges picks up all the former crossings with uv and so that ux and vy cross. Now move x around v and y around u so they pick up all crossings with edges incident to those vertices. See Figure 12.3. $\qquad\square$

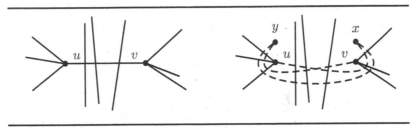

FIGURE 12.3: Edge uv before *(left)* and after *(right)* being replaced by a path $uxyv$.

Proof of Theorem 12.2. We already saw that $\text{max-cr}(C_n) \leq n(n-3)/2$ for all n, and that the bound is attainable for odd n in the proof of Theorem 12.1. We also know how to draw C_4 with one crossing, and we saw that that is the maximum in a good drawing in the proof of Lemma 1.5. So $\text{max-cr}(C_4) = 1$. Figure 12.4 shows that $\text{max-cr}(C_6) \geq 9$ which agrees with the upper bound.

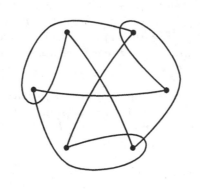

FIGURE 12.4: A good drawing of C_6 with nine crossings.

We can now inductively show that there is a drawing of C_n for n even and $n \geq 6$ in which every edge crosses every edge that it is not adjacent to. Figure 12.4 is the base case, and Lemma 12.3 allows us to add two vertices to a cycle, so that the new (and old) edges cross every edge they are not adjacent to. This attains the bound $n(n-3)/2$. □

We are not aware of a more direct construction for max-cr(C_n) that works for all even n.[1]

The result on max-$\overline{\mathrm{cr}}(C_n)$ can be extended to find max-$\overline{\mathrm{cr}}(W_n)$ for the wheel graph W_n; for example, if n is odd, then adding the center vertex anywhere on the convex boundary and connecting it via straight-line segments to the vertices of the C_n adds $(n^2 - 2n - 1)/2$ crossings, resulting in $(n^2 - 2n - 1)/2 + n(n-3)/2 = n^2 - 2.5n - 0.5$ crossings, and a similar construction works for even n. Establishing the matching lower bounds is harder though, see Feder [133].

Theorem 12.4. *We have*

$$\text{max-}\overline{\mathrm{cr}}(W_n) = \begin{cases} n^2 - 2.5n - 0.5 & \textit{if } n \textit{ is odd} \\ n^2 - 3n + 1 & \textit{if } n \textit{ is even} \end{cases}$$

There are further results on generalized wheel graphs [30], and we also know that asymptotically, max-cr$(W_n) \sim 5n^2/4$ [182], see Exercise (12.6).

---

[1]For $n \equiv 2 \bmod 4$ see [176, Figure 11].

12.2 Complete Graphs

Compared to cycles, complete n-partite graphs are easily dealt with.

Theorem 12.5 (Harborth [175]). *We have*

$$\text{max-cr}(K_{x_1,\ldots,x_k}) = \text{max-}\overline{\text{cr}}(K_{x_1,\ldots,x_n}) = \binom{n}{4} - \sum_{i=1}^{k}\binom{x_i}{4} - \sum_{i=1}^{k}\binom{x_i}{3}(n-x_i),$$

where $n = \sum_{i=1}^{k} x_i$.

In particular,

$$\text{max-cr}(K_n) = \text{max-}\overline{\text{cr}}(K_n) = \binom{n}{4}, \text{and}$$

$$\text{max-cr}(K_{m,n}) = \text{max-}\overline{\text{cr}}(K_{m,n}) = \binom{m}{2}\binom{n}{2}.$$

The following, slightly simplified proof, is based on [30].

Proof. To see the lower bound, consider convex rectilinear drawings of $K_{x_1,\ldots,x_k}$ in which the x_i vertices of the ith partition are consecutive along the boundary. Then if we pick four vertices from at least three different partitions, they contain a C_4 with a crossing as a subgraph. To count the number of ways in which the four vertices can be picked, it is easier to count all ways four vertices can be picked, and subtract the number of ways they can be chosen from less than 3 partitions, that is, one partition, $\sum_{i=1}^{k}\binom{x_i}{4}$, or two partitions, $\sum_{i=1}^{k}\binom{x_i}{3}(n-x_i)$.

For the upper bound, note that $\text{max-cr}(K_4) \leq 1$, since any two pairs of independent edges in a K_4 form a C_4 which can have at most one crossing in a good drawing. So any four vertices, in an arbitrary good drawing of $K_{x_1,\ldots,x_k}$, can induce at most one crossings, and they can only do so if they belong to at least three different partitions. The count then is exactly as above. □

12.3 Thrackles

12.3.1 The Thrackle Bound

In a good drawing, crossing edges cannot be adjacent, giving us a simple upper bound on max-cr, the *thrackle bound*.

Lemma 12.6 (Thrackle Bound. Piazza, Ringeisen, and Stueckle [281]).

$$\text{max-cr}(G) \leq \vartheta(G) \quad := \quad \frac{1}{2} \sum_{uv \in E(G)} (m - \deg(u) - \deg(v) + 1)$$

$$= \quad = \binom{m}{2} - \sum_{u \in V(G)} \binom{\deg(u)}{2},$$

where $m = |E(G)|$.

Proof. Edge uv can only cross edges not adjacent to it, and there are $m - \deg(u) - \deg(v) + 1$ of those. □

The lemma is a convenient first upper bound for max-cr, and max-$\overline{\text{cr}}$, and often it is sharp, as in the following example.

Theorem 12.7 (Piazza, Ringeisen, and Stueckle [281]). *If G is a forest, then* max-cr$(G) = \vartheta(G)$.

Graphs achieving $\vartheta(G)$, that is, max-cr$(G) = \vartheta(G)$, are known as *thrackles* or *thrackleable*; so the theorem states that forests are thrackles.

Proof. We start with an observation: if G is a thrackle, and G contains a vertex v of degree 1, then any graph obtained by adding edges incident to v is still a thrackle. To see this, take a thrackle-drawing of G, and let u be the unique neighbor of v. Any new edge added to v can be drawn following uv closely on one side from v to u, picking up crossings with all edges of G except edges incident to u, and then turning around u to pick up crossings with all edges incident to u, stopping short of uv on the other side.

This observation allows us to prove the result by induction. If we cannot apply the observation to a forest G, then G must be a matching (any isolated vertices we can remove, they do not affect the result), and we can easily draw a matching so that every two edges cross. □

The proof of Theorem 12.7 requires edges to bend (to turn around a vertex), and this is necessary; in Theorem 12.22 we will see a characterization of which forests satisfy max-$\overline{\text{cr}}(G) = \vartheta(G)$.

In all the cases we have studied so far, max-$\overline{\text{cr}}$ was achieved by a convex rectilinear drawing; this suggests defining max-$\overline{\text{cr}}°(G)$ as the largest number of crossings in a convex rectilinear drawing of G, the *convex maximum rectilinear crossing number*. It has been conjectured [28] that max-$\overline{\text{cr}}(G) = $ max-$\overline{\text{cr}}°(G)$, but a counterexample to this conjecture has been announced [90]; it is still possible though that the conjecture holds for bipartite graphs G.

Theorem 12.8 (Verbitsky [344]). *We have*

$$\vartheta(G)/3 \leq \text{max-}\overline{\text{cr}}°(G) \leq \vartheta(G).$$

Proof. The upper bound follows from Lemma 12.6. For the lower bound, place the vertices of G in a random permutation along a circle, and draw edges as straight-line segment. There are potentially $\vartheta(G)$ pairs of edges that may cross. Any particular pair of independent edges crosses with probability $1/3$ (of the 24 permutations of the 4 endpoints, 8 result in a crossing). Hence, the expected number of crossings is $\vartheta(G)/3$, which implies that there is a permutation that achieves max-$\overline{\mathrm{cr}}^\circ(G) \geq \vartheta(G)$. □

Since max-$\overline{\mathrm{cr}}^\circ(G) \leq$ max-$\overline{\mathrm{cr}}(G) \leq$ max-cr$(G) \leq \vartheta(G)$, Theorem 12.8 means that we can always approximate any of the maximum crossing numbers to within a factor of 3 using $\vartheta(G)$. The proof shows that we can even find a drawing realizing that bound using an efficient randomized construction. Bald, Johnson, and Liu [41] show that the algorithm can be derandomized, so there is an efficient approximation algorithm for all the maximum crossing number notions we have seen so far.

12.3.2 Conway's Thrackle Conjecture

We introduced thrackles as graphs which have a good drawing in which every pair of independent edges crosses once (or, equivalently, every pair of edges intersects once). In this language, Theorem 12.1 states that C_n is a thrackle for $n = 3$ and $n \geq 5$, and fails to be thrackleable for $n = 4$. The main driving force behind the study of thrackles has been Conway's conjecture.

Conjecture 12.9 (Conway's Thrackle Conjecture). If G is a thrackle with m edges and n vertices, then $m \leq n$.

This conjecture has been open for many decades, but there has been some partial progress.

Theorem 12.10 (Woodall [351]). *The thrackle conjecture is true if and only if no graph consisting of two even-length cycles sharing a vertex can be thrackled.*

Let us say an edge is *thrackled* if it crosses every non-adjacent edge exactly once. In the proof, we will make use of a basic tool called *Conway doubling*. Given a drawing of a graph containing a cycle C_n we can replace C_n in the drawing as shown in Figure 12.5.

Figure 12.6 shows the results of applying Conway for C_3 and C_4.

Conway doubling a cycle of odd length doubles the length of the cycle; Conway doubling a cycle of even length results in two cycles of the original length as illustrated in the drawings. The point of Conway doubling is that if the edges of the original cycle were thrackled, then so are the edges of the new cycle(s), and any other edges which were thrackled remain thrackled. Hence, if the original graph was a thrackle, then so is the new graph. Applying the construction to a triangle explains the thrackled drawing of C_6 we saw earlier.

Proof of Theorem 12.10. Let G be a minimal counterexample to the thrackle

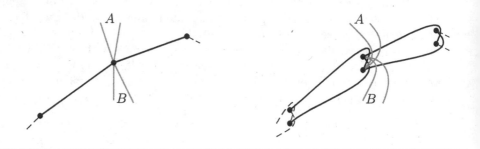

FIGURE 12.5: Conway doubling at a vertex. *Left:* cycle (black), separating rotation at vertex into two parts, A and B. *Right:* Incidences after Conway doubling.

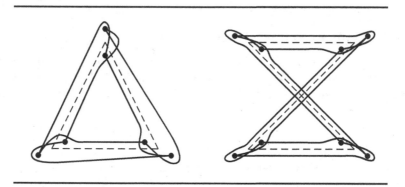

FIGURE 12.6: *Left:* Conway doubling C_3 (dashed). *Right:* Conway doubling C_4 (dashed).

conjecture, with $m = |E(G)|$ and $n = |V(G)|$. G is connected, since otherwise it could be split into two disjoint graphs, both of which are thrackleable, and at least one of which has more edges than vertices, so a smaller counterexample. We can assume that $m = n+1$, since if $m > n+1$ and G is thrackleable, then removing any edge will leave n unchanged, and result in another thrackleable graph with one fewer edge. There cannot be any vertices of degree 1, since removing that vertex (and its incident edge) would yield a smaller thrackleable graph. So, we are left with a connected graph, in which all vertices must have degree 2 except for two vertices of degree 3, or one vertex of degree 4. In other words, two cycles connected by a (possibly empty) path (these graphs are sometimes called *bar-bells*, or *figure-8* graphs, if the path is empty), or three paths between two vertices (θ-*graphs*). If there are two vertices of degree 3, pick a shortest path between them. We can shorten this path, by redrawing one of the edges incident to a degree-3 vertex as shown in Figure 12.7, eventually

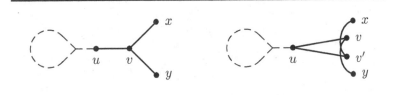

FIGURE 12.7: Shortening a path at a degree 3-vertex. *Left:* Vertex v with three neighbors. *Right:* After the split u is the new degree 3-vertex, and the path has been shortened.

leaving us with a single vertex of degree 4, so a figure-8 graph. If either or both of the two cycles in the figure-8 graph has odd length, we apply Conway-doubling to it. We end up with a thrackleable figure-8 graph consisting of two even cycles. □

Theorem 12.11 (Woodall [351]). *Assuming the truth of Conway's thrackle conjecture, a thrackle contains at most one odd-length cycle, contains no C_4, and every connected component contains at most one cycle.*

Proof. There cannot be two disjoint cycles of odd length: they would have to cross each other an odd number of times, but two closed simple curves in the plane cross an even number of times (as usual, we can remove self-crossings to obtain simple curves), and we already saw that no C_4 is thrackleable. Two cycles in the same component implies the presence of a bar-bell or figure-8 graph, which violates the edge bound (here we are using the assumed truth of the conjecture). This proves necessity of the classification. It remains to show that any graph satisfying these conditions is a thrackle (we will not need the truth of the conjecture for that part).

It is sufficient to prove this result for a collection of cycles at most one of which is odd, and none of them a C_4: to any thrackle drawing we can add an edge uv incident to an existing vertex, say u, and a new vertex v using the same idea we saw in Theorem 12.7. Pick an edge uw, route uv close to one side of uw (picking up the same crossings as uw), and wrap it once around v so that it crosses all edges incident to v except uw. Then uv is a thrackled edge. This construction allows us to remove all tree-parts from each component. Lemma 12.3 now implies that we can assume that all cycles are 6-cycles, except possibly one, which is a C_3. Using Conway doubling reduces this to the case of one C_6, and, possibly, a C_3. Since $C_6 \cup C_3$ is thrackleable, as shown in Figure 12.8, this proves sufficiency. □

12.3.3 Generalized Thrackles

A *generalized thrackle drawing* of a graph is a drawing in which every pair of edges intersects an odd number of times (so every pair of independent edges

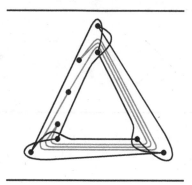

FIGURE 12.8: A thrackle-drawing of $C_6 \cup C_3$, edges of C_3 are gray.

crosses an odd number of times, while adjacent edges cross an even number of times, since the common endpoint counts as an intersection). A graph which has a generalized thrackle drawing is called a *generalized thrackle*. Any thrackle is a generalized thrackle, but the converse is not true, as witnessed by C_4 which is a generalized thrackle, but not a thrackle.

Theorem 12.12 (Cairns and Nikolayevsky [82], Pelsmajer, Schaefer, and Štefankovič [277]). *A bipartite graph has a generalized thrackle drawing in some surface, if and only if it can be embedded in that surface.*

Recall that an (e, v)-move in a drawing changes the parity of crossing between e and any edge incident to v, see Figure 11.1. Flipping the rotation at a vertex changes the parity of crossing between any two edges incident to the vertex, see Figure 12.9.

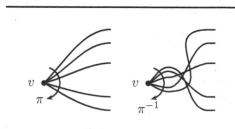

FIGURE 12.9: Flipping the rotation at vertex v.

Proof. Let U, W be the partitions of the graph and fix an arbitrary order $<$ on W. Given a drawing D of the graph in the surface, we perform a (uw, w')-move for all $u \in U$, $w, w' \in W$, for which $w < w'$, and flip all the rotations of vertices in U. The result of these moves is that the parity of *crossing* changes for every pair of independent edges. This is a consequence of the (e, v)-moves we perform; however, these moves also change the parity of crossing between any two edges sharing a vertex in U, which we correct by flipping the rotations of all vertices in U. Let D' be the new drawing. If D was an embedding of the graph, then D' is a generalized thrackle drawing, and if D was a generalized thrackle drawing, then D' is a drawing in which every pair of edges crosses an even number of times, so by Theorem 11.1 it corresponds to an embedding. □

Corollary 12.13 (Cairns and Nikolayevsky [82]). *If a bipartite graph is a generalized thrackle on some surface with Euler genus* eg, *then* $m \leq 2(n +$ eg $-2)$.

Proof. This immediately follows from the Euler-Poincaré formula, see Corollary A.17, and Theorem 12.12. □

Theorem 12.14 (Cairns and Nikolayevsky [82]). *In the plane,*

(i) $m \leq 3/2(n - 1)$ *for thrackles, and*

(ii) $m \leq 2(n - 1)$ *for generalized thrackles.*

Proof. We start with item (ii). Suppose G is a generalized thrackle; if G is bipartite, then the bound immediately follows from Corollary 12.13. So G contains an odd cycle C. We can apply Conway doubling to C, close to the simple closed curve γ obtained from C by removing self-crossings. We can reattach edges to vertices on C so that the resulting graph G' is a generalized thrackle, as we saw in Figure 12.5. G' is bipartite: every edge crosses γ an odd number of times (since the drawing is a generalized thrackle drawing), so every odd cycle closes γ an odd number of times, but that is not possible, since two cycles in the plane cross an even number of times. So G' is bipartite, with $n' = |V(G')| = n + k$, $m' = |E(G')| = m + k$, where k is the length of C. By Theorem 12.12, G' has an embedding in the plane with the same rotation system. In particular, noting the new rotations at the doubled C, it bounds a face in the embedding. If C has length k, this means, G' has a face of size $2k$. So $2m' \geq 4(f' - 1) + 2k$, where f' is the number of faces in the embedding of G'. Then

$$
\begin{aligned}
m = m' - k &\geq 2(f' - 1) + k - k \\
&= 2(f' - 1) \\
&= 2(2 - n' + m' - 1) \\
&= 2 - 2n + 2m,
\end{aligned}
$$

so $m \leq 2(n - 1)$.

To see (i), note that since C_4 is not thrackleable, no thrackle can contain a C_4. We can then treat G as a generalized thrackle without a C_4. In both the bipartite case (G contains no C_4), and the non-bipartite case (G' contains no C_4) this allows us to improve the bound on m to $3/2(n - 1)$. □

The bound on generalized thrackles is tight, as witnessed by the generalized thrackle drawing in Figure 12.10.

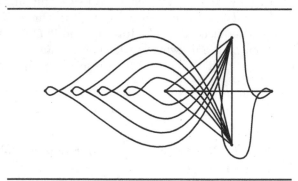

FIGURE 12.10: A generalized thrackle with n vertices, and $2(n-1)$ edges [82].

Pinchasi [283] observed that the result on generalized thrackles implies an earlier result on rectilinear drawings of graphs. Call two edges in a rectilinear drawing *parallel* if they do not cross and their endpoints are in convex position, so they are opposite sides of convex quadrilateral.

Corollary 12.15 (Katchalski and Last [203]; Valtr [342]; Pinchasei [283]). *If there are no parallel edges in a rectilinear drawing of a graph on m vertices and n edges, then $m \leq 2(n-1)$.*

Proof. Fix the rectilinear drawing of the graph; for every edge $e = uv$ pick one of the two half-spaces with e on its boundary, and perform (e, w)-moves for every vertex w in the interior of the halfspace. Then take the ends of e at u and v and—within the half-space—move them to the opposite side of where they used to attach. Figure 12.11 shows the effect on two adjacent edges e and f.

Consider two edges e and f in the original drawing. Since e and f are not parallel, they must either be (*i*) adjacent, (*ii*) cross, or (*iii*) their convex hull is a triangle. If e and f are adjacent, the (e, w)-moves we perform change the parity of crossing between e and f if f lies within the half-space we chose for e. This is compensated for by moving the end of e past f in the half-space. If e and f cross, we perform one (e, w)- and one (f, w')-move, where w is an endpoint of f, and w' an endpoint of e, so the parity of crossing between e and f does not change. If the convex hull of $e \cup f$ is a triangle, then the parity of crossing between e and f changes, so it becomes odd, since one of the two edges has only one endpoint of the other in its chosen half-space.

We have obtained a drawing of the graph in which all edges cross oddly, except pairs of adjacent edges, which cross evenly. Flipping the rotation at every vertex gives us a generalized thrackle drawing of the graph, which by Theorem 12.14 implies that $m \leq 2(n-1)$. □

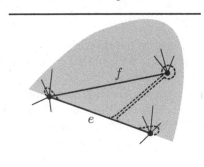

FIGURE 12.11: Modifying edge e in its halfspace (gray side): performing (e, w) moves, and adjusting ends of e.

Pinchasi's proof is slightly different: he visualizes the moves as taking a geodesic drawing on the sphere and replacing each edge with the remaining part of the great cycle it lies on.

Theorem 12.12 can be generalized to non-bipartite graphs; for that we need a special type of embedding. A graph has a *parity embedding (drawing)* in a surface if it can be embedded (drawn) so that all cycles of even length are two-sided curves, and all cycles of odd length are one-sided.

Theorem 12.16 (Cairns and Nikolayevsky [82]). *A graph has a generalized thrackle drawing in an orientable surface, if and only if it has a parity embedding in the surface with an additional crosscap.*

This result extends to non-orientable surfaces, using a different notion of parity embedding [277]. We prove the forward direction of the theorem for the sphere, following the approach in [277].

Proof. In a generalized thrackle drawing of G on the sphere, every pair of edges intersects an odd number of times. By flipping the rotation at each vertex, we can ensure that every pair of edges *crosses* an odd number of times. Pick a point on the sphere (not part of the drawing), and redraw each edge so it passes through that point, each edge at a different angle. Every pair of edges crosses an even number of times in this (non-normal) drawing. Replacing the point with a crosscap, we conclude that G has a drawing in the projective plane in which every pair of edges crosses an even number of times. By Theorem 11.1, G has an embedding in the projective plane with the same embedding scheme. Since every edge passed through the crosscap once in the original drawing, the number of times a cycle passes through the crosscap is the same, modulo 2, as the length of the cycle, so we have a parity-embedding of G in the projective plane. □

12.3.4 Superthrackles

When defining a thrackle, we required that every pair of edges intersects once. What happens if instead we require that every pair of edges *crosses* once, including adjacent edges? Archdeacon and Stor [36] introduced the term *superthrackle* for this variant. Every thrackle is a superthrackle: locally flip the rotation at each vertex in a thrackle drawing (we did this in Theorem 12.12 to get a superthrackle drawing), and the example of C_4 shows that the two notions differ: the generalized thrackle drawing of C_4 is a superthrackle drawing.

On the other hand, every superthrackle is a generalized thrackle (flip the rotation at each vertex, every pair of adjacent edges will then cross evenly, so intersect oddly). Surprisingly, Archdeacon and Stor showed that the two classes are the same.

Theorem 12.17 (Archdeacon and Stor [36]). *A graph is a generalized thrackle if and only if it is a superthrackle.*

We first illustrate the proof in a special case, which also explains the name superthrackle.

Lemma 12.18. *If G is a bipartite generalized thrackle, then it has a superthrackle drawing with a single crossing.*

Note that this drawing is not normal (which we typically require), since more than two edges cross in a point.

Proof. If G is a bipartite generalized thrackle, it is planar by Theorem 12.12. Fix a plane embedding of G. The dual of the embedding of G is Eulerian (since every face has even size), so, by Theorem A.18, contains an Eulerian circuit without self-crossings. Moving apart the points at which the circuit touches itself, we obtain a closed curve, let us call it γ, which crosses every edge of G exactly once. Cut the surface along γ, flip one of the resulting disks, and merge the disks connecting them with an annulus. The order of the severed edges along the boundaries of the disks is reversed, so as we connect them along the annulus we can do so by passing all of them through the same crossing point. □

Proof of Theorem 12.17. We already saw that a superthrackle can be drawn as a generalized thrackle, so we only have to show the reverse direction.

Since the boundary of a face must be two-sided, all faces have even size, hence the dual graph of the embedding is Eulerian, and, by Theorem A.18 contains an Eulerian circuit without self-crossings. Moving apart the points at which the circuit touches itself, we obtain a closed curve, let us call it γ, which crosses every edge of G exactly once. If γ is contractible, then G must be bipartite (since every edge crosses γ once, there can be no odd cycles). By Theorem 12.12, G is planar, and Lemma 12.18 deals with this case. Hence γ must be one-sided. Cutting the surface along γ results in a sphere with a hole.

Fill in the hole with a disk. The ends of the edges along the boundary of the disk are in reverse order, so we can connect them by curves which cross in a single point of the disk. □

The proof implies that every generalized thrackle has a (non-normal) superthrackle drawing with a single crossing point, so all generalized thrackles have degenerate crossing number at most 1.

12.3.5 A Better Bound

In Theorem 12.14 we saw that every thrackle on n vertices has at most $1.5(n-1)$ edges. This bound can be improved slightly, by studying the embedding of the thrackle in the plane or projective plane.

Theorem 12.19 (Goddyn and Xu [153]). *If G is a thrackle, then $m \leq 1.4(n-1)$, where $m = |E(G)|$ and $n = |V(G)|$.*

The result builds on previous work that shows that certain graphs are not thrackleable. We state these results without proof.

Theorem 12.20. *The following graphs are not thrackleable:*

- *Two C_5 overlapping in a single vertex (Misereh and Nikolayevsky [244]).*

- *Two C_6 overlapping in a single vertex (Fulek and Pach [141]).*

Lemma 12.21. *A graph consisting of two overlapping C_6 is thrackleable if and only if it is C_6 itself.*

Proof. If the two C_6 share an edge, and they are not the same, then there must be an edge belonging to both C_6 which is incident at one end to two edges, one belonging to each of the C_6. Applying the redrawing illustrated in Figure 12.7 (in Theorem 12.10), we can split the shared edge to obtain a thrackle drawing of two C_6 which share one fewer edge. We can therefore assume that the two C_6 do not share any edges at all.

Let us use C and C' for the two C_6-graphs, and consider C. No edge of C' can be an edge in C (as we just saw), and it cannot be a chord of C, since that would form a C_4, which is not possible, or a C_3 and a C_5 sharing a single edge, see the left of Figure 12.12.

We can eliminate the shared edge as before (Figure 12.7) to obtain a thrackle-drawing of a C_3 and a C_5 sharing a point. Applying Lemma 12.3 to an edge of the C_3 which does not belong to C_5 we obtain a thrackle-drawing of two C_5 sharing a vertex, which we know is not possible by Theorem 12.20. We conclude that C' does not contain a chord of C. A similar argument shows that there cannot be a path of length 2 in C' which connects two vertices of C. It would either form a C_4, two C_5 overlapping in a path of length 2 (after using redrawing of Figure 12.7 on both edges, we obtain a thrackle-drawing of two C_5 sharing a vertex), or a C_5 and a C_3 sharing a single edge (which

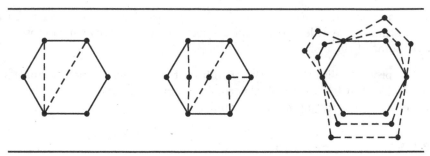

FIGURE 12.12: Edges of C' (dashed) connecting vertices of C by paths of lengths 1, on the left, 2, in the middle, and 3, on the right.

we already dealt with). Hence any C' path between two C-vertices must have length at least 3, and therefore, there must be exactly two such paths of length 3, attaching to the same vertices of C. This either leads to a C_4, two C_5 sharing a path of length 2 (which we already dealt with), or three paths of length 3 between two vertices. Unravelling one of these paths (as in Figure 12.7) gives us two C_6 sharing a vertex, which we know to be impossible. □

Proof of Theorem 12.19. We need to show that $m \leq 1.4(n-1)$ for a thrackle G. Suppose not, and let G be a minimal counterexample. Then G is 2-connected, since otherwise $G = G_1 \cup G_2$, where $|V(G_1) \cap V(G_2)| \leq 1$. Setting $n_1 := |V(G_1)|$ and $n_2 := |V(G_2)|$, we have $n_1 + n_2 \leq n + 1$, so $|E(G)| = |E(G_1)| + |E(G_2)| \leq 1.4(n_1 - 1) + 1.4(n_2 - 1) = 1.4(n_1 + n_2 - 2) \leq 1.4(n-1)$, a contradiction.

By Theorem 12.16, G, as a thrackle, has a parity embedding in the projective plane. Since the boundary of a face is two-sided, every face of the embedding must be a cycle of even length (here we use that G is 2-connected: boundaries are cycles in the graph). Since there are no 4-faces, all faces have length at least 6. Call an edge e *bad* if it is incident to a 6-face. By Lemma 12.21, a bad edge can be incident to at most one 6-face, so it is incident to a 6-face, and an (even) face of length at least 8. An 8-face can have at most 6 bad edges, since 7 bad edges form a single block along the face, and all these edges would have to belong to the same 6-face, which is not possible. On the other hand, it is possible for an 8-face to have 6 bad edges. They form two blocks around the boundary, so correspond to two 6-faces, see Figure 12.13. We can now complete the proof with a discharging argument. Initially assign to each face the length of the face as its weight. We then use one discharging rule: for any bad edge of the drawing, transfer weight 1/6 from the face of length at least 8 to the 6-face incident to the bad edge.

Since each 6-face is incident to 6 bad edges, the final weight of a 6-face is 7. An 8-face loses at most $6 \cdot 1/6 = 1$ weight, so also has weight at least 7 after the discharging. Finally, any face of length $\ell \geq 10$ has final weight $\ell(1 - 1/6) \geq 10 \cdot 5/6 > 7$. We conclude that all faces have weight at least

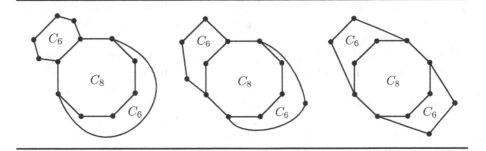

FIGURE 12.13: All possible embeddings of an 8-face with 6 bad edges.

7 after the discharge. Since the total weight was $2m$ to start with, and did not change, we conclude that there are at most $2/7m$ faces in the drawing. The Euler-Poincare formula for the projective plane then implies that $1 \leq n - m + f \leq n - m + 2/7m$, which implies $m \leq 1.4(n-1)$, which is what we had to show. $\qquad\qquad\square$

12.3.6 Geometric and Monotone Thrackles

A *geometric thrackle* is a thrackle which can be realized by a rectilinear thrackle drawing. There is a full characterization of geometric thrackles.

Theorem 12.22 (Woodall [351]). *A graph is a geometric thrackle if and only if removing all degree-1 vertices yields a disjoint union of paths, or a cycle of odd length.*

In other words, Conway's conjecture is true for geometric thrackles. *Caterpillars* are graphs for which removal of all degree-1 vertices results in a path. Equivalently, caterpillars are trees which do not contain a T_2-subgraph (a $K_{1,3}$ in which every edge has been subdivided once). The first case of the theorem corresponds to the graph being a disjoint union of caterpillars. The theorem implies that a forest is a geometric thrackle, if and only if it is a caterpillar.

Proof. T_2 is not a geometric thrackle: if cuv is one of the paths of length 2 from the center vertex c, then for uv to cross the other two edges incident to c, the left or right angle at cu must be greater than π, but this cannot be true for all three edges incident to u. Hence, any geometric thrackle which is a tree must be a disjoint union of caterpillars (since caterpillars are the connected trees not containing T_2). So, suppose the graph contains a cycle. Using Theorem 12.1 and the proof of Theorem 12.10 we conclude that there can be at most one cycle C_n and it must have odd length. There cannot be a path of length 2 attached to the C_n, since this would result in a T_2 if $n \geq 5$, and, for $n = 3$ would require that the outer edge of that path crosses every edge of the triangle, which is impossible. So removing all degree-1 vertices results in the cycle.

This shows necessity of the classification, and already implies the truth of Conway's conjecture for geometric thrackles, since for each of the two types of graphs we have $m \leq n$.

We are left with showing sufficiency, that is, the two classes of graphs are realizable as geometric thrackles. Caterpillars have planar bipartite rectilinear drawings, that is, their vertices can be placed on two parallel lines so that edges only occur between the two lines, and the drawing is planar (traverse the dominating path, place all neighbors of the current vertex on the other line, with the next vertex on the path last, and continue alternating between the lines like this). Reversing the order of the vertices on one of the lines yields a rectilinear drawing in which every pair of independent edges cross, and all edges lie between the two lines. Moving the two lines apart, and the vertices close to each other, we can make each caterpillar thrackle drawing arbitrarily close to a line. If the original graph is the union of k caterpillars, take a drawing of k line segments which intersect pairwise, and replace each with the caterpillar thrackle drawings we created above.

In the second case, we start with the geometric thrackle drawing of the odd-length cycle we produced in Theorem 12.1, placing the vertices on a circle. We can now add additional edges incident to some vertex u of the cycle close to one of the cycle edges uv, and just to the right of v on the circle so that it crosses any other edges incident to v. See Figure 12.14. □

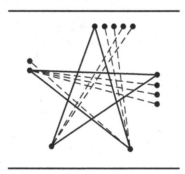

FIGURE 12.14: Adding thrackled edges to an odd thrackled cycle.

There is a direct argument establishing the thrackle conjecture for geometric graphs, due to Perles, as reported in [260]. The basic idea for a proof can be found in a paper by Erdős [125], see Exercise (12.17).

Theorem 12.23 (Erdős [125] and Woodall [351]). *Any geometric thrackle has at most as many edges as vertices.*

This is the base case of Theorem 5.35 on pairwise disjoint edges in a rectilinear drawing.

Proof. Call a vertex v *pointed* if all edges incident to v lie in a halfspace with v

on the boundary. We claim that removing the leftmost edge incident to every pointed vertex removes all edges of the graph, implying the result. By leftmost we mean the edge just before the halfspace in the rotation at v. Suppose for a contradiction that we are left with an edge uv. Since uv was not the leftmost edge at either u or v, the leftmost edges at u and v must lie on opposite sides of uv, but then they cannot cross. $\square$

The condition that the thrackle-drawing be rectilinear can be relaxed to x-monotonicity (which implies the same result for pseudolinear thrackles, since by Lemma 5.10 every pseudolinear drawing is isomorphic to an x=monotone drawing).

Theorem 12.24 (Pach and Sterling [260]). *Any x-monotone thrackle has at most as many edges as vertices.*

Proof. Fix an x-monotone thrackle drawing. By perturbing vertices slightly, we can assume that they are in general position. The projections to the x-axis of any two edges intersect in a closed set, so, by Helly's theorem, there is a vertical line ℓ which intersects every edge, possibly in an endpoint.

If ℓ does not pass through a vertex, then the graph is bipartite (since every edge intersects ℓ), so, unless it is a tree, it contains a cycle of even length $k \geq 6$ (since we already know that C_4 is not thrackleable). Consider a path $tuvw$ of length 3, so tu and vw cross in c. Triangle ucw either contains both t and w (what we called a B-configuration in Chapter 6), or neither, let us call that a $\ltimes$-configuration. Suppose it contains both, then t lies between u and v (with respect to the x-axis), so the edge st preceding t must cross uv so that $stuv$ forms a $\ltimes$ configuration. So, we can always assume that t and w lie outside the triangle ucv. The triangle is empty, that is, it does not contain any vertices, since any edge incident to that vertex would have to cross uv, and then would not be able to cross both tu and vw. So any edge crossing the boundary of ucw must cross it exactly twice, and since every edge other than tu, uv, and vw must cross uv, all these edges cross exactly one of uc or cv. See Figure 12.15. This implies that all these edges cross exactly one of ct or cw, with the exception of st, the edge preceding t on the cycle, and the edge after w. So we can remove edges tu, uv, and vw and replace them with a new edge tw routed as tcw. We obtain an x-monotone thrackle drawing of C_{k-2}, which, recursively, yields a thrackle drawing of C_4 which we know not to be possible. (This argument fails for the thrackle-drawing of C_6, since all its paths of lengths three are B-configurations, there are no $\ltimes$-configurations).

So the graph must be a tree, and the number of edges is strictly less than the number of vertices.

If ℓ does pass through a vertex v, we can modify the drawing as follows: split v into two vertices, one incident with the edges to the left, and the other incident with the edges to the right of ℓ. We can ensure that every left edge intersects every right edge, by placing the two copies of v appropriately, while keeping the drawing x-monotone, see Figure 12.16. Since ℓ no longer passes

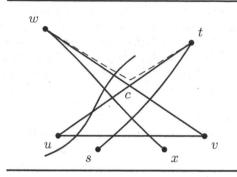

FIGURE 12.15: The triangle ucv.

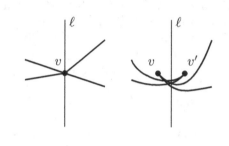

FIGURE 12.16: Intersecting left and right edges.

through a vertex, we can apply the argument from the first case. The upper bound in the first case is strict, so this compensates for increasing the number of vertices by one, and we conclude that the number of edges is at most the number of vertices. $\qquad\square$

12.3.7 The Subthrackle Bound

The thrackle bound max-cr$(G) \leq \vartheta(G)$ can be improved by taking into account the fact that any C_4 can cause at most one crossing. Let $\vartheta'(G) := \vartheta(G) - c + k$, where c is the number of C_4-subgraphs of G and k the number of K_4-subgraphs.

Lemma 12.25 (Subthrackle Bound. Ringeisen, Stueckle, and Piazza [288]). *We have*

$$\text{max-cr}(G) \leq \vartheta'(G).$$

Graphs for which max-cr$(G) = \vartheta'(G)$ are *subthrackles* or *subthrackleable*. The subthrackle bound is not optimal, for example if G is the *butterfly* graph, two triangles with a shared vertex, then max-cr$(G) = 4$, while $\vartheta'(G) = 5$. So the butterfly graph is not a subthrackle.

Proof. Recall that $\vartheta(G)$ counts the number of pairs of independent edges. There are three graphs on 4 vertices which contain a C_4: C_4, $C_4 + e$, and K_4. All three of these contribute at most one crossing to max-cr. The first two contain two pairs of independent edges, and a single C_4-subgraph, while K_4 contains three C_4-subgraphs, and three pairs of independent edges. In each case, max-cr$(G) \leq \vartheta(G) - c + k$. □

Let us see an example where the bound is sharp: the *domino* graph, that is, a C_6 with a chord connecting two vertices of distance 3.

Lemma 12.26. *The maximum crossing number of the domino graph is 9.*

Proof. We saw earlier that max-cr$(C_6) = 9$, and, as it turns out in the three non-isomorphic cr-maximal good drawings of C_6 every two vertices of distance 3 lie in the same face, so we can add an edge without introducing any crossings. To see that this is optimal, we calculate $\vartheta' = \vartheta - c + k = ((\binom{7}{2}) - 10) - 2 + 0 = 9$. □

Why did we not argue that the domino graph, containing a C_6, must have a drawing with at least 9 crossings? The reason is that, as far as we know, max-cr is not monotone. If G is a subgraph of H we do not know whether max-cr$(G) \leq$ max-cr(H). We study the subgraph problem in more detail in the next section.

We add one final observation, which to some degree explains why for many families of graphs a $\overline{cr}$-maximal drawing of the graph is convex. Call G a *geometric subthrackle* if max-$\overline{cr}(G) = \vartheta'(G)$.

Theorem 12.27 (Stueckle, Piazza, and Ringeisen [323]). *If G is a geometric subthrackle, then max-$\overline{cr}(G) =$ max-$\overline{cr}°(G)$.*

Proof. If G is a geometric subthrackle, then $\vartheta'(G) = \vartheta(G) - c + k =$ max-$\overline{cr}(G)$, where c is the number of C_4- and k the number of K_4-subgraphs of G. This means that if two non-adjacent edges in a $\overline{cr}$-maximal drawing of G do not cross, they must be part of a C_4 whose remaining two edges cross. Suppose there is a $\overline{cr}$-maximal drawing of G with a vertex v in the interior of the convex hull of the drawing. If v is isolated, we can move it to the boundary of the convex hull, so v has a neighbor u. The ray from u through v must cross an edge xy on the boundary of the convex hull (if it hits a vertex on the convex hull, we perturb that vertex slightly). Now xy and uv are two edges which do not cross, and $\{u, v, x, y\}$ induce a subgraph of a plane K_4 contradicting our argument that two of the other edges must cross. Hence, all vertices lie on the boundary of the convex hull. □

Stueckle, Piazza, and Ringeisen [323] show that the connected geometric subthrackles are exactly the complements of proper circular-arc graphs.

12.4   The Subgraph Monotonicity Problem

There is another open question associated with the maximum crossing number which, while not having garnered as much attention as Conway's conjecture, seems evenly appealing.

Question 12.28 (Ringeisen, Stueckle, and Piazza [288]). Is is true that if G is a subgraph of H, then max-cr$(G) \leq$ max-cr(H)? What if G is an induced subgraph of H?

When looking at the domino-graph example, in Lemma 12.26, we mentioned that monotonicity of max-cr is not immediately obvious. The problem is that a cr-maximal drawing of G may not extend to any good drawing of H: it is easy to construct examples of good drawings of a graph G in which two vertices cannot be connected by an edge (while keeping the drawing good). The two obstacles are having to cross an edge twice, or crossing an adjacent edge. For rectilinear drawings these problems can be dealt with by perturbing vertices.

Theorem 12.29 (Ringeisen, Stueckle, and Piazza [288]). *The maximum rectilinear crossing number is monotone.*

Proof. Suppose G is a subgraph of H. Fix a rectilinear drawing of G with max-$\overline{\mathrm{cr}}(G)$ crossings. Extend this to a rectilinear drawing of H. By perturbing vertices of H if necessary, we can ensure that no edge passes through a vertex; since the drawing is rectilinear, it is automatically good, and contains at least max-$\overline{\mathrm{cr}}(G)$ crossings, so max-$\overline{\mathrm{cr}}(H) \geq$ max-$\overline{\mathrm{cr}}(G)$. $\qquad\square$

If H is a thrackle, and G a subgraph of H, then obviously max-cr$(G) \leq \vartheta(G) \leq \vartheta(H) =$ max-cr(H), so the maximum crossing number is monotone for thrackles. Cottingham managed to extend this result to subthrackles.

Theorem 12.30 (Cottingham [96]). *If G is a subgraph of H, and H is subthrackleable, then* max-cr$(G) \leq$ max-cr(H).

Proof. We claim that if $G = H - \{e\}$ for some edge $e \in E(H)$, then $\vartheta'(G) \leq \vartheta'(H)$. The theorem then follows by induction (removing isolated vertices does not affect either ϑ' or max-cr). To see that the claim is true, we consider $\vartheta'(H) = \vartheta(H) - c + k$, where c and k are the number of C_4 and K_4-subgraphs of H. To compare this count to $\vartheta'(G)$, let n_e be the number of edges of H not adjacent to e, and k_e the number of K_4-subgraphs of H containing e. Finally, let c_e be the number of C_4-subgraphs of H which contain e *and which are not* K_4. Then the total number of C_4-subgraphs e belongs to is $c_e + 2k_e$. We therefore have $\vartheta'(G) = \vartheta'(H) - n_e + (c_e + 2k_e) - k_e = \vartheta'(H) - (n_e - c_e - k_e)$. Since $n_e \geq c_e + k_e$, the result follows. $\qquad\square$

While this is a good start, we would prefer results that imply max-cr(G) $\leq$ max-cr(H) based on properties of G rather than H, and we already have an inkling of how this could work. Using the same idea as in Theorem 12.29 we can obtain the following result.

Corollary 12.31. *If* max-cr(G) = max-$\overline{\text{cr}}$(G) *and G is a subgraph of H, then* max-cr(G) $\leq$ max-cr(H).

The reason this works is that we can start with a cr-maximal drawing of G which is rectilinear; this allows us to extend the drawing to H without encountering any of the obstacles we mentioned earlier. It immediately follows that we have subgraph monotonicity for G being a complete graph (n-partite or not), paths (indeed caterpillars), and cycles and wheels C_n and W_n for which n is even.

How can we weaken the hypothesis so that it covers a wider variety of graphs? Here is a sketch of an idea that turns out to work for *induced* subgraphs G of H: let G' be the graph obtained by identifying all vertices in $V(H) - V(G)$ with each other. Then G' contains G and is a subgraph of the join $G + v$ which is why we call it a *partial join* of G, and, in this case, more specifically, the partial join of G with respect to H.

If max-cr(G') $\geq$ max-cr(G), then max-cr(H) $\geq$ max-cr(G) by replacing the vertex joined to G with a drawing of $H - G$; the method was introduced by Ringeisen, Stueckle, and Piazza [288] as the *aura technique*, named for the annulus-shaped region they used to reconnect G to $H - G$. We take a slightly different approach here.

Theorem 12.32. *If G is an induced subgraph of H, and* max-cr(G) $\leq$ max-cr(G') *for the partial join G' of G with respect to H, then* max-cr(G) $\leq$ max-cr(H).

Proof. Fix a good drawing of G' with at least max-cr(G) crossings. Let v be the partial join vertex we added to G to obtain G'. We create $n' = V(H) - V(G)$ copies of the star centered at v with the new endpoints close to v, and connect their centers (the duplicate copies of v) with a $K_{n'}$. We can ensure that this drawing is good (imagine the original star with center v drawn so that v lies on a line, and all edges incident to it are drawn as straight-line segments below the line. Perturb v along the line to create duplicates, and so that the edges incident to the duplicates cross the same edges as the original edges, apart from the other duplicate edges, of course).

The resulting drawing contains a drawing of H, which in turn contains a drawing of G. Since every join edge of G' corresponds to an edge connecting G to $H - V(G)$, this drawing of H has at least max-cr(G') $\geq$ max-cr(G) crossings. $\qquad\square$

If we do not know how H extends G, we can anticipate all such ways, which is the content of the following corollary.

Corollary 12.33. *If* max-cr(G) $\leq$ max-cr(G') *for every partial join G' of G, then* max-cr(G) $\leq$ max-cr(H) *for all graphs H of which G is an induced subgraph.*

The construction in Theorem 12.32 does not work if there is an edge of H between two vertices of G, so we require G to be an induced subgraph of H. We can adopt the method to the non-induced case by anticipating all possible ways such edges could be added.

Corollary 12.34. *If* max-cr(G) $\leq$ max-cr(G') *for every partial join G' of G and all of its supergraphs on the same vertex set, then* max-cr(G) $\leq$ max-cr(H) *for all graphs H of which G is a subgraph.*

We note that the conditions in both corollaries are necessary, not just sufficient, so we do have an improved line of attack on the original subgraph problem.

Theorem 12.35 (Ringeisen, Stueckle, and Piazza [288]). *If H contains an induced cycle of length $n \geq 5$, then* max-cr(H) $\geq n(n-3)/2$.

Proof. We already know this to be true for odd n, since in that case max-cr(C_n) = max-$\overline{\mathrm{cr}}(C_n)$. By Corollary 12.33 it is sufficient to show that every partial join of C_n, with n even, has a good drawing with at least $n(n-3)/2$ crossings. We proceed as in the proof of Theorem 12.2. Figure 12.17 shows that there is a good drawing of W_6 which contains a drawing of C_6 with max-cr$(C_6) = 9$ crossings. So the lower bound of 9 remains true if any of the edges incident to the join vertex are deleted, showing that the theorem is true for C_6. We now apply Lemma 12.3 to extend the C_6 to a C_8 with max-cr(C_8)

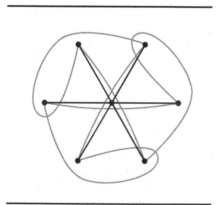

FIGURE 12.17: A good drawing of W_6 extending a drawing of C_6 (in gray) with max-cr$(C_6) = 9$ crossings.

crossings; the redrawing is local, and we can add two spokes to the wheel to connect the two new vertices to the center of the wheel. Again, we have a good drawing of W_n which contains a drawing of C_n with max-cr(C_n) crossings, so we can apply Corollary 12.33 to get the result for C_8, and recursively, any even n which completes the proof. $\square$

So one way to use Corollary 12.33 is to show that a cr-maximal good drawing of G can be extended to a good drawing of $G+v$. This has been done for several other classes of graphs including families of C_4 sharing a vertex, and families of C_4 sharing an edge [157], see Exercises (12.21), (12.23), and (12.24). Open cases include grid graphs $P_n \square P_n$ and prism graphs $C_n \square K_2$ [157].

We consider one final variant, which follows immediately from Corollary 12.34.

Corollary 12.36. *Let G be a graph on n vertices. If there is a good drawing of K_{n+1} which contains a cr-maximal drawing of G, then max-cr$(G) \leq$ max-cr(H) for all graphs H of which G is a subgraph.*

The corollary, together with Figure 12.18, then establishes the case $n = 6$ of the following theorem. We leave the remaining cases to Exercise (12.26).

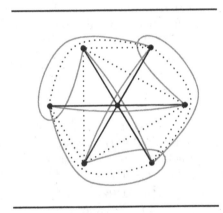

FIGURE 12.18: A cr-maximal drawing of C_6 (gray) as part of a good drawing of K_7 (edges are dotted, except for black edges incident to new vertex).

Theorem 12.37. *If C_n is a subgraph of H, then max-cr$(C_n) \leq$ max-cr(H).*

Using the same idea, one can prove the following theorem, see Exercise (12.28).

Theorem 12.38. *If G is a forest, and G is a subgraph of H, then max-cr$(G) \leq$ max-cr(H).*

12.5   Hypercubes

Theorem 12.39 (Harborth [178]). *We have*

$$\text{max-cr}(Q_3) = 34.$$

We do not include a proof of this result, but note that it is interesting, because $\vartheta'(Q_3) = 36$, so it is an example of a graph which is not a subthrackle, and, as Piazza, Ringeisten and Stueckle [282] observed it is edge- and vertex-critical with that property.

Theorem 12.40 (Alpert, Feder, Harborth, and Klein [29]). *For $n \geq 0$ we have*

$$\text{max-}\overline{\text{cr}}(Q_n) \geq 2^{n-3}(2^n n^2 - 2n^2 - 2^{n+1}n + 3 \cdot 2^n - 2).$$

Proof. We recursively construct a bipartite drawing of Q_n. For Q_1 this is immediate. To draw Q_n take a recursively constructed copy of Q_{n-1} and move its left vertices so that they lie above all the right vertices. Now mirror that drawing along the midline between the two partition lines to get a second drawing of a Q_{n-1}. Connect the 2^{n-2} vertices in the top half to the right 2^{n-2} vertices in the top half in reverse order, causing $\binom{2^{n-2}}{2}$ crossings between these matching edges. Do the same for the bottom half.

Let v_n be the number of crossings in the drawing. Then v_n needs to count the $2v_{n-1}$ crossings in the Q_{n-1}, the $((n-1)2^{n-2})^2$ crossings between the two Q_{n-1} (every two such edges cross), the $2\binom{2^{n-2}}{2}$ crossings between the edges matching the two Q_{n-1}, and, finally, the crossings between the Q_{n-1}-edges and the matching edges. Each matching edge crosses $(n-1)(2^{n-2}-1)$ edges in the two Q_{n-1}, for a total of $(n-1)2^{n-1}(2^{n-2}-1)$ such crossings.

In summary, $v_n = 2v_{n-1} + ((n-1)2^{n-2})^2 + 2\binom{2^{n-2}}{2} + (n-1)2^{n-1}(2^{n-2} - 1) = 2v_{n-1} + 2^{n-3}(2^{n-1}n^2 - 4n + 2)$ and $v_1 = v_2 = 0$, which solves as $v_n = 2^{n-3}(2^n n^2 - 2n^2 - 2^{n+1}n + 3 \cdot 2^n - 2)$. $\square$

Alpert, Feder, Harborth, and Klein [29] conjecture that the bound in the Theorem is sharp, and manage to prove so for the first non-trivial case, Q_3.

Corollary 12.41 (Alpert, Feder, Harborth, and Klein [29]). *We have*

$$\text{max-}\overline{\text{cr}}(Q_3) = 28.$$

Proof. The lower bound follows from Theorem 12.40. To see that there cannot be more than 28 crossings, let D be a rectilinear drawing of Q_3. Two crossing edges either belong to a C_4 or belong to an induced C_6 in which they are opposite of each other or are incident to a common edge. Each C_4 can contribute at most 1 crossing; each induced C_6 can contribute 7 crossings as we saw in Theorem 12.1. Each pair of opposite edges occurs in two induced C_6-subgraphs, so the total number of crossings is at most $6 + 4 \cdot 7 - 6 = 28$. $\square$

Theorem 12.42 (Alpert, Feder, Harborth, and Klein [29]). *For $n \geq 0$ we have*

$$\text{max-}\overline{\text{cr}}(Q_n) \leq n2^{n-3}((2^n - 1)n - \frac{1}{3}(4n^2 - 1)).$$

For the proof we note that Q_i occurs $\binom{n}{i}2^{n-i}$ times as a subgraph of Q_n.

Proof. Lemma 12.25 gives us an upper bound of

$$\binom{n2^{n-1}}{2} - 2^n\binom{n}{2} - \binom{n}{2}2^{n-2} = n2^{n-3}(2^n n - 5n + 3).$$

We improve this bound by using $\text{max-}\overline{\text{cr}}(Q_3) = 28$ from Corollary 12.41. Each Q_3 subgraph contains $\vartheta(Q_3) = 42$ pairs of independent edges. Of these we have already counted the 6 pairs of edges belonging to C_4-subgraphs, leaving 36 pairs of independent edges. Since $\text{max-}\overline{\text{cr}}(Q_3) = 28$, there are at least 8 pairs of independent edges in each Q_3 which cannot cross each other, and which we have not counted yet. Moreover, since two different Q_3 overlap in a C_4 or one of its subgraphs, none of these 8 crossings gets counted in more than one Q_3. Hence, we can remove $8\binom{n}{3}2^{n-3} = \binom{n}{3}2^n$ crossings from our earlier count to get a bound of

$$n2^{n-3}(2^n n - 5n + 3) - \binom{n}{3}2^n = n2^{n-3}((2^n - 1)n - \frac{1}{3}(4n^2 - 1)).$$

$\square$

12.6 Complexity

We already mentioned the conjecture that $\text{max-}\overline{\text{cr}}(G) = \text{max-}\overline{\text{cr}}^\circ(G)$ for bipartite graphs. The truth of the conjecture would tell us something surprising about the nature of $\text{max-}\overline{\text{cr}}$ for bipartite graphs: it would be a combinatorial problem, its complexity in **NP**, while the best upper bound on $\text{max-}\overline{\text{cr}}$ at this point is $\exists\mathbb{R}$.

Theorem 12.43 (Bald, Johnson, and Liu [41]). *Computing $\text{max-}\overline{\text{cr}}$ is **NP**-hard, computing $\text{max-}\overline{\text{cr}}^\circ$ is **NP**-complete.*

For the proof we need the **NP**-complete *Maximum Cut Problem*: does G contain a subset $U \subseteq V(G)$ of vertices so that there are at least k edges between U and $V(G) - U$. The number k is the size of the cut.

Proof. Let (G, k) be a graph and an integer k. Let G' be G together with $t := \binom{n}{4} + 1$ disjoint edges M. If G has a maximum cut of size k, then G' has a convex rectilinear drawing with $kt + \binom{t}{2}$ crossings: let $(U, V(G') - U)$ be a cut of size k; place U above the x-axis, $V(G') - U$ below it, and add the t edges in M close to the x-axis, so they cross each other pairwise, as well as all k edges of the cut. So max-$\overline{\mathrm{cr}}^{\circ}(G') \geq kt + \binom{t}{2}$.

On the other hand, suppose that max-$\overline{\mathrm{cr}}(G') \geq kt + \binom{t}{2}$. We can assume that the t edges in M cross each other in a drawing of G' realizing max-$\overline{\mathrm{cr}}(G')$: if not, take an edge in M that incurs most crossings with edges of G and draw the remaining $t - 1$ edges of M close to it, so they cross the same set F of edges of G and each other pairwise. Since edges of G cross less than t times with each other, and edges of M cross each other $\binom{t}{2}$ times, there must be at least $kt - (t - 1) = k(t - 1) + 1$ crossings between M and F. So F must have size at least k (otherwise, there are at most $t(k - 1) < k(t - 1) + 1$ crossings between M and F). Take an edge $e \in M$ and extend it to a line; all crossings with the line occur on e, otherwise we could have extended e to increase the number of crossings (here we use that the drawing is rectilinear, so no two edges can cross twice). So the line splits the vertices of G into two sets, U and $V(G) - U$, and there are at least k edges between U and $V(G) - U$, namely the edges in F. Therefore, G has a maximum cut of size at least k.

We conclude that the maximum cut problem reduces to both max-$\overline{\mathrm{cr}}$ and max-$\overline{\mathrm{cr}}^{\circ}$, so both problems are **NP**-hard. It is easy to see that max-$\overline{\mathrm{cr}}^{\circ}$ lies in **NP**, we only need to guess the order along the convex hull. For max-$\overline{\mathrm{cr}}$, the best upper bound we know is $\exists \mathbb{R}$. □

It was recently shown that max-cr is **NP**-complete [90].

12.7 Notes

Baltzer's [44] polygon problem may well be the oldest crossing number problem in the literature; a complete solution was not found until 1923 when Steinitz published his paper [321] proving Theorem 12.1, but this is not the end of the story; for a survey of Baltzer's problem, see Grünbaum [164].

It was also Grünbaum [163] who introduced the maximum rectilinear crossing number, though other special cases, like max-$\overline{\mathrm{cr}}(nP_2)$, had often occurred as puzzles. The only other value of max-$\overline{\mathrm{cr}}$ which we did not present in this chapter is for the Petersen graph, which has maximum rectilinear crossing number 49, a result by Feder, Harborth, Herzberg and Klein [134].

Grünbaum [163] also discusses max-cr, apparently unaware, that Ringel [290] had introduced that notion earlier, and calculated its values for complete graphs. Harborth and Zahn [182] calculate max-cr for all graphs of order at most 6, and use that to identify all thrackles in that group.

One of the first published mentions of Conway's conjecture occurs in a collection of open combinatorial problems by Guy [167] in 1970. There is no single survey covering the state of the art, unfortunately, but the current chapter should be a good starting point to find additional research on thrackles.

12.8 Exercises

(12.1) Prove Theorem 12.5, namely show that max-cr(K_n) = max-$\overline{\mathrm{cr}}(K_n)$ = $\binom{n}{4}$, and max-cr$(K_{m,n})$ = max-$\overline{\mathrm{cr}}(K_{m,n})$ = $\binom{m}{2}\binom{n}{2}$. *Hint:* C_4 holds the key.

(12.2) Show that max-cr$(K_{n_1,\dots,n_k})$ = $\binom{n}{4} - \sum_{i=1}^{k}\left(\binom{n_i}{4} + (n - n_i)\binom{n_i}{3}\right)$.

For the following two problems let $\mathrm{spec}_\gamma(G) := \{\gamma(D) : D \text{ is a good}$ drawing of $G\}$, the *spectrum of G under* γ. We write $\mathrm{spec}(G)$ for $\mathrm{spec}_{\mathrm{cr}}(G)$.

(12.3) Show that $\mathrm{spec}(C_n) = \{0,\dots,n(n-3)/2)\}$.

(12.4) Show that

$$\mathrm{spec}_{\text{max-}\overline{\mathrm{cr}}}(C_n) = \begin{cases} \{0,\dots,n(n-3)/2\} - \{n(n-3)/2 - 1\} & \text{if } n \text{ is odd} \\ \{0,\dots,n(n-4)/2 + 1\} & \text{if } n \text{ is even.} \end{cases}$$

(12.5) Show that the maximum monotone crossing number of C_n equals max-$\overline{\mathrm{cr}}(C_n)$.

(12.6) Show that max-cr$(W_n) \sim 5/4n^2$. *Hint:* For the lower bound extend the subthrackle bound to also account for graphs consisting of two triangles sharing a vertex.

(12.7) Show that max-$\overline{\mathrm{cr}}(tK_4) = 20\binom{t}{2} + t$.

(12.8) Show that a planar graph G with minimum degree at least 2 satisfies max-cr$(G) = \Theta(n^2)$. For the upper bound show that max-cr$(G) < 3n^2$.

(12.9) Show that (i) every subdrawing of a K_n in a cr-maximal drawing of $K_{n'}$, $n' \geq n$, is cr-maximal. (ii) The crossing-free edges in a cr-maximal drawing of K_n form a Hamiltonian cycle, or a spanning set of disjoint paths.

(12.10) Show that the only graphs for which cr(G) = max-cr(G) are triangles and stars.

(12.11) Show that for every n there is an N so that any good drawing of a K_N contains a cr-maximal subdrawing of K_n. *Hint:* N will be huge. It is open whether the result remains true for arbitrary graphs G instead of K_n.

(12.12) Every graph is thrackleable on an orientable surface of genus at most $(m - n + \vartheta(G) + 1)/2$, where $\vartheta(G)$ is the thrackle bound for graph G.

(12.13) If we had defined a generalized thrackle drawing as a drawing in which every pair of edges *crosses* an odd number of times, does this lead to a new notion of generalized thrackle? Prove or disprove.

(12.14) Show that the wheel with an odd number of spokes is a generalized thrackle.

(12.15) Show that to prove the thrackle conjecture, it is sufficient to show that no θ-graph (a graph consisting of three disjoint paths between two vertices) is thrackleable.

(12.16) Show that to prove the thrackle conjecture, it is sufficient to show that no bar-bell graph (a graph consisting of two cycles connected by a non-empty path) is thrackleable.

(12.17) Prove the geometric thrackle conjecture adapting Erdős' proof in "On sets of distances of n points" [125]. *Hint:* Finding the relevant passage is part of the exercise.

(12.18) Show that the result in Corollary 12.15 is tight: there are graphs with n vertices and $2(n - 1)$ edges which can be drawn without parallel edges.

(12.19) Show that $\vartheta(C_m \square C_n) - \vartheta'(C_m \square C_n)$ is unbounded.

(12.20) Show that any non-planar graph has maximum crossing number at least 5.

(12.21) Let G be the graph consisting of n copies of C_4 with one common vertex. Show that if G is an induced subgraph of H, then $\mathrm{cr}(H) \geq 6n^2 - 5n$.

(12.22) Show that if G is a subgraph of H, then $\text{max-cr}(G) \leq 3\,\text{max-cr}(H)$.

(12.23) Let G be the graph consisting of n copies of C_4 with one common edge. Show that if G is an induced subgraph of H, then $\mathrm{cr}(H) \geq (7n^2 - 5n)/2$.

(12.24) Show that if G is a forest which occurs as an induced subgraph in H, then $\vartheta(G) \leq \text{max-cr}(H)$.

(12.25) Determine the maximum rectilinear crossing number of a disjoint union of cycles. *Hint:* Start with a pair of cycles; the solution will scale.

(12.26) Show that if H contains a cycle C_n, $n \geq 3$, then $\text{max-cr}(C_n) \leq \text{max-cr}(H)$. *Hint:* The base is covered, extending it requires some care.

(12.27) Show that if G is an *apex graph*, that is, there is a vertex which is adjacent to all other vertices, and G is an induced subgraph of H, then $\text{max-cr}(G) \leq \text{max-cr}(H)$.

(12.28) Show that if H contains a forest G as a subgraph, then max-cr$(G) \leq$ max-cr(H).

(12.29) Show that if max-$\overline{\text{cr}}$ or max-$\overline{\text{cr}}°$ can be approximated to within a constant factor in polynomial time, then $\mathbf{P} = \mathbf{NP}$. *Hint:* A small modification in the proof of Theorem 12.43 will be sufficient.

Part III

Appendices

Appendix A

Basics of Topological Graph Theory

A.1 Curves

A *curve* is the image of a continuous function from $[0, 1]$ to the plane (or surface). The curve is *simple* if the function is injective, and it is *closed* if its two endpoints are the same. We use the term *simple closed curve* to denote a closed curve which is simple except for its two endpoints, which are the same. Equivalently, a simple closed curve is a curve which is homeomorphic to the unit circle. While curves in general can behave rather awkwardly (they can be space-filling, for example), simple curves are more well-behaved as witnessed by the following two theorems. A subset of the plane or surface is *(arc-) connected* if any two points in the set can be connected by a curve belonging to the set. A maximal connected subset is called a *region*.

Theorem A.1 (Jordan Curve Theorem [249]). *The complement of any simple closed curve in the plane consists of two connected regions.*

The Jordan Curve Theorem implies that any two simple closed curves cross an even number of times in the plane (assuming they intersect finitely often). We can exploit this fact to prove the following "homological" version of the Jordan curve theorem.

Lemma A.2. *The complement of a closed curve with finitely many self-intersections can be two-colored so that colors on opposite sides of the boundary are different.*

Proof. Add a vertex to each face of the complement, and connect two vertices if they share a boundary. The resulting graph is bipartite, since every cycle of the graph corresponds to a simple closed curve that crosses the given closed curve an even number of times. We can therefore 2-color the faces. □

We occasionally use a much stronger form of the Jordan Curve Theorem, known as the Jordan-Schoenflies Theorem.

Theorem A.3 (Jordan-Schoenflies Theorem [249]). *Any homeomorphism of a simple closed curve to the unit circle can be extended to a homeomorphism of the plane.*

Theorem A.3 is often used to simplify visual arguments: suppose we have a drawing of a graph containing an edge e. We can then use the theorem to redraw the graph so that e is a straight-line segment (for example, we do so in the proof of Lemma 9.2). See Exercise (A.3).

A.2   Embeddings and Planar Graphs

A *drawing* of a graph $G = (V, E)$ on a plane or surface Σ assigns to each vertex in V a distinct point in Σ, and to each edge uv a simple curve (often called *arc*) in Σ connecting the points assigned to u and v. We typically assume that the drawn edge uv does not contain any other points assigned to vertices (though we may occasionally violate this condition). If G has a drawing on Σ in which the interiors of all arcs are disjoint from each other and do not contain any points associated to vertices, we say G is *embeddable in Σ* and we call the drawing an *embedding* of G.[1] A graph together with a drawing of the graph in a surface is called a *topological* graph. Often the graph, its drawing, and the topological graph are not clearly distinguished.

A graph embeddable in the plane is *planar*, and the embedding is often called a *plane* (topological) graph. The regions of a plane embedding are known as *faces*. All but one face of the embedding will be bounded, these are known as the *inner* faces, while one face, the *outer* face is unbounded.

Euler's formula relates the number of vertices, edges, and faces in a plane embedding.

Theorem A.4 (Euler's Formula). *For a plane graph G we have $n - m + f = 2 + c$, where $n = |V(G)|$, $m = |E(G)|$, f is the number of faces of the embedding, and c is the number of connected components of G.*

Proof. To show this result we will switch to multigraphs, that is, we allow loops and multiple edges; this simplifies the argument. We prove the result by induction on m. If $m = 0$, then G consists of n isolated vertices, so $f = 1$ and $n = c$, and the equality holds. If $m > 0$ pick any edge e of the graph. If e is a loop, removing it does not change $-m + f$ so we can apply induction. If e connects two distinct vertices, we can contract e. This leaves $n - m$ unchanged, and, again, the result follows by induction. $\qquad\square$

Corollary A.5. *If a plane graph G can be embedded in the plane so that the smallest face size is g, then*

$$m \leq \frac{g}{g-2}(n-2),$$

---
[1] The term *representation* is often used in this case, and embedding reserved for equivalence classes of drawings under homeomorphisms of the surface, but this view will not be useful to us.

where $n = |V(G)| \geq 3$ and $m = |E(G)|$.

The *girth* of a graph is the length of a shortest cycle. Corollary A.5 implies that if a planar graph has girth g, then $m \leq g/(g-2)\,(n-2)$, where we let $g = 3$ in case G is a forest (a forest has $m \leq n \leq 3(n-2)$ edges, for $n \geq 3$).

Proof. Every face has size at least g, and since every edge is incident to at most two faces (one, if it is a cut-edge), we have $gf \leq 2m$ where f is the number of faces of the embedding. By Euler's Formula, we have $n - m + f = 2$, so $2g = gn - gm + gf \leq gn + m(2-g)$, so $m \geq g/(g-2)\,(n-2)$. $\qquad\square$

We conclude that any planar graph has at most $3(n-2)$ edges, and any planar bipartite graph has at most $2(n-2)$ edges. This immediately implies that K_5 and $K_{3,3}$ are *non-planar*; they cannot be embedded in the plane. Kuratowski's theorem shows that these two graphs (and their subdivisions) are the only obstructions to planarity.

Theorem A.6 (Kuratowski [249, Section 2.3]). *A graph is planar if and only if it does not contain a subdivision of K_5 or $K_{3,3}$.*

Wagner later strengthened Kuratowski's theorem by showing that a graph is planar if and only if it does not contain K_5 or $K_{3,3}$ as a minor [249, Theorem 2.4.1], where H is a *minor* of G if it can be obtained from G by a sequence of vertex and edge deletions, as well as edge contractions.

A graph is *outerplanar* if it has an embedding in which all vertices are incident on the outer face. Corollary A.5 implies that an n-vertex outerplanar graph can have at most $2n - 3$ edges (add a vertex connected to each of the n vertices, and apply the corollary). The only obstruction to outer-planarity is a K_4-subdivision (or minor), see Exercise (A.5).

The edge bounds of $3(n-2)$ edges for planar graphs and $2(n-2)$ edges for planar bipartite graphs are sharp, as witnessed by triangulations. A *triangulation* is a plane graph in which every face is a triangle. Triangulations have exactly $3(n-2)$ edges, and there are various ways to construct bipartite graphs with $\lfloor 2(n-2) \rfloor$ edges, see Exercise (A.6). Mader showed that any graph with at least $3n - 5$ edges contains a subdivision of K_5 [249, p.42].

Any plane graph can be extended to a triangulation.

Lemma A.7. *Every plane embedding on at least three vertices can be extended to a plane triangulation by adding edges.*

Proof. Add edges to the given plane embedding as long as this is possible without introducing crossings or parallel edges. Every face of the resulting embedding is incident to at most three vertices: if a face were incident to more than three vertices, say a, b, c, d, then at least one of the edges between two of these vertices must be absent (otherwise we would have an outerplanar embedding of K_4, which is ruled out by Exercise (A.5)), so we could add this edge to the face without creating a parallel edge or destroying planarity. Since we assumed that we cannot add any more edges, this is impossible, and every face is incident to at most three vertices. On the other hand, every face is incident to at least three vertices (two vertices cannot bound a face if there are no parallel edges), each face is incident to exactly three vertices. Since we can connect the three vertices by edges inside the faces, those edges are present in the embedding (the only other reason not to add them would be an existing, parallel edge in the graph; but then the face would be incident to additional vertices). Therefore, each face is bounded by the three edges between its incident vertices. □

We often use Lemma A.7 to extend a plane embedding to a triangulation, since this gives us a 3-connected graph whose embedding is essentially unique.

Lemma A.8. *Every plane triangulation on at least three vertices is 3-connected.*

Two drawings of a graph are *isomorphic* if there is a homeomorphism of the plane (or surface) that turns one into the other. Sufficiently connected graphs, such as triangulations, have unique embeddings, up to isomorphism.

Theorem A.9 (Whitney's Theorem). *Any two embeddings of a subdivision of a 3-connected planar graph are isomorphic on the sphere.*

Consequently, in the plane, embeddings of subdivisions of 3-connected graphs are unique (up to homeomorphisms of the plane) once an outer face has been chosen.

The *topological dual* G^* of a plane graph G is the multigraph which has a vertex for every face of the plane graph, and an edge between two vertices for any edge in G incident to both faces.

Theorem A.10 (Bern and Gilbert [59]). *Every plane rectilinear triangulation has a plane rectilinear dual.*

Proof. Place the vertices of the dual at the *incenter* of each triangle, that is, the point where the angle bisectors meet. □

Planar graphs have excellent separation properties, an ingredient of many algorithmic results for planar graphs. We say a set $S \subseteq V(G)$ *separates* $A, B \subseteq V(G)$, if any A-B path intersects S. We also say that S is a separator for A, B in G.

Theorem A.11 (Planar Separator Theorem. Lipton and Tarjan [230]). *The vertices of a weighted n-vertex planar graph can be partitioned into three sets A, B and S so that S separates A and B, and both A and B are at most 2/3 of the total weight, and S is of size at most $\sqrt{8n}$. The partition can be found in linear time.*

We say an edge set $T \subseteq E(G)$ *separates* $A, B \subseteq V(G)$, if any A-B path contains an edge in T.

Theorem A.12 (Planar Edge Separator Theorem. Gazit and Miller [147]). *Every weighted planar graph contains an edge separator T of size at most $1.58(\sum_{v \in V(G)} \deg_G^2(v))^{1/2}$, separating two sets A and B each of weight at most 2/3 of the total weight. The separator can be found in polynomial time.*

A.3 Bigons

Hass and Scott [183] proved various results about substructures of self-intersecting curves. A *bigon* consists of two subarcs of a curve (or two curves) only intersecting in the two distinct endpoints of the two arcs. Of interest to us is one result that implies that two simple curves with the same endpoints form an empty bigon. A bigon is *empty* if the two arcs do not cross any other part of the curve (or the two curves), and the region bounded by the bigon does not contain any vertices (including vertices belonging to other curves). The existence of an empty bigon is not immediately obvious, for example, Figure A.1 shows that the requirement that the endpoints of the curves are the same is necessary.

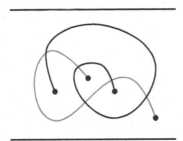

FIGURE A.1: Two simple curves crossing without forming an empty bigon.

We need a slightly stronger result.

Lemma A.13. *Two simple curves with the same endpoints form at least two empty bigons.*

Proof. Suppose we have two curves γ_1 and γ_2 crossing each other $k \geq 0$ times. Replacing each crossing with a dummy vertex results in a plane graph with two vertices of degree 2, k vertices of degree 4, and $2k + 2$ edges. Let f_2 be the number bigons, and $f_{\geq 4}$ the number of faces of size at least 4; note that all faces have even size, since the two underlying curves are simple. By Euler's formula, Theorem A.4, the total number of faces, $f = f_2 + f_{\geq 4} = 2 - (k + 2) + 2k + 2 = k + 2$. So if $f_2 \leq 1$, then $f_{\geq 4} \geq k + 1$. However, we know that $2k + 2 \geq f_2 + 2f_{\geq 4}$ (since each face of size ℓ is bounded by ℓ edges, and each edge bounds at most two faces). So $2k + 2 \geq k + 2 + f_{\geq 4}$, and $f_{\geq 4} \leq k$, contradicting the earlier bound. So $f_2 \geq 2$, and we have at least two (empty) bigons. $\qquad\square$

Corollary A.14. *If two simple curves with the same endpoints cross at least twice, then there are two crossing-free subarcs connecting two crossings.*

Proof. By Lemma A.13 there are two empty bigons; if either of the two bigons is incident to two crossings, we are done; therefore, each bigon is incident on at least one endpoint. It is not possible that one of the bigons is incident on both endpoints, because then there are no crossings, so each bigon is incident on exactly one endpoint. We can then suppress both endpoints and their incident bigons, giving us two new endpoints (the former crossings incident to the suppressed bigons), and two curves. Applying Lemma A.13 gives us a bigon for the new curves. For the original curves, this bigon corresponds to two subarcs connecting two crossings, which is otherwise free of crossings. $\qquad\square$

A.4   Graphs on Surfaces

Technically, a *surface* is a compact, closed 2-manifold; intuitively, it is something that locally looks like a disk. The classification theorem for surfaces [249] shows that every surface can be obtained from a sphere by adding handles and crosscaps. A surface Σ constructed from the sphere by adding h handles and c crosscaps has *Euler genus* $\mathrm{eg}(\Sigma) := 2h + c$. If $c = 0$, we write S_h for the surface of *genus h*. If $h = 0$, we say that the surface has *non-orientable genus c*, and write that surface as N_c. Finally, the *Euler characteristic* of Σ is $2 - \mathrm{eg}(\Sigma)$, but we will mostly avoid this notion. Table A.1 gives an overview of surfaces with small Euler genus.

surface	genus	non-orientable genus	Euler characteristic	Euler genus
sphere (S_0)	0	na	2	0
projective plane (N_1)	na	1	1	1
torus (S_1)	1	na	0	2
Klein Bottle (N_2)	na	2	0	2
Dyck surface (N_3)	na	3	-1	3
Double Torus (S_2)	2	na	-2	4

TABLE A.1: Some surfaces and their genera.

We already defined what it means for a graph to be drawn or embedded in a surface. We say G is *2-cell embedded* in Σ if every face of the embedding is homeomorphic to a disk. Minimum genus embeddings are 2-cell embeddings [249, Section 3.4].

Euler's formula generalizes to arbitrary surfaces.

Theorem A.15 (Euler-Poincaré Formula). *If G is embedded on a surface of Euler characteristic eg, then $n - m + f \geq 2 -$ eg, where $n = |V(G)|$, $m = |E(G)|$, f is the number of faces of the embedding, and c is the number of connected components of G. If G is 2-cell embedded, we have $n - m + f = 2 -$ eg.*

Theorem A.15 gives us a bound on the minimum genus of an orientable surface into which a graph can be embedded.

Theorem A.16. *Every graph has a 2-cell embedding on an orientable surface of genus at most $(m - n + 1)/2$.*

We conclude that every graph has Euler genus at most $m - n + 1$.

Proof. Every graph does have a 2-cell embedding on some orientable surface (take a minimum genus embedding). On the other hand, if the graph does have a 2-cell embedding on a surface of genus g, then by the Euler-Poincaré Formula, Theorem A.15, we have $2 - 2g = n - m + f$, where n, m, and f are the number of vertices, edges and faces of the embedding. Hence $g \leq (m - n + 1)/2$, which is what we needed. $\qquad\square$

The Euler-Poincaré Formula implies the following result, using essentially the same proof we saw in Corollary A.5.

Corollary A.17. *If a multigraph G can be embedded in a surface with Euler genus eg so that the smallest face size is g, then*

$$m \leq \frac{g}{g - 2}(n + \text{eg} - 2),$$

where $n = |V(G)| \geq 3$, $m = |E(G)|$.

As earlier, we can conclude that a graph of girth g satisfies $m \leq \frac{g}{g-2}(n + \text{eg} - 2)$, where we let $g = 3$ if G is a forest.

Euler (partially) proved that a graph contains what we now call an *Eulerian circuit*, a closed walk that traverses every edge exactly once, if and only if every vertex has even degree. We call a graph *Eulerian* if all vertices have even degree. If the graph is embedded in a surface, we can find an Eulerian circuit that does not cross itself (considering the circuit as a closed curve on the surface, self-intersections are allowed).

Theorem A.18 (Belyĭ [57]). *Every connected Eulerian graph embedded in a surface contains an Eulerian circuit without self-crossings.*

Proof. Let D be the embedding of a connected Eulerian graph G in some surface. If G has a vertex v of degree greater than 2, let e, f, and g be three consecutive edges at v. We try splitting v into two copies, one adjacent to e and f, the other adjacent to the remaining edges incident to v. If the resulting graph is connected, then it is Eulerian, by induction, it contains an Eulerian circuit without self-crossings, which also is an Eulerian circuit without self-crossings in the original graph (after identifying the two copies of v). If the resulting graph is not connected, we have two Eulerian subgraphs of G, one of which contains e and f. Hence, if instead of working with e and f, we split f and g off v, the graph remains connected, and we can apply induction as above. $\square$

In the plane any two simple closed curves are isomorphic, by the Jordan-Schoenflies theorem. In a surface Σ, the situation is more complex; let γ be a simple closed curve. Then γ can be *one-sided* or *two-sided* depending on how many sides it has (to be one-sided, γ must be drawn on a non-orientable surface). If $\Sigma - \gamma$ is connected, then γ is *non-separating*, otherwise it is *separating*. Cutting the surface along a non-separating curve, and filling in any boundary holes by disks reduces the Euler genus of the surface. A separating curve is called *contractible* if it is isomorphic to the boundary of a disk (it can be contracted to a point within the surface).

A cycle in an embedded graph traces a simple closed curve in a surface, so we can extend all the terms we introduced for simple closed curves to cycles. The following two lemmas tell us something abound embedding a non-planar graph in a surface.

Lemma A.19. *If a graph can be embedded in a surface so that all cycles of the graph are contractible, then the graph is planar.*

Therefore the embedding of a non-planar graph in a surface must contain a non-contractible cycle. The next lemma strengthens this result slightly.

Lemma A.20. *An embedding of a non-planar graph in a surface contains a non-separating cycle.*

Proof. Suppose G is embedded in a surface Σ. If the drawing of G contains a separating cycle with underling curve γ and so that all of G lies in one of the two components of $\Sigma - \gamma$, we can replace (if necessary) the other component with a disk, with its boundary glued in along γ. In other words, we can assume that all separating cycles are contractible. If all cycles were separating, they would all be contractible, which would imply that the graph is planar, a contradiction. Therefore at least one of the cycles must be non-separating. $\square$

A.5   Rotations and Embedding Schemes

Embeddings in surfaces can be hard to work with, and we sometimes switch to a combinatorial way of describing them. For embeddings in an orientable surface, this can be done using rotation systems. The *rotation* at a vertex is a cyclic permutation ρ of the edges incident to the vertex. For an actual embedding, $\rho(e) = f$ means that f is the next edge after e in a clockwise direction around the vertex.[2] A *rotation system* assigns a rotation to every vertex of the graph. A drawing of a graph *realizes* the rotation system if the rotation at each vertex of the drawing is as described by the rotation system. For an (abstract) graph and a rotation system, we can construct a surface on which the graph embeds realizing the given rotation system (use the rotation system to trace all faces and glue the faces together to form a surface on which the graph embeds [249, Theorem 3.2.4]).

Theorem A.21. *In linear time we can compute the smallest genus of an orientable surface into which a graph with a given rotation system can be embedded.*

Proof. Fix a minimum genus orientable surface Σ on which the graph can be embedded realizing the given rotation system. Then this embedding is a 2-cell embedding, so, by the Euler-Poincaré Formula, we can calculate the genus of the surface by counting vertices, edges, faces, and connected components. To count the number of faces, we construct the set of faces, by tracing them via the rotation system. Since each edge will be traversed twice, this takes linear time, as do all other counts. So g can be calculated in linear time. $\qquad\square$

For non-orientable surfaces we need more information. Suppose we have a graph embedded in a non-orientable surface. At each vertex we have a rotation. Since the surface is not orientable, we cannot distinguish clockwise and anticlockwise, so we arbitrarily label each rotation with one of the two directions. If an edge uv has rotations at u and v with opposite directions, we label uv with -1, and 1 otherwise. This is the *signature* of the edge. A rotation system together with a signature (a mapping from $E(G)$ to $\{-1, 1\}$) is called an *embedding scheme*. As we saw an actual embedding can be described by an embedding scheme, and in the reverse direction, an embedding scheme uniquely identifies an embedding on a surface up to homeomorphism [249, Theorem 3.3.1]. Note that if we flip the direction of rotation at a vertex, and negate the signatures of all edges incident to the vertex, we are still describing the same embedding. We therefore call two embedding schemes *equivalent* if they only differ by a set of such direction flips. Signatures allow us to determine the sidedness of a cycle: a cycle is one-sided if and only if the product of the signatures along the cycle's edges is -1. We refer to [249] for details on how to work with embedding schemes.

---

[2]In the presence of loops we need to work with permutations of half-edges.

A.6 Crossings in Drawings

Here we present two results on how to make drawings with crossings "nicer". A drawing is *polygonal* if every edge is drawn as a polygonal arc, that is, a sequence of straight-line segments. The common point of two straight-line segments is called a *bend*.

Lemma A.22. *For every drawing of a graph, there is an isomorphic polygonal drawing of the graph even if the locations of vertices are fixed.*

Proof. Apply a homeomorphism of the plane to ensure that the locations of vertices in the drawing D as are prescribed. Replace each crossing in D with a vertex. Let $\varepsilon > 0$ be smaller than the distance between any two non-incident objects (vertices and edges) of the graph and so that any circle around a vertex v of radius ε intersects each edge incident to v exactly once. ε exists because of compactness. Inside each such circle, we can replace every edge by a straight-line segment (connecting v to the intersection on the boundary), and outside the circles, we can replace each edge by a polygonal arc in a ε neighborhood without introducing crossings (using a compactness argument). □

Note that this proof gives no bound on the number of bends or the size of the drawing, a consequence of using compactness. One can use a result due to Pach and Wenger [271] to show that the number of bends in each edge can be assumed to be linear. Since we only use the lemma for theoretical arguments, we leave the details to Exercise (A.8).

If more than two edges intersect in a point, that point may be the result of various crossing and touching pairs. The following lemma tells us that we can separate the crossing points and remove touching points altogether.

Lemma A.23. *Suppose we are given a drawing of a graph in which more than two edges intersect in a common point, and so that k pairs of edges cross in that point. We can then perturb the edges involved to replace the common intersection point with k crossing points (and no further intersections). If the edges are polygonal, with a bend in the common point, this redrawing can be performed without increasing the number of bends.*

Proof. Let p be the common point; pick an edge e whose segments ending at p are closest in the rotation at p in the sense that no other edge has both its segments in the clockwise wedge formed by the segments of e. Move the bend of e at p away from p along any ray in the wedge. This creates proper crossings of e with any of the edges having segments in its wedge, which are exactly the edges e crossed to start with, so we are not increasing the number of crossings. □

A.7   Notes

For basic background on surfaces, we refer the reader to Diestel [114]. Standard references for topological graph theory include "Graphs on Surfaces" by Mohar and Thomassen [249] and "Topological Graph Theory" by Gross and Tucker [162]; for a shorter survey, see Archdeacon's "Topological Graph Theory" [33]. For a more topological and geometric point of view, we recommend Stillwell's eminently readable "Classical topology and combinatorial group theory" [322].

The most exhaustive (and modern) treatment of rotation systems and embedding schemes can be found in [249, Chapter 3].

A.8   Exercises

(A.1) Show that any two simple closed curves in the plane that intersect finitely often, intersect an even number of times. *Hint:* Use the Jordan Curve Theorem, Theorem A.1.

(A.2) Show that if a contractible curve in a surface intersects another simple closed curve in the surface a finite number of times, the two curves intersect an even number of times.

(A.3) Let e be an edge in a plane drawing of a graph G. Show that there is an isomorphic drawing of G in which e is drawn as a straight-line segment. *Hint:* Use Theorem A.3.

(A.4) Prove Lemma A.8.

(A.5) Show that a graph is outer-planar if and only if it does not contain a subdivision of K_4. *Hint:* Use Kuratowski's theorem.

(A.6) Show that for every $n \geq 4$, there is a bipartite n-vertex graph with $\lfloor 2(n-2) \rfloor$ edges.

(A.7) Prove Lemma A.19. *Hint:* Contract edges of a spanning tree, maintaining rotation system.

(A.8) Prove Lemma A.22 with the additional constraint that every edge has at most a linear number of bends. *Hint:* Use a result by Pach and Wenger [271]. Work with a planar skeleton, then add edges; if two edges cross, remove the crossing by rerouting. Multiple crossings can be removed using the method described earlier.

Appendix B

Basics of Complexity

B.1 Algorithms, Time, and Space

When talking about solving a problem algorithmically, it is (reasonably) safe to think about a program in your favorite programming language (Python, Java, etc.). A formalization would require a formal machine model, such as RAM machines[1], which we will try to avoid. There is danger in being too informal though: the short (Python) code fragment

```
x = 2
for i in range(n):
    x = x*x
print(x)
```

calculates 2^{2^n} in $O(n)$ steps. So while it looks like printing x would take a single step, that is misleading, since it would really take 2^n steps to print all digits of x. The solution is to restrict the programming language, only allowing integer variables and the list and dictionary datatype (which is sufficient to encode graphs), and disallowing multiplication.[2]

The *time* taken by a program on some input is measured in the number of steps (commands) performed by the algorithm. The space used by an algorithm is measured in the total number of bits of data stored by the algorithm in variables. Consider running the following code on input $G = 1 : [2, 3, 4], 2 : [3, 4], 3 : [4]$.

```
edge_count = 0
for v in G:
    for u in G[v]:
        edge_count += 1
return edge_count//2
```

If we measure the size of an input x by its *bitlength*, written $|x|$, the number of bits it takes to write down, then this program takes $O(|G|)$ steps to calculate

---

[1]Traditional Turing machines are less appropriate, since their restriction to a single, non-random access data structure (the tape) distorts an analysis of running times), see [273].

[2]In this restricted language, we can simulate multiplication with integers with polynomially many bits by using repeated addition.

309

the number of edges in the graph G. We say the algorithm takes *linear time*.[3] How much space does the algorithm use? We could say linear space, but it often makes sense to exclude input and output variables from the accounting, so this program only requires logarithmic space to store variables u and v.

B.2   Computational Complexity

Strictly speaking, the computational complexity of a problem like crossing number is not well-defined: we need to specify an encoding. We will always encode our problems in binary. Therefore, a *computational (decision) problem* is a subset of $\{0,1\}^*$, the set of finite binary strings, consisting of the positive instances of the problem. For example, the crossing number problem would be encoded as the set

$$\mathrm{CNP} := \{\langle G, k \rangle : \mathrm{cr}(G) \leq k\},$$

where $\langle \cdot \rangle$ is a (succinct) binary encoding mechanism.[4] So $(K_5, 0) \notin \mathrm{CNP}$, and $(K_5, 1) \in \mathrm{CNP}$.

Computational complexity theory begins by comparing computational problems and trying to classify them by hardness. As comparison tools we use translations between the problems: if problem A can be efficiently translated into problem B, and problem B can be solved easily, then so can A. There are different ways to make precise the notion of efficient translation, but here we will use polynomial-time many-one reductions.

Definition B.1. We say a function $f : \{0,1\}^* \to \{0,1\}^*$ can be *computed in polynomial time* if there is an algorithm which calculates $f(x)$ for every $x \in \{0,1\}^*$ in time $O(|x|^k)$ for some fixed $k \in \mathbb{N}$. A computational problem A *(polynomial-time many-one) reduces* to a computational problem B if there is a polynomial time computable function f, the *(polynomial-time many-one) reduction*, so that $x \in A$ if and only if $f(x) \in B$ for all $x \in \{0,1\}^*$.

There are two ways to look at a reduction: if A reduces to B, and B is easy, then so is A; or, if A is hard, then so is B.[5] Let us make the first claim precise. We say a computational decision problem A can be *decided in*

---

[3]There is a bit of subtlety here. Why did we not replace the inner loop by edge_count += len(G[v])? The answer is that this hides the fact that computing the size (length) of G[v] takes time proportionally to the size, and not a single step.

[4]Imagine writing the details of G and k using Unicode and then converting to binary. The actual encoding itself does not matter as long as it is succinct. In particular, we generally do not encode integers in unary: if we write k as 1^k, this can distort the computational complexity analysis.

[5]The expression "A reduces to B" is somewhat unfortunate, in that it may (wrongly) suggest that B is "smaller", or "easier" in some sense than A, but the opposite is the case.

polynomial time, if $\chi_A(x) := \begin{cases} 1 & x \in A \\ 0 & \text{else} \end{cases}$, the *characteristic function* of A can be computed in polynomial time. We say that A is *decidable* if χ_A can be computed by an algorithm (without any bounds).

Lemma B.2. *If A reduces to B and B can be decided in polynomial time, then A can be decided in polynomial time.*

Polynomial time will be the base of our computational complexity investigations, it is what we consider computationally easy. We write $\mathbf{P} := \{A \subseteq \mathcal{P}(\{0,1\}^*) : \chi_A$ can be decided in polynomial-time$\}$, the class of all problems solvable in polynomial time. Many natural graph algorithm problems can be solved in polynomial time, for example planarity, which even has a *linear time* algorithm, that is an algorithm running in time $O(n)$. This highly non-trivial result is due to Hopcroft and Tarjan [194].

A *complexity class* $\mathcal{C} \subseteq \mathcal{P}(\{0,1\}^*)$ is a set of computational decision problems which is downward closed under reductions. That is, if $B \in \mathcal{C}$, and A reduces to B, we must also have $A \in \mathcal{C}$. A problem B is *hard* for $\mathcal{C}$, or $\mathcal{C}$-*hard* if A reduces to B for every $A \in \mathcal{C}$. If, moreover, $B \in \mathcal{C}$, we say B is $\mathcal{C}$-*complete* or *complete* for $\mathcal{C}$. By definition, $\mathbf{P} = \mathcal{C}(\{0\})$ (see Exercise (B.3)).

We can use complete problems to define complexity classes: if A is a computational (decision) problem, then $\mathcal{C}(A) := \{B \subseteq \mathcal{P}(\{0,1\}^*)$ so that B reduces to $A\}$, is a complexity class with A as a complete problem.

Using this idea, we can define $\mathbf{NP}$, *non-deterministic polynomial time* using any of its complete problems, for example, SAT $:= \{\varphi : \varphi$ is a satisfiable Boolean formula$\}$, the *(Boolean) satisfiability problem*. With that, $\mathbf{NP} := \mathcal{C}(\text{SAT})$. Typically, $\mathbf{NP}$ is defined using a machine model (non-deterministic machines), and then SAT is shown $\mathbf{NP}$-complete for $\mathbf{NP}$. We avoid the machine model, but we do need a version of it to simplify showing that certain problems are in $\mathbf{NP}$.

Theorem B.3 (Cook [146]). *Let $A = \{(x_i, w_i) : i \in \mathbb{N}\}$ be a decision problem decidable in polynomial time, and so that $|w_i| \le p(|x_i|)$ for some polynomial p. Then $A_1 := \{x :$ there is a w so that $(x, w) \in A\}$, the projection of A onto its first coordinate, lies in $\mathbf{NP}$.*

We often call a w for which $(x, w) \in A$ a *witness* for $x \in A_1$. Intuitively, Cook's theorem says that problems for which membership can be verified in polynomial time if one knows a polynomial size witnesses belong to $\mathbf{NP}$. This will allow us to easily place problems in $\mathbf{NP}$. For $\mathbf{NP}$-hardness proofs we now use the second aspect of a reduction we mentioned earlier.

Lemma B.4. *If A reduces to B and A is $\mathcal{C}$-complete, then B is $\mathcal{C}$-hard. If $B \in \mathcal{C}$, then it is $\mathcal{C}$-complete.*

There is a large number of known $\mathbf{NP}$-complete problems we can use for $\mathbf{NP}$-hardness reductions, see [146] for the classical treatment of this subject.

Other traditional complexity classes include $\mathbf{EXP}$, exponential time, the class of problems solvable in exponential time, and **PSPACE**, the class of problems solvable in polynomial space. If we want to define these via complete problems, we can define **PSPACE** through any of its complete problems, such as generalized geography, Sokoban, or quantified Boolean formula, and **EXP**, via generalized chess, or the bounded halting problem. The complexity classes relate to each other as follows:

$$\mathbf{P} \subseteq \mathbf{NP} \subseteq \mathbf{PSPACE} \subseteq \mathbf{EXP}.$$

It is known that $\mathbf{P} \neq \mathbf{EXP}$, but none of the other inclusions are known to be proper.

For graph drawing, there is another complexity class which turns out to be useful. Let ETR := $\{\varphi : \varphi$ is a true, existentially quantified sentence in the theory of the reals$\}$. The theory of the reals has constants 0, 1, functions $+$, $*$, comparisons $<$, $=$, and all the Boolean operations. For example,

$$(\exists x) : x * x = 1 + 1$$

is a sentence expressing the existence of $\sqrt{2}$, and thus belongs to ETR; similarly,

$$(\exists x_1, x_2, \ldots, x_n) : x_1 * x_1 < 1 \wedge x_2 * x_2 < x_1 \wedge \cdots \wedge x_n * x_n < x_{n-1} \wedge x_n > 0,$$

expresses that $0 < x_n < 2^{2^n}$. In spite of the high precision required to express a solution, it is known that membership in ETR can be decided in polynomial space, a result due to Canny [83]. So if we define $\exists \mathbb{R} := \mathcal{C}(\text{ETR})$, then $\exists \mathbb{R}$, the computational complexity class of the existential theory of the reals, contains **NP**, and lies in polynomial space, **PSPACE**, that is $\mathbf{NP} \subseteq \exists \mathbb{R} \subseteq \mathbf{PSPACE}$. $\exists \mathbb{R}$ captures the complexity of many problems such as pseudoline stretchability, and the rectilinear crossing number problem.

B.3 Notes

For an introduction to complexity theory, and **NP**-completeness in particular, Garey and Johnson's book is still the standard reference [146]. For material on the existential theory of the reals, see [237, 298].

B.4 Exercises

(B.1) Show Lemma B.2.

(B.2) Show Lemma B.4.

(B.3) Show that $\mathbf{P} = \mathcal{C}(\{0\})$. Why can we not define $\mathbf{P}$ as $\mathcal{C}(\emptyset)$?

(B.4) Show that $\{(G, k) : G$ contains a clique of size at least $k\}$ lies in $\mathbf{NP}$.

(B.5) In the *sum of square root problems* you are given a list of integers $(k_1, \ldots, k_n, k)$, and you need to decide whether $\sum_{i=1}^{n} k_i^{1/2} < k$. Show that this problem belongs to $\exists \mathbb{R}$.

Bibliography

[1] Bernardo M Ábrego, Oswin Aichholzer, Silvia Fernández-Merchant, Thomas Hackl, Jürgen Pammer, Alexander Pilz, Pedro Ramos, Gelasio Salazar, and Birgit Vogtenhuber. All good drawings of small complete graphs. In *EuroCG15*, pages 57–60, 2015. Available at `http://eurocg15.fri.uni-lj.si/pub/eurocg15-book-of-abstracts.pdf` (last accessed 10/8/2017).

[2] Bernardo M. Ábrego, Oswin Aichholzer, Silvia Fernández-Merchant, Dan McQuillan, Bojan Mohar, Petra Mutzel, Pedro Ramos, R. Bruce Richter, and Birgit Vogtenhuber. Bishellable drawings of K_n. *CoRR*, abs/1510.00549, 2015.

[3] Bernardo M. Ábrego, Oswin Aichholzer, Silvia Fernández-Merchant, Pedro Ramos, and Gelasio Salazar. The 2-page crossing number of K_n. *Discrete Comput. Geom.*, 49(4):747–777, 2013.

[4] Bernardo M. Ábrego, Oswin Aichholzer, Silvia Fernández-Merchant, Pedro Ramos, and Gelasio Salazar. More on the crossing number of K_n: Monotone drawings. *Electronic Notes in Discrete Mathematics*, 44:411–414, 2013.

[5] Bernardo M. Ábrego, Oswin Aichholzer, Silvia Fernández-Merchant, Pedro Ramos, and Gelasio Salazar. Shellable drawings and the cylindrical crossing number of K_n. *Discrete Comput. Geom.*, 52(4):743–753, 2014.

[6] Bernardo M. Ábrego, Mario Cetina, Silvia Fernández-Merchant, Jesús Leaños, and Gelasio Salazar. 3-symmetric and 3-decomposable geometric drawings of K_n. *Discrete Appl. Math.*, 158(12):1240–1458, 2010.

[7] Bernardo M. Ábrego, Mario Cetina, Silvia Fernández-Merchant, Jesús Leaños, and Gelasio Salazar. On $\leq k$-edges, crossings, and halving lines of geometric drawings of K_n. *Discrete Comput. Geom.*, 48(1):192–215, 2012.

[8] Bernardo M. Ábrego and Silvia Fernández-Merchant. A lower bound for the rectilinear crossing number. *Graphs Combin.*, 21(3):293–300, 2005.

[9] Bernardo M. Ábrego and Silvia Fernández-Merchant. Geometric drawings of K_n with few crossings. *J. Combin. Theory Ser. A*, 114(2):373–379, 2007.

315

[10] Bernardo M. Ábrego and Silvia Fernández-Merchant. The rectilinear local crossing number of K_n. *ArXiv e-prints*, August 2015.

[11] Bernardo M. Ábrego, Silvia Fernández-Merchant, and Gelasio Salazar. The rectilinear crossing number of K_n: closing in (or are we?). In *Thirty essays on geometric graph theory*, pages 5–18. Springer, New York, 2013.

[12] Eyal Ackerman. On the maximum number of edges in topological graphs with no four pairwise crossing edges. *Discrete Comput. Geom.*, 41(3):365–375, 2009.

[13] Eyal Ackerman. On topological graphs with at most four crossings per edge. *ArXiv e-prints*, September 2015.

[14] Eyal Ackerman, Balázs Keszegh, and Mate Vizer. *On the Size of Planarly Connected Crossing Graphs*, pages 311–320. Springer International Publishing, Cham, 2016.

[15] Eyal Ackerman and Marcus Schaefer. A crossing lemma for the pair-crossing number. In *Graph drawing*, volume 8871 of *Lecture Notes in Comput. Sci.*, pages 222–233. Springer, Heidelberg, 2014.

[16] Pankaj K. Agarwal, Boris Aronov, János Pach, Richard Pollack, and Micha Sharir. Quasi-planar graphs have a linear number of edges. *Combinatorica*, 17(1):1–9, 1997.

[17] Oswin Aichholzer. Oswin aichholzer's homepage. Available online at `http://www.ist.tugraz.at/staff/aichholzer/research/rp/triangulations/crossing/` (last accessed 2/3/2017).

[18] Oswin Aichholzer, Ruy Fabila-Monroy, Hernán González-Aguilar, Thomas Hackl, Marco A. Heredia, Clemens Huemer, Jorge Urrutia, Pavel Valtr, and Birgit Vogtenhuber. On k-gons and k-holes in point sets. *Comput. Geom.*, 48(7):528–537, 2015.

[19] Oswin Aichholzer, Jesús García, David Orden, and Pedro Ramos. New lower bounds for the number of $(\leq k)$-edges and the rectilinear crossing number of K_n. *Discrete Comput. Geom.*, 38(1):1–14, 2007.

[20] Oswin Aichholzer, Thomas Hackl, Alexander Pilz, Pedro Ramos, Vera Sacristán, and Birgit Vogtenhuber. Empty triangles in good drawings of the complete graph. *Graphs Combin.*, 31(2):335–345, 2015.

[21] Oswin Aichholzer, Thomas Hackl, Alexander Pilz, Gelasio Salazar, and Birgit Vogtenhuber. Deciding monotonicity of good drawings of the complete graph, 2015. Abstracts available at `http://dccg.upc.edu/egc15/en/program/` (last accessed 3/25/2017).

[22] Martin Aigner and Günter M. Ziegler. *Proofs from The Book*. Springer-Verlag, Berlin, fifth edition, 2014.

[23] Miklós Ajtai, Vašek Chvátal, Monroe M. Newborn, and Endre Sze-merédi. Crossing-free subgraphs. In *Theory and practice of combina-torics*, volume 60 of *North-Holland Math. Stud.*, pages 9–12. North-Holland, Amsterdam, 1982.

[24] Michael O. Albertson. Chromatic number, independence ratio, and crossing number. *Ars Math. Contemp.*, 1(1):1–6, 2008.

[25] Michael O. Albertson, Daniel W. Cranston, and Jacob Fox. Crossings, colorings, and cliques. *Electron. J. Combin.*, 16(1):Research Paper 45, 11, 2009.

[26] V. B. Alekseev and V. S. Gončakov. The thickness of an arbitrary complete graph. *Mat. Sb. (N.S.)*, 101(143)(2):212–230, 1976.

[27] Noga Alon and Paul Erdős. Disjoint edges in geometric graphs. *Discrete Comput. Geom.*, 4(4):287–290, 1989.

[28] Matthew Alpert, Elie Feder, and Heiko Harborth. The maximum of the maximum rectilinear crossing numbers of d-regular graphs of order n. *Electron. J. Combin.*, 16(1):Research Paper 54, 16, 2009.

[29] Matthew Alpert, Elie Feder, Heiko Harborth, and Sheldon Klein. The maximum rectilinear crossing number of the n dimensional cube graph. In *Proceedings of the Fortieth Southeastern International Conference on Combinatorics, Graph Theory and Computing*, volume 195, pages 147–158, 2009.

[30] Matthew Alpert, Elie Feder, and Yehuda Isseroff. The maximum recti-linear crossing number of generalized wheel graphs. In *Proceedings of the Forty-Second Southeastern International Conference on Combinatorics, Graph Theory and Computing*, volume 209, pages 33–46, 2011.

[31] Patrizio Angelini, Michael A. Bekos, Franz J. Brandenburg, Giordano Da Lozzo, Giuseppe Di Battista, Walter Didimo, Giuseppe Liotta, Fab-rizio Montecchiani, and Ignaz Rutter. On the Relationship between k-Planar and k-Quasi Planar Graphs. *ArXiv e-prints*, February 2017.

[32] Dan Archdeacon. A combinatorial generalization of drawing the complete graph, 1995. Available online at http://www.cems.uvm.edu/TopologicalGraphTheoryProblems/combcrkn.htm (last accessed 6/22/2017).

[33] Dan Archdeacon. Topological graph theory: a survey. *Congr. Numer.*, 115:5–54, 1996.

[34] Dan Archdeacon, C. Paul Bonnington, and Jozef Širáň. Trading cross-ings for handles and crosscaps. *J. Graph Theory*, 38(4):230–243, 2001.

[35] Dan Archdeacon and R. Bruce Richter. On the parity of crossing numbers. *J. Graph Theory*, 12(3):307–310, 1988.

[36] Dan Archdeacon and Kirsten Stor. Superthrackles. *Australas. J. Combin.*, 67:145–158, 2017.

[37] Boris Aronov, Paul Erdős, Wayne Goddard, Daniel J. Kleitman, Michael Klugerman, J'anos Pach, and Leonard J. Schulman. Crossing families. *Combinatorica*, 14(2):127–134, 1994.

[38] Alan Arroyo, Dan McQuillan, R. Bruce Richter, and Gelasio Salazar. Drawings of K_n with the same rotation scheme are the same up to triangle-flips (Gioans Theorem). *Australas. J. Combin.*, 67:131–144, 2017.

[39] Kouhei Asano. The crossing number of $K_{1,3,n}$ and $K_{2,3,n}$. *J. Graph Theory*, 10(1):1–8, 1986.

[40] Gail Adele Atneosen. *On the Embeddability of Compacta in N-Books: Intrinsic and Extrinsic Properties*. PhD thesis, Michigan State University, 1968.

[41] Samuel Bald, Matthew P. Johnson, and Ou Liu. Approximating the maximum rectilinear crossing number. In Thang N. Dinh and My T. Thai, editors, *Computing and Combinatorics - 22nd International Conference, COCOON 2016, Ho Chi Minh City, Vietnam, August 2-4, 2016, Proceedings*, volume 9797 of *Lecture Notes in Computer Science*, pages 455–467. Springer, 2016.

[42] Martin Balko, Radoslav Fulek, and Jan Kynčl. Crossing numbers and combinatorial characterization of monotone drawings of K_n. *Discrete Comput. Geom.*, 53(1):107–143, 2015.

[43] József Balogh, Jesús Leaños, Shengjun Pan, R. Bruce Richter, and Gelasio Salazar. The convex hull of every optimal pseudolinear drawing of K_n is a triangle. *Australas. J. Combin.*, 38:155–162, 2007.

[44] R Baltzer. Eine Erinnerung an Möbius und seinen Freund Weiske. In *Berichte über die Verhandlungen der Königlich Sächsischen Gesellschaft der Wissenschaften zu Leipzig. Mathematisch-Physische Classe.*, volume 37, pages 1–6. S. Hirzel, Leipzig, 1885.

[45] Imre Bárány and Zoltán Füredi. Empty simplices in Euclidean space. *Canad. Math. Bull.*, 30(4):436–445, 1987.

[46] Imre Bárány and Pavel Valtr. Planar point sets with a small number of empty convex polygons. *Studia Sci. Math. Hungar.*, 41(2):243–266, 2004.

[47] János Barát and Géza Tóth. Towards the Albertson conjecture. *Electron. J. Combin.*, 17(1):Research Paper 73, 15, 2010.

[48] Luis Barba, Frank Duque, Ruy Fabila Monroy, and Carlos Hidalgo-Toscano. Drawing the horton set in an integer grid of minimun size. In *Proceedings of the 26th Canadian Conference on Computational Geometry, CCCG 2014, Halifax, Nova Scotia, Canada, 2014*. Carleton University, Ottawa, Canada, 2014.

[49] Calvin Pascal Barton. *The Rectilinear Crossing Number For Complete Simple Graphs in E_2*. PhD thesis, The University of Texas at Austin, 1970.

[50] Joseph Battle, Frank Harary, and Yukihiro Kodama. Every planar graph with nine points has a nonplanar complement. *Bull. Amer. Math. Soc.*, 68:569–571, 1962.

[51] Laurent Beaudou, Antoine Gerbaud, Roland Grappe, and Frédéric Palesi. Drawing disconnected graphs on the Klein bottle. *Graphs Combin.*, 26(4):471–481, 2010.

[52] Lowell Beineke and Robin Wilson. The early history of the brick factory problem. *Math. Intelligencer*, 32(2):41–48, 2010.

[53] Lowell W. Beineke and Frank Harary. The thickness of the complete graph. *Canad. J. Math.*, 17:850–859, 1965.

[54] Lowell W. Beineke, Frank Harary, and John W. Moon. On the thickness of the complete bipartite graph. *Proc. Cambridge Philos. Soc.*, 60:1–5, 1964.

[55] Lowell W. Beineke and Richard D. Ringeisen. On the crossing numbers of products of cycles and graphs of order four. *J. Graph Theory*, 4(2):145–155, 1980.

[56] Michael A. Bekos, Till Bruckdorfer, Michael Kaufmann, and Chrysanthi Raftopoulou. 1-planar-graphs have constant book thickness. In *Algorithms—ESA 2015*, volume 9294 of *Lecture Notes in Comput. Sci.*, pages 130–141. Springer, Heidelberg, 2015.

[57] S. B. Bclyĭ. Self-nonintersecting and nonintersecting chains. *Mat. Notes*, 34(4):802–804, 1983.

[58] Seymour Benzer. On the topology of the genetic fine structure. *Proceedings of the National Academy of Science*, 45:1607–1620, 1959.

[59] Marshall Bern and John R. Gilbert. Drawing the planar dual. *Inform. Process. Lett.*, 43(1):7–13, 1992.

[60] Frank Bernhart and Paul C. Kainen. The book thickness of a graph. *J. Combin. Theory Ser. B*, 27(3):320–331, 1979.

[61] Sandeep N. Bhatt and Frank Thomson Leighton. A framework for solving VLSI graph layout problems. *J. Comput. System Sci.*, 28(2):300–343, 1984.

[62] Therese Biedl and Martin Derka. 1-string B_2-VPG representation of planar graphs. *J. Comput. Geom.*, 7(2):191–215, 2016.

[63] Daniel Bienstock. Some provably hard crossing number problems. *Discrete Comput. Geom.*, 6(5):443–459, 1991.

[64] Daniel Bienstock and Nathaniel Dean. New results on rectilinear crossing numbers and plane embeddings. *J. Graph Theory*, 16(5):389–398, 1992.

[65] Daniel Bienstock and Nathaniel Dean. Bounds for rectilinear crossing numbers. *J. Graph Theory*, 17(3):333–348, 1993.

[66] Thomas Bilski. Embedding graphs in books: a survey. *IEEE Proceedings (Computers and Digital Techniques)*, 139:134–138(4), March 1992.

[67] Jaroslav Blažek and Milan Koman. A minimal problem concerning complete plane graphs. In *Theory of Graphs and its Applications (Proc. Sympos. Smolenice, 1963)*, pages 113–117. Publ. House Czechoslovak Acad. Sci., Prague, 1964.

[68] Rainer Bodendiek, Heinz Schumacher, and Klaus Wagner. Bemerkungen zu einem Sechsfarbenproblem von G. Ringel. *Abh. Math. Sem. Univ. Hamburg*, 53:41–52, 1983.

[69] Drago Bokal. On the crossing numbers of Cartesian products with paths. *J. Combin. Theory Ser. B*, 97(3):381–384, 2007.

[70] Drago Bokal. On the crossing numbers of Cartesian products with trees. *J. Graph Theory*, 56(4):287–300, 2007.

[71] Drago Bokal. Infinite families of crossing-critical graphs with prescribed average degree and crossing number. *J. Graph Theory*, 65(2):139–162, 2010.

[72] Drago Bokal, Éva Czabarka, László A. Székely, and Imrich Vrt'o. General lower bounds for the minor crossing number of graphs. *Discrete Comput. Geom.*, 44(2):463–483, 2010.

[73] Drago Bokal, Gašper Fijavž, and Bojan Mohar. The minor crossing number. *SIAM J. Discrete Math.*, 20(2):344–356, 2006.

[74] Oleg V. Borodin. A new proof of the 6 color theorem. *J. Graph Theory*, 19(4):507–521, 1995.

[75] Alex Brodsky, Stephane Durocher, and Ellen Gethner. The rectilinear crossing number of K_{10} is 62. *Electron. J. Combin.*, 8(1):Research Paper 23, 30, 2001.

[76] Christoph Buchheim, Markus Chimani, Carsten Gutwenger, Michael Jünger, and Petra Mutzel. Crossings and planarization. In Tamassia [331], pages 43–85.

[77] Jonathan F. Buss and Peter W. Shor. On the pagenumber of planar graphs. In *Proceedings of the sixteenth annual ACM symposium on Theory of computing*, STOC '84, pages 98–100. ACM, New York, NY, USA, 1984.

[78] Sergio Cabello. Hardness of approximation for crossing number. *Discrete Comput. Geom.*, 49(2):348–358, 2013.

[79] Sergio Cabello, Éric Colin de Verdière, and Francis Lazarus. Algorithms for the edge-width of an embedded graph. *Comput. Geom.*, 45(5-6):215–224, 2012.

[80] Sergio Cabello and Bojan Mohar. Adding one edge to planar graphs makes crossing number and 1-planarity hard. *SIAM J. Comput.*, 42(5):1803–1829, 2013.

[81] Sergio Cabello, Bojan Mohar, and Robert Sámal. Drawing a disconnected graph on the torus (extended abstract). *Electronic Notes in Discrete Mathematics*, 49:779–786, 2015.

[82] Grant Cairns and Yury Nikolayevsky. Bounds for generalized thrackles. *Discrete Comput. Geom.*, 23(2):191–206, 2000.

[83] John Canny. *The complexity of robot motion planning*, volume 1987 of *ACM Doctoral Dissertation Awards*. MIT Press, Cambridge, MA, 1988.

[84] Vasilis Capoyleas and János Pach. A Turán-type theorem on chords of a convex polygon. *J. Combin. Theory Ser. B*, 56(1):9–15, 1992.

[85] Jean Cardinal and Stefan Felsner. Topological drawings of complete bipartite graphs. In *Graph drawing and network visualization*, volume 9801 of *Lecture Notes in Comput. Sci.*, pages 441–453. Springer, Cham, 2016.

[86] Mario Cetina, Cesar Hernández-Vélez, Jesús Leaños, and Cristóbal Villalobos. Point sets that minimize ($\leq k$)-edges, 3-decomposable drawings, and the rectilinear crossing number of K_{30}. *Discrete Math.*, 311(16):1646–1657, 2011.

[87] Jérémie Chalopin and Daniel Gonçalves. Every planar graph is the intersection graph of segments in the plane: extended abstract. In Michael

Mitzenmacher, editor, *Proceedings of the 41st Annual ACM Symposium on Theory of Computing, STOC 2009, Bethesda, MD, USA, May 31 - June 2, 2009*, pages 631–638. ACM, 2009.

[88] Jérémie Chalopin, Daniel Gonçalves, and Pascal Ochem. Planar graphs have 1-string representations. *Discrete Comput. Geom.*, 43(3):626–647, 2010.

[89] Chandra Chekuri and Anastasios Sidiropoulos. Approximation algorithms for Euler genus and related problems. *CoRR*, abs/1304.2416, 2013.

[90] Markus Chimani, Stefan Felsner, Stephen G. Kobourov, Torsten Ueckerdt, Pavel Valtr, and Alexander Wolff. On the maximum crossing number. *CoRR*, abs/1705.05176, 2017.

[91] Markus Chimani and Petr Hliněný. A tighter insertion-based approximation of the crossing number. *J. Comb. Optim.*, 33(4):1183–1225, 2017.

[92] Chaim Chojnacki (Haim Hanani). Über wesentlich unplättbare Kurven im drei-dimensionalen Raume. *Fundamenta Mathematicae*, 23:135–142, 1934.

[93] Robin Christian, R. Bruce Richter, and Gelasio Salazar. Zarankiewicz's conjecture is finite for each fixed m. *J. Combin. Theory Ser. B*, 103(2):237–247, 2013.

[94] George Chrystal. *Algebra: An elementary text-book for the higher classes of secondary schools and for colleges*. Adam and Charles Black, Edinburgh, 1889.

[95] Julia Chuzhoy, Yury Makarychev, and Anastasios Sidiropoulos. On graph crossing number and edge planarization. In *Proceedings of the Twenty-Second Annual ACM-SIAM Symposium on Discrete Algorithms*, pages 1050–1069. SIAM, Philadelphia, PA, 2011.

[96] Judith Elaine Cottingham. *Thrackles, surfaces, and maximum drawings of graphs*. ProQuest LLC, Ann Arbor, MI, 1993. Thesis (Ph.D.)– Clemson University.

[97] Éva Czabarka, Ondrej Sýkora, László A. Székely, and Imrich Vrt'o. Biplanar crossing numbers. I. A survey of results and problems. In *More sets, graphs and numbers*, volume 15 of *Bolyai Soc. Math. Stud.*, pages 57–77. Springer, Berlin, 2006.

[98] Éva Czabarka, Ondrej Sýkora, László A. Székely, and Imrich Vrt'o. Biplanar crossing numbers. II. Comparing crossing numbers and biplanar crossing numbers using the probabilistic method. *Random Structures Algorithms*, 33(4):480–496, 2008.

[99] Elias Dahlhaus, David S. Johnson, Christos H. Papadimitriou, Paul D. Seymour, and Mihalis Yannakakis. The complexity of multiterminal cuts. *SIAM J. Comput.*, 23(4):864–894, 1994.

[100] Maurits de Graaf and Alexander Schrijver. Grid minors of graphs on the torus. *J. Combin. Theory Ser. B*, 61(1):57–62, 1994.

[101] Etienne de Klerk, J. Maharry, Dmitrii V. Pasechnik, R. Bruce Richter, and Gelasio Salazar. Improved bounds for the crossing numbers of $K_{m,n}$ and K_n. *SIAM J. Discrete Math.*, 20(1):189–202, 2006.

[102] Etienne de Klerk and Dmitrii V. Pasechnik. Improved lower bounds for the 2-page crossing numbers of $K_{m,n}$ and K_n via semidefinite programming. *SIAM J. Optim.*, 22(2):581–595, 2012.

[103] Etienne de Klerk, Dmitrii V. Pasechnik, and Gelasio Salazar. Improved lower bounds on book crossing numbers of complete graphs. *SIAM J. Discrete Math.*, 27(2):619–633, 2013.

[104] Etienne de Klerk, Dmitrii V. Pasechnik, and Gelasio Salazar. Book drawings of complete bipartite graphs. *Discrete Appl. Math.*, 167:80–93, 2014.

[105] Etienne de Klerk, Dmitrii V. Pasechnik, and Alexander Schrijver. Reduction of symmetric semidefinite programs using the regular *-representation. *Math. Program.*, 109(2-3, Ser. B):613–624, 2007.

[106] Éric C. de Verdière, Vojtěch Kaluža, Pavel Paták, Zuzana Patáková, and Martin Tancer. A Direct Proof of the Strong Hanani-Tutte Theorem on the Projective Plane. *ArXiv e-prints*, August 2016.

[107] Alice M. Dean, Joan P. Hutchinson, and Edward R. Scheinerman. On the thickness and arboricity of a graph. *J. Combin. Theory Ser. B*, 52(1):147–151, 1991.

[108] Alice M. Dean and R. Bruce Richter. The crossing number of $C_4 \times C_4$. *J. Graph Theory*, 19(1):125–129, 1995.

[109] Knut Dehnhardt. *Leere konvexe Vielecke in ebenen Punktmengen*. PhD thesis, Technische Universität Braunschweig, 1987.

[110] Matt DeVos, Bojan Mohar, and Robert Šámal. Unexpected behaviour of crossing sequences. *J. Combin. Theory Ser. B*, 101(6):448–463, 2011.

[111] Giuseppe Di Battista, Roberto Tamassia, and Ioannis G. Tollis. Area requirement and symmetry display of planar upward drawings. *Discrete Comput. Geom.*, 7(4):381–401, 1992.

[112] Walter Didimo. Density of straight-line 1-planar graph drawings. *Inform. Process. Lett.*, 113(7):236–240, 2013.

[113] Reinhard Diestel. *Graph theory*, volume 173 of *Graduate Texts in Mathematics*. Springer, Heidelberg, fourth edition, 2010.

[114] Reinhard Diestel. *Graph theory*, volume 173 of *Graduate Texts in Mathematics*. Springer, Heidelberg, fifth edition, 2016.

[115] Hristo N. Djidjev and Shankar M. Venkatesan. Planarization of graphs embedded on surfaces. In *Graph-theoretic concepts in computer science (Aachen, 1995)*, volume 1017 of *Lecture Notes in Comput. Sci.*, pages 62–72. Springer, Berlin, 1995.

[116] Hristo N. Djidjev and Imrich Vrt'o. Crossing numbers and cutwidths. *J. Graph Algorithms Appl.*, 7(3):245–251, 2003.

[117] Hristo N. Djidjev and Imrich Vrt'o. Planar crossing numbers of graphs of bounded genus. *Discrete Comput. Geom.*, 48(2):393–415, 2012.

[118] Vida Dujmović, David Eppstein, and David R. Wood. Structure of graphs with locally restricted crossings. *SIAM J. Discrete Math.*, 31(2):805–824, 2017.

[119] Stephane Durocher, Ellen Gethner, and Debajyoti Mondal. On the biplanar crossing number of k_n. In Thomas C. Shermer, editor, *Proceedings of the 28th Canadian Conference on Computational Geometry, CCCG 2016, August 3-5, 2016, Simon Fraser University, Vancouver, British Columbia, Canada*, pages 93–100. Simon Fraser University, Vancouver, British Columbia, Canada, 2016.

[120] Roger B. Eggleton. *Crossing Numbers of graphs*. PhD thesis, University of Calgary, Calgary, 1973.

[121] Roger B. Eggleton and Richard K. Guy. The crossing number of the n-cube, June 1970.

[122] György Elekes. On the number of sums and products. *Acta Arith.*, 81(4):365–367, 1997.

[123] Hikoe Enomoto, Tomoki Nakamigawa, and Katsuhiro Ota. On the pagenumber of complete bipartite graphs. *J. Combin. Theory Ser. B*, 71(1):111–120, 1997.

[124] P. Erdős and Richard K. Guy. Crossing number problems. *Amer. Math. Monthly*, 80:52–58, 1973.

[125] Paul Erdős. On sets of distances of n points. *Amer. Math. Monthly*, 53:248–250, 1946.

[126] Paul Erdős and Alfréd Rényi. On the evolution of random graphs. *Magyar Tud. Akad. Mat. Kutató Int. Közl.*, 5:17–61, 1960.

[127] Paul Erdős and George Szekeres. A combinatorial problem in geometry. *Compositio Math.*, 2:463–470, 1935.

[128] Paul Erdős and George Szekeres. On some extremum problems in elementary geometry. *Ann. Univ. Sci. Budapest. Eötvös Sect. Math.*, 3–4:53–62, 1960/1961.

[129] Guy Even, Sudipto Guha, and Baruch Schieber. Improved approximations of crossings in graph drawings and VLSI layout areas. *SIAM J. Comput.*, 32(1):231–252, 2002.

[130] Luerbio Faria, Celina M. H. de Figueiredo, R. Bruce Richter, and Imrich Vrt'o. The same upper bound for both: the 2-page and the rectilinear crossing numbers of the n-cube. *J. Graph Theory*, 83(1):19–33, 2016.

[131] Luerbio Faria, Celina Miraglia Herrera de Figueiredo, Ondrej Sýkora, and Imrich Vrt'o. An improved upper bound on the crossing number of the hypercube. *J. Graph Theory*, 59(2):145–161, 2008.

[132] István Fáry. On straight line representation of planar graphs. *Acta Univ. Szeged. Sect. Sci. Math.*, 11:229–233, 1948.

[133] Elie Feder. The maximum rectilinear crossing number of the wheel graph. In *Proceedings of the Forty-Second Southeastern International Conference on Combinatorics, Graph Theory and Computing*, volume 210, pages 21–32, 2011.

[134] Elie Feder, Heiko Harborth, Steven Herzberg, and Sheldon Klein. The maximum rectilinear crossing number of the Petersen graph. In *Proceedings of the Forty-First Southeastern International Conference on Combinatorics, Graph Theory and Computing*, volume 206, pages 31–40, 2010.

[135] Stefan Felsner. On the number of arrangements of pseudolines. *Discrete Comput. Geom.*, 18(3):257–267, 1997. ACM Symposium on Computational Geometry (Philadelphia, PA, 1996).

[136] Stefan Felsner. *Geometric graphs and arrangements*. Advanced Lectures in Mathematics. Friedr. Vieweg & Sohn, Wiesbaden, 2004.

[137] Jacob Fox and János Pach. A separator theorem for string graphs and its applications. *Combin. Probab. Comput.*, 19(3):371–390, 2010.

[138] Jacob Fox, János Pach, and Andrew Suk. Approximating the rectilinear crossing number. In *Graph drawing and network visualization*, volume 9801 of *Lecture Notes in Comput. Sci.*, pages 413–426. Springer, Cham, 2016.

[139] Jacob Fox and Benny Sudakov. Density theorems for bipartite graphs and related Ramsey-type results. *Combinatorica*, 29(2):153–196, 2009.

[140] Radoslav Fulek. Estimating the number of disjoint edges in simple topological graphs via cylindrical drawings. *SIAM J. Discrete Math.*, 28(1):116–121, 2014.

[141] Radoslav Fulek and János Pach. A computational approach to Conway's thrackle conjecture. *Comput. Geom.*, 44(6-7):345–355, 2011.

[142] Radoslav Fulek, Michael J. Pelsmajer, Marcus Schaefer, and Daniel Štefankovič. Adjacent crossings do matter. *J. Graph Algorithms Appl.*, 16(3):759–782, 2012.

[143] Radoslav Fulek, Michael J. Pelsmajer, Marcus Schaefer, and Daniel Štefankovič. Hanani-Tutte, monotone drawings, and level-planarity. In *Thirty essays on geometric graph theory*, pages 263–287. Springer, New York, 2013.

[144] W. H. Furry and Daniel J. Kleitman. Maximal rectilinear crossing of cycles. *Studies in Appl. Math.*, 56(2):159–167, 1977.

[145] Enrique Garcia-Moreno and Gelasio Salazar. Bounding the crossing number of a graph in terms of the crossing number of a minor with small maximum degree. *J. Graph Theory*, 36(3):168–173, 2001.

[146] Michael R. Garey and David S. Johnson. Crossing number is NP-complete. *SIAM J. Algebraic Discrete Methods*, 4(3):312–316, 1983.

[147] Hillel Gazit and Gary L. Miller. Planar separators and the Euclidean norm. In *Algorithms (Tokyo, 1990)*, volume 450 of *Lecture Notes in Comput. Sci.*, pages 338–347. Springer, Berlin, 1990.

[148] Tobias Gerken. Empty convex hexagons in planar point sets. *Discrete Comput. Geom.*, 39(1-3):239–272, 2008.

[149] Ellen Gethner, Leslie Hogben, Bernard Lidický, Florian Pfender, Amanda Ruiz, and Michael Young. On crossing numbers of complete tripartite and balanced complete multipartite graphs. *J. Graph Theory*, 84(4):552–565, 2017.

[150] Emeric Gioan. Complete graph drawings up to triangle mutations. In *Graph-theoretic concepts in computer science*, volume 3787 of *Lecture Notes in Comput. Sci.*, pages 139–150. Springer, Berlin, 2005.

[151] Isidoro Gitler, Petr Hliněný, Jesús Leaños, and Gelasio Salazar. The crossing number of a projective graph is quadratic in the face-width. *Electron. J. Combin.*, 15(1):Research paper 46, 8, 2008.

[152] Lev Yu. Glebsky and Gelasio Salazar. The crossing number of $C_m \times C_n$ is as conjectured for $n \geq m(m+1)$. *J. Graph Theory*, 47(1):53–72, 2004.

[153] Luis Goddyn and Yian Xu. On the bounds of Conway's thrackles. *Discrete Comput. Geom.*, pages 1–7, 2017.

[154] Jacob E. Goodman and Richard Pollack. Proof of Grünbaum's conjecture on the stretchability of certain arrangements of pseudolines. *J. Combin. Theory Ser. A*, 29(3):385–390, 1980.

[155] Jacob E. Goodman and Richard Pollack. Upper bounds for configurations and polytopes in $\mathbf{R}^d$. *Discrete Comput. Geom.*, 1(3):219–227, 1986.

[156] Jacob E. Goodman, Richard Pollack, and Bernd Sturmfels. The intrinsic spread of a configuration in $\mathbf{R}^d$. *J. Amer. Math. Soc.*, 3(3):639–651, 1990.

[157] J. E. Green and Richard D. Ringeisen. Lower bounds for the maximum crossing number using certain subgraphs. In *Proceedings of the Twenty-third Southeastern International Conference on Combinatorics, Graph Theory, and Computing (Boca Raton, FL, 1992)*, volume 90, pages 193–203, 1992.

[158] D. Yu. Grigor'ev and Nikolaĭ N. Vorobjov, Jr. Solving systems of polynomial inequalities in subexponential time. *J. Symbolic Comput.*, 5(1-2):37–64, 1988.

[159] Alexander Grigoriev and Hans L. Bodlaender. Algorithms for graphs embeddable with few crossings per edge. *Algorithmica*, 49(1):1–11, 2007.

[160] Martin Grohe. Computing crossing numbers in quadratic time. *J. Comput. System Sci.*, 68(2):285–302, 2004.

[161] Hans-Dietrich O. F. Gronau and Heiko Harborth. Numbers of nonisomorphic drawings for small graphs. In *Proceedings of the Twentieth Southeastern Conference on Combinatorics, Graph Theory, and Computing (Boca Raton, FL, 1989)*, volume 71, pages 105–114, 1990.

[162] Jonathan L. Gross and Thomas W. Tucker. *Topological graph theory*. Dover Publications, Inc., Mineola, NY, 2001.

[163] Branko Grünbaum. *Arrangements and spreads*. Conference Board of the Mathematical Sciences Regional Conference Series in Mathematics, No. 10. American Mathematical Society Providence, R.I., 1972.

[164] Branko Grünbaum. Selfintersections of polygons. *Geombinatorics*, 8(2):37–45, 1998.

[165] Richard K. Guy. A combinatorial problem. *Nabla (Bull. Malayan Math. Soc)*, 7:68–72, 1960.

[166] Richard K. Guy. The decline and fall of Zarankiewicz's theorem. In *Proof Techniques in Graph Theory (Proc. Second Ann Arbor Graph Theory Conf., Ann Arbor, Mich., 1968)*, pages 63–69. Academic Press, New York, 1969.

[167] Richard K. Guy. Twenty odd questions in combinatorics. In *Proc. Second Chapel Hill Conf. on Combinatorial Mathematics and its Applications (Univ. North Carolina, Chapel Hill, N.C., 1970)*, pages 209–237. Univ. North Carolina, Chapel Hill, N.C., 1970.

[168] Richard K. Guy. Crossing numbers of graphs. In *Graph theory and applications (Proc. Conf., Western Michigan Univ., Kalamazoo, Mich., 1972; dedicated to the memory of J. W. T. Youngs)*, pages 111–124. Lecture Notes in Math., Vol. 303, Springer, Berlin, 1972.

[169] Richard K. Guy. The slimming number and genus of graphs. *Canad. Math. Bull.*, 15:195–200, 1972.

[170] Richard K. Guy, Tom Jenkyns, and Jonathan Schaer. The toroidal crossing number of the complete graph. *J. Combinatorial Theory*, 4:376–390, 1968.

[171] Frank Harary. *Graph theory*. Addison-Wesley Publishing Co., Reading, Mass., 1969.

[172] Frank Harary and Anthony Hill. On the number of crossings in a complete graph. *Proc. Edinburgh Math. Soc. (2)*, 13:333–338, 1962/1963.

[173] Frank Harary, Paul C. Kainen, and Allen J. Schwenk. Toroidal graphs with arbitrarily high crossing numbers. *Nanta Math.*, 6(1):58–67, 1973.

[174] Heiko Harborth. über die Kreuzungszahl vollständiger, n-geteilter Graphen. *Math. Nachr.*, 48:179–188, 1971.

[175] Heiko Harborth. Parity of numbers of crossings for complete n-partite graphs. *Math. Slovaca*, 26(2):77–95, 1976.

[176] Heiko Harborth. Drawings of graphs and multiple crossings. In *Graph theory with applications to algorithms and computer science (Kalamazoo, Mich., 1984)*, pages 413–421. Wiley, New York, 1985.

[177] Heiko Harborth. Drawings of the cycle graph. In *Nineteenth Southeastern Conference on Combinatorics, Graph Theory, and Computing (Baton Rouge, LA, 1988)*, volume 66, pages 15–22, 1988.

[178] Heiko Harborth. Maximum number of crossings for the cube graph. In *Proceedings of the Twenty-second Southeastern Conference on Combinatorics, Graph Theory, and Computing (Baton Rouge, LA, 1991)*, volume 82, pages 117–122, 1991.

[179] Heiko Harborth. Empty triangles in drawings of the complete graph. *Discrete Math.*, 191(1-3):109–111, 1998.

[180] Heiko Harborth. Special numbers of crossings for complete graphs. *Discrete Math.*, 244(1-3):95–102, 2002.

[181] Heiko Harborth and Ingrid Mengersen. Edges without crossings in drawings of complete graphs. *J. Combinatorial Theory Ser. B*, 17:299–311, 1974.

[182] Heiko Harborth and Sophie Zahn. Maximum number of crossings in drawings of small graphs. In *Graph theory, combinatorics, and algorithms, Vol. 1, 2 (Kalamazoo, MI, 1992)*, pages 485–495. Wiley, New York, 1995.

[183] Joel Hass and Peter Scott. Intersections of curves on surfaces. *Israel J. Math.*, 51(1-2):90–120, 1985.

[184] César Hernández-Vélez, Jesús Leaños, and Gelasio Salazar. On the pseudolinear crossing number. *Journal of Graph Theory*, 84(3):297–310, 2017.

[185] César Hernández-Vélez, Carolina Medina, and Gelasio Salazar. The optimal drawings of $K_{5,n}$. *Electron. J. Combin.*, 21(4):Paper 4.1, 29, 2014.

[186] Petr Hliněný and Markus Chimani. Approximating the crossing number of graphs embeddable in any orientable surface. In *Proceedings of the Twenty-First Annual ACM-SIAM Symposium on Discrete Algorithms*, pages 918–927. SIAM, Philadelphia, PA, 2010.

[187] Petr Hliněný. Crossing-number critical graphs have bounded pathwidth. *J. Combin. Theory Ser. B*, 88(2):347–367, 2003.

[188] Petr Hliněný. Crossing number is hard for cubic graphs. *J. Combin. Theory Ser. B*, 96(4):455–471, 2006.

[189] Pak Tung Ho. The crossing number of $K_{1,m,n}$. *Discrete Math.*, 308(24):5996–6002, 2008.

[190] Michael Hoffmann and Csaba D. Tóth. Two-Planar Graphs Are Quasiplanar. *ArXiv e-prints*, May 2017.

[191] Seok-Hee Hong, Peter Eades, Giuseppe Liotta, and Sheung-Hung Poon. Fáry's theorem for 1-planar graphs. In Joachim Gudmundsson, Julián Mestre, and Taso Viglas, editors, *Computing and Combinatorics - 18th Annual International Conference, COCOON 2012, Sydney, Australia, August 20-22, 2012. Proceedings*, volume 7434 of *Lecture Notes in Computer Science*, pages 335–346. Springer, 2012.

[192] Seok-Hee Hong and Hiroshi Nagamochi. Re-embedding a 1-plane graph into a straight-line drawing in linear time. In Hu and Nöllenburg [196], pages 321–334.

[193] Shlomo Hoory, Nathan Linial, and Avi Wigderson. Expander graphs and their applications. *Bull. Amer. Math. Soc. (N.S.)*, 43(4):439–561, 2006.

[194] John Hopcroft and Robert Tarjan. Efficient planarity testing. *J. Assoc. Comput. Mach.*, 21:549–568, 1974.

[195] Joseph D. Horton. Sets with no empty convex 7-gons. *Canad. Math. Bull.*, 26(4):482–484, 1983.

[196] Yifan Hu and Martin Nöllenburg, editors. *Graph Drawing and Network Visualization - 24th International Symposium, GD 2016, Athens, Greece, September 19-21, 2016, Revised Selected Papers*, volume 9801 of *Lecture Notes in Computer Science*. Springer, 2016.

[197] Mark R. Jerrum. The complexity of finding minimum-length generator sequences. *Theoret. Comput. Sci.*, 36(2-3):265–289, 1985.

[198] Hector A. Juarez and Gelasio Salazar. Drawings of $C_m \times C_n$ with one disjoint family. II. *J. Combin. Theory Ser. B*, 82(1):161–165, 2001.

[199] Mark Jungerman. The non-orientable genus of the n-cube. *Pacific J. Math.*, 76(2):443–451, 1978.

[200] Paul C. Kainen. On a problem of P. Erdős. *J. Combinatorial Theory*, 5:374–377, 1968.

[201] Paul C. Kainen. A lower bound for crossing numbers of graphs with applications to K_n, $K_{p,q}$, and $Q(d)$. *J. Combinatorial Theory Ser. B*, 12:287–298, 1972.

[202] Paul C. Kainen and Arthur T. White. On stable crossing numbers. *J. Graph Theory*, 2(3):181–187, 1978.

[203] M. Katchalski and H. Last. On geometric graphs with no two edges in convex position. *Discrete Comput. Geom.*, 19(3, Special Issue):399–404, 1998.

[204] Meir Katchalski and Amram Meir. On empty triangles determined by points in the plane. *Acta Math. Hungar.*, 51(3-4):323–328, 1988.

[205] Ken-ichi Kawarabayashi and Bruce Reed. Computing crossing number in linear time. In *STOC'07—Proceedings of the 39th Annual ACM Symposium on Theory of Computing*, pages 382–390. ACM, New York, 2007.

[206] Michael Kleinert. Die Dicke des n-dimensionalen Würfel-Graphen. *J. Combinatorial Theory*, 3:10–15, 1967.

[207] Daniel J. Kleitman. The crossing number of $K_{5,n}$. *J. Combinatorial Theory*, 9:315–323, 1970.

[208] Daniel J. Kleitman. A note on the parity of the number of crossings of a graph. *J. Combinatorial Theory Ser. B*, 21(1):88–89, 1976.

[209] Marián Klešč, Daniela Kravecová, and Jana Petrillová. On the crossing numbers of Cartesian products of paths with special graphs. *Carpathian J. Math.*, 30(3):317–325, 2014.

[210] Stephen G. Kobourov, Giuseppe Liotta, and Fabrizio Montecchiani. An annotated bibliography on 1-planarity. *CoRR*, abs/1703.02261, 2017.

[211] Martin Kochol. Construction of crossing-critical graphs. *Discrete Math.*, 66(3):311–313, 1987.

[212] Martin Kochol. Linear jump of crossing number for non-Kuratowski edge of a graph. *Rad. Mat.*, 7(1):177–184, 1991.

[213] Petr Kolman and Jiří Matoušek. Crossing number, pair-crossing number, and expansion. *J. Combin. Theory Ser. B*, 92(1):99–113, 2004.

[214] Vladimir P. Korzhik and Bojan Mohar. Minimal obstructions for 1-immersions and hardness of 1-planarity testing. *J. Graph Theory*, 72(1):30–71, 2013.

[215] Anton Kotzig. On certain decompositions of a graph. *Mat.-Fyz. Časopis. Slovensk. Akad. Vied*, 5:144–151, 1955.

[216] Jan Kratochvíl. String graphs. II. Recognizing string graphs is NP-hard. *J. Combin. Theory Ser. B*, 52(1):67–78, 1991.

[217] Jan Kratochvíl. Crossing number of abstract topological graphs. In *Graph drawing (Montréal, QC, 1998)*, volume 1547 of *Lecture Notes in Comput. Sci.*, pages 238–245. Springer, Berlin, 1998.

[218] Jan Kratochvíl and Jiří Matoušek. String graphs requiring exponential representations. *J. Combin. Theory Ser. B*, 53(1):1–4, 1991.

[219] Jan Kratochvíl and Jiří Matoušek. NP-hardness results for intersection graphs. *Comment. Math. Univ. Carolin.*, 30(4):761–773, 1989.

[220] Yaakov Shimeon Kupitz. On pairs of disjoint segments in convex position in the plane. In *Convexity and graph theory (Jerusalem, 1981)*, volume 87 of *North-Holland Math. Stud.*, pages 203–208. North-Holland, Amsterdam, 1984.

[221] Jan Kynčl. Enumeration of simple complete topological graphs. *European J. Combin.*, 30(7):1676–1685, 2009.

[222] Jan Kynčl. Simple realizability of complete abstract topological graphs in P. *Discrete Comput. Geom.*, 45(3):383–399, 2011.

[223] Jan Kynčl. *Simple Realizability of Complete Abstract Topological Graphs Simplified*, pages 309–320. Springer International Publishing, Cham, 2015.

[224] James R. Lee. Separators in region intersection graphs. *ArXiv e-prints*, August 2016.

[225] Frank Thomson Leighton. New lower bound techniques for VLSI. *Math. Systems Theory*, 17(1):47–70, 1984.

[226] Tom Leighton and Satish Rao. Multicommodity max-flow min-cut theorems and their use in designing approximation algorithms. *J. ACM*, 46(6):787–832, 1999.

[227] Friedrich Levi. Die teilung der projektiven Ebene durch Gerade oder Pseudogerade. *Ber. Math.-Phys. Kl. Schs. Akad. Wiss. Leipzig*, 78:256–267, July 1926.

[228] Annegret Liebers. Planarizing graphs—a survey and annotated bibliography. *J. Graph Algorithms Appl.*, 5:no. 1, 74, 2001.

[229] Xuemin Lin and Peter Eades. Towards area requirements for drawing hierarchically planar graphs. *Theoret. Comput. Sci.*, 292(3):679–695, 2003.

[230] Richard J. Lipton and Robert E. Tarjan. A separator theorem for planar graphs. In *Proceedings of a Conference on Theoretical Computer Science (Univ. Waterloo, Waterloo, Ont., 1977)*, pages 1–10. Comput. Sci. Dept., Univ. Waterloo, Waterloo, Ont., 1978.

[231] Peter C. Liu and R. C. Geldmacher. On the deletion of nonplanar edges of a graph. In *Proceedings of the Tenth Southeastern Conference on Combinatorics, Graph Theory and Computing (Florida Atlantic Univ., Boca Raton, Fla., 1979)*, Congress. Numer., XXIII–XXIV, pages 727–738. Utilitas Math., Winnipeg, Man., 1979.

[232] László Lovász, Katalin Vesztergombi, Uli Wagner, and Emo Welzl. Convex quadrilaterals and k-sets. In *Towards a theory of geometric graphs*, volume 342 of *Contemp. Math.*, pages 139–148. Amer. Math. Soc., Providence, RI, 2004.

[233] Sam Loyd. *Sam Loyd and his puzzles*. Barse & Co, New York, 1928.

[234] Tom Madej. Bounds for the crossing number of the N-cube. *J. Graph Theory*, 15(1):81–97, 1991.

[235] Seth M. Malitz. Graphs with E edges have pagenumber $O(\sqrt{E})$. *J. Algorithms*, 17(1):71–84, 1994.

[236] Anthony Mansfield. Determining the thickness of graphs is NP-hard. *Math. Proc. Cambridge Philos. Soc.*, 93(1):9–23, 1983.

[237] Jiří Matoušek. Intersection graphs of segments and $\exists\mathbb{R}$. *ArXiv e-prints*, June 2014.

[238] Jiří Matoušek. String graphs and separators. In *Geometry, structure and randomness in combinatorics*, volume 18 of *CRM Series*, pages 61–97. Ed. Norm., Pisa, 2015.

[239] Dan McQuillan, Shengjun Pan, and R. Bruce Richter. On the crossing number of K_{13}. *J. Combin. Theory Ser. B*, 115:224–235, 2015.

[240] Dan McQuillan and R. Bruce Richter. A parity theorem for drawings of complete and complete bipartite graphs. *Amer. Math. Monthly*, 117(3):267–273, 2010.

[241] Dan McQuillan and R. Bruce Richter. On the crossing number of K_n without computer assistance. *J. Graph Theory*, 82(4):387–432, 2016.

[242] Ingrid Mengersen. *Kreuzungsfreie Kanten in vollständigen n-geteilten Graphen*. PhD thesis, Technische Universität Braunschweig, 1975.

[243] Ingrid Mengersen. Die Maximalzahl von kreuzungsfreien Kanten in Darstellungen von vollständigen n-geteilten Graphen. *Math. Nachr.*, 85:131–139, 1978.

[244] Grace Misereh and Yuri Nikolayevsky. Thrackles containing a standard musquash. *ArXiv e-prints*, January 2016.

[245] Nikolai E. Mnëv. The universality theorems on the classification problem of configuration varieties and convex polytopes varieties. In *Topology and geometry—Rohlin Seminar*, volume 1346 of *Lecture Notes in Math.*, pages 527–543. Springer, Berlin, 1988.

[246] August Ferdinand Möbius. *Der barycentrische Calcul*. Johann Ambrosuis Barth, Leipzig, 1827. Available online at `http://sites.mathdoc.fr/cgi-bin/oeitem?id=OE_MOBIUS__1_1_0` (last accessed 9/27/2016).

[247] Bojan Mohar. A linear time algorithm for embedding graphs in an arbitrary surface. *SIAM J. Discrete Math.*, 12(1):6–26, 1999.

[248] Bojan Mohar. The genus crossing number. *Ars Math. Contemp.*, 2(2):157–162, 2009.

[249] Bojan Mohar and Carsten Thomassen. *Graphs on surfaces*. Johns Hopkins Studies in the Mathematical Sciences. Johns Hopkins. University Press, Baltimore, MD, 2001.

[250] Ruy Fabila Monroy and Jorge López. Computational search of small point sets with small rectilinear crossing number. *J. Graph Algorithms Appl.*, 18(3):393–399, 2014.

[251] John W. Moon. On the distribution of crossings in random complete graphs. *J. Soc. Indust. Appl. Math.*, 13:506–510, 1965.

[252] Walter Morris and V. Soltan. The Erdős-Szekeres problem on points in convex position—a survey. *Bull. Amer. Math. Soc. (N.S.)*, 37(4):437–458, 2000.

[253] Petra Mutzel, Thomas Odenthal, and Mark Scharbrodt. The thickness of graphs: a survey. *Graphs Combin.*, 14(1):59–73, 1998.

[254] Bogdan Oporowski and David Zhao. Coloring graphs with crossings. *Discrete Math.*, 309(9):2948–2951, 2009.

[255] Zhangdong Ouyang, Jing Wang, and Yuanqiu Huang. Two recursive inequalities for crossing numbers of graphs. *Frontiers of Mathematics in China*, pages 1–7, 2016.

[256] János Pach, Radoš Radoičić, Gábor Tardos, and Géza Tóth. Improving the crossing lemma by finding more crossings in sparse graphs. *Discrete Comput. Geom.*, 36(4):527–552, 2006.

[257] János Pach, Radoš Radoičić, and Géza Tóth. Relaxing planarity for topological graphs. In *More sets, graphs and numbers*, volume 15 of *Bolyai Soc. Math. Stud.*, pages 285–300. Springer, Berlin, 2006.

[258] János Pach, Farhad Shahrokhi, and Mario Szegedy. Applications of the crossing number. *Algorithmica*, 16(1):111–117, 1996.

[259] János Pach, Joel Spencer, and Géza Tóth. New bounds on crossing numbers. *Discrete Comput. Geom.*, 24(4):623–644, 2000. ACM Symposium on Computational Geometry (Miami, FL, 1999).

[260] János Pach and Ethan Sterling. Conway's conjecture for monotone thrackles. *Amer. Math. Monthly*, 118(6):544–548, 2011.

[261] János Pach, László A. Székely, Csaba D. Tóth, and G'eza Tóth. Note on k-planar crossing numbers. *ArXiv e-prints*, November 2016.

[262] János Pach and Jenő Törőcsik. Some geometric applications of Dilworth's theorem. *Discrete Comput. Geom.*, 12(1):1–7, 1994.

[263] János Pach and Géza Tóth. Graphs drawn with few crossings per edge. *Combinatorica*, 17(3):427–439, 1997.

[264] János Pach and Géza Tóth. Thirteen problems on crossing numbers. *Geombinatorics*, 9(4):194–207, 2000.

[265] János Pach and Géza Tóth. Which crossing number is it anyway? *J. Combin. Theory Ser. B*, 80(2):225–246, 2000.

[266] János Pach and Géza Tóth. Monotone drawings of planar graphs. *J. Graph Theory*, 46(1):39–47, 2004.

[267] János Pach and Géza Tóth. Crossing number of toroidal graphs. In *Topics in discrete mathematics*, volume 26 of *Algorithms Combin.*, pages 581–590. Springer, Berlin, 2006.

[268] János Pach and Géza Tóth. How many ways can one draw a graph? *Combinatorica*, 26(5):559–576, 2006.

[269] János Pach and Géza Tóth. Disjoint edges in topological graphs. *J. Comb.*, 1(3-4):335–344, 2010.

[270] János Pach and Géza Tóth. Monotone crossing number. *Mosc. J. Comb. Number Theory*, 2(3):18–33, 2012.

[271] János Pach and Rephael Wenger. Embedding planar graphs at fixed vertex locations. In *Graph drawing (Montréal, QC, 1998)*, volume 1547 of *Lecture Notes in Comput. Sci.*, pages 263–274. Springer, Berlin, 1998.

[272] Shengjun Pan and R. Bruce Richter. The crossing number of K_{11} is 100. *J. Graph Theory*, 56(2):128–134, 2007.

[273] Christos H. Papadimitriou. *Computational complexity*. Addison-Wesley Publishing Company, Reading, MA, 1994.

[274] Ed Pegg and Geoffrey Exoo. Crossing number graphs. beyond sudoku. *Mathematica*, 11, 2009. Available at http://www.mathematica-journal.com/2009/11/crossing-number-graphs/#more-10860 (Last accessed 2/6/2017).

[275] Michael J. Pelsmajer, Marcus Schaefer, and Despina Stasi. Strong Hanani-Tutte on the projective plane. *SIAM J. Discrete Math.*, 23(3):1317–1323, 2009.

[276] Michael J. Pelsmajer, Marcus Schaefer, and Daniel Štefankovič. Odd crossing number and crossing number are not the same. *Discrete Comput. Geom.*, 39(1-3):442–454, 2008.

[277] Michael J. Pelsmajer, Marcus Schaefer, and Daniel Štefankovič. Removing even crossings on surfaces. *European J. Combin.*, 30(7):1704–1717, 2009.

[278] Michael J. Pelsmajer, Marcus Schaefer, and Daniel Štefankovič. Removing independently even crossings. *SIAM J. Discrete Math.*, 24(2):379–393, 2010.

[279] Michael J. Pelsmajer, Marcus Schaefer, and Daniel Štefankovič. Crossing numbers of graphs with rotation systems. *Algorithmica*, 60(3):679–702, 2011.

[280] Richard E. Pfiefer. The historical development of J. J. Sylvester's four point problem. *Math. Mag.*, 62(5):309–317, 1989.

[281] Barry L. Piazza, Richard D. Ringeisen, and Samuel K. Stueckle. Properties of nonminimum crossings for some classes of graphs. In *Graph theory, combinatorics, and applications. Vol. 2*, pages 975–989. Wiley, New York, 1991.

[282] Barry L. Piazza, Richard D. Ringeisen, and Samuel K. Stueckle. Subthrackleable graphs and four cycles. *Discrete Math.*, 127(1-3):265–276, 1994.

[283] Rom Pinchasi. Geometric graphs with no two parallel edges. *Combinatorica*, 28(1):127–130, 2008.

[284] R. Bruce Richter and Gelasio Salazar. Crossing numbers. In *Topics in topological graph theory*, volume 128 of *Encyclopedia Math. Appl.*, pages 133–150. Cambridge Univ. Press, Cambridge, 2009.

[285] R. Bruce Richter and Carsten Thomassen. Minimal graphs with crossing number at least k. *J. Combin. Theory Ser. B*, 58(2):217–224, 1993.

[286] R. Bruce Richter and Carsten Thomassen. Relations between crossing numbers of complete and complete bipartite graphs. *Amer. Math. Monthly*, 104(2):131–137, 1997.

[287] Richard D. Ringeisen and Lowell W. Beineke. The crossing number of $C_3 \times C_n$. *J. Combin. Theory Ser. B*, 24(2):134–136, 1978.

[288] Richard D. Ringeisen, Samuel K. Stueckle, and Barry L. Piazza. Subgraphs and bounds on maximum crossings. *Bull. Inst. Combin. Appl.*, 2:33–46, 1991.

[289] Gerhard Ringel. Teilungen der Ebene durch Geraden oder topologische Geraden. *Math. Z.*, 64:79–102 (1956), 1955.

[290] Gerhard Ringel. Extremal problems in the theory of graphs. In *Theory of Graphs and its Applications (Proc. Sympos. Smolenice, 1963)*, pages 85–90. Publ. House Czechoslovak Acad. Sci., Prague, 1964.

[291] Gerhard Ringel. Ein Sechsfarbenproblem auf der Kugel. *Abh. Math. Sem. Univ. Hamburg*, 29:107–117, 1965.

[292] Adrian Riskin. On the outerplanar crossing numbers of $K_{m,n}$. *Bull. Inst. Combin. Appl.*, 39:16–20, 2003.

[293] Neil Robertson and Paul Seymour. Excluding a graph with one crossing. In *Graph structure theory*, volume 147 of *Contemp. Math.*, pages 669–675. Amer. Math. Soc., Providence, RI, 1993.

[294] Jianping Roth and Wendy Myrvold. Simpler projective plane embedding. *Ars Combin.*, 75:135–155, 2005.

[295] Andres J. Ruiz-Vargas. Empty triangles in complete topological graphs. *Discrete Comput. Geom.*, 53(4):703–712, 2015.

[296] Thomas L. Saaty. Two theorems on the minimum number of intersections for complete graphs. *J. Combinatorial Theory*, 2:571–584, 1967.

[297] Gelasio Salazar and Edgardo Ugalde. An improved bound for the crossing number of $C_m \times C_n$: a self-contained proof using mostly combinatorial arguments. *Graphs Combin.*, 20(2):247–253, 2004.

[298] Marcus Schaefer. Complexity of some geometric and topological problems. In *Graph drawing*, volume 5849 of *Lecture Notes in Comput. Sci.*, pages 334–344. Springer, Berlin, 2010.

[299] Marcus Schaefer. Hanani-Tutte and related results. In *Geometry—intuitive, discrete, and convex*, volume 24 of *Bolyai Soc. Math. Stud.*, pages 259–299. János Bolyai Math. Soc., Budapest, 2013.

[300] Marcus Schaefer, Eric Sedgwick, and Daniel Štefankovič. Recognizing string graphs in NP. *J. Comput. System Sci.*, 67(2):365–380, 2003.

[301] Marcus Schaefer and Daniel Štefankovič. Decidability of string graphs. *J. Comput. System Sci.*, 68(2):319–334, 2004.

[302] Marcus Schaefer and Daniel Štefankovič. Fixed points, Nash equilibria, and the existential theory of the reals. *Theory Comput. Syst.*, 60(2):172–193, 2017.

[303] Edward R. Scheinerman. *Intersection Classes and Multiple Intersection Parameters of Graphs*. ProQuest LLC, Ann Arbor, MI, 1984. Thesis (Ph.D.)–Princeton University.

[304] Edward R. Scheinerman and Herbert S. Wilf. The rectilinear crossing number of a complete graph and Sylvester's "four point problem" of geometric probability. *Amer. Math. Monthly*, 101(10):939–943, 1994.

[305] Alexander Schrijver. *Combinatorial optimization. Polyhedra and efficiency. Vol. C*, volume 24 of *Algorithms and Combinatorics*. Springer-Verlag, Berlin, 2003. Disjoint paths, hypergraphs, Chapters 70–83.

[306] Heinz Schumacher. Ein 7-Farbensatz 1-einbettbarer Graphen auf der projektiven Ebene. *Abh. Math. Sem. Univ. Hamburg*, 54:5–14, 1984.

[307] Heinz Schumacher. On 2-embeddable graphs. In *Topics in combinatorics and graph theory (Oberwolfach, 1990)*, pages 651–661. Physica, Heidelberg, 1990.

[308] Farhad Shahrokhi, Ondrej Sýkora, László A. Székely, and Imrich Vrt'o. Book embeddings and crossing numbers. In *Graph-theoretic concepts in computer science*, volume 903 of *Lecture Notes in Comput. Sci.*, pages 256–268. Springer, Berlin, 1995.

[309] Farhad Shahrokhi, Ondrej Sýkora, László A. Székely, and Imrich Vrt'o. Drawings of graphs on surfaces with few crossings. *Algorithmica*, 16(1):118–131, 1996.

[310] Farhad Shahrokhi, Ondrej Sýkora, László A. Székely, and Imrich Vrt'o. Intersection of curves and crossing number of $C_m \times C_n$ on surfaces. *Discrete Comput. Geom.*, 19(2):237–247, 1998.

[311] Farhad Shahrokhi, Ondrej Sýkora, László A. Székely, and Imrich Vrt'o. The gap between crossing numbers and convex crossing numbers. In *Towards a theory of geometric graphs*, volume 342 of *Contemp. Math.*, pages 249–258. Amer. Math. Soc., Providence, RI, 2004.

[312] Farhad Shahrokhi, Ondrej Sýkora, László A. Székely, and Imrich Vrt'o. On k-planar crossing numbers. *Discrete Appl. Math.*, 155(9):1106–1115, 2007.

[313] Farhad Shahrokhi, Ondrej Sýkora, Lász;ó A. Székely, and Imrich Vrt'o. Crossing numbers: bounds and applications. In *Intuitive geometry (Budapest, 1995)*, volume 6 of *Bolyai Soc. Math. Stud.*, pages 179–206. János Bolyai Math. Soc., Budapest, 1997.

[314] Peter W. Shor. Stretchability of pseudolines is NP-hard. In *Applied geometry and discrete mathematics*, volume 4 of *DIMACS Ser. Discrete Math. Theoret. Comput. Sci.*, pages 531–554. Amer. Math. Soc., Providence, RI, 1991.

[315] F. W. Sinden. Topology of thin film RC circuits. *The Bell System Technical Journal*, 45(9):1639–1662, 1966.

[316] David A. Singer. The rectilinear crossing number of certain graphs, 1971. Unpublished manuscript, available at author's webpage at `http://www.cwru.edu/artsci/math/singer/publish/Rectilinear_crossings.pdf` (Accessed 3/7/2017).

[317] Jozef Širáň. The crossing function of a graph. *Abh. Math. Sem. Univ. Hamburg*, 53:131–133, 1983.

[318] Joel Spencer. The biplanar crossing number of the random graph. In *Towards a theory of geometric graphs*, volume 342 of *Contemp. Math.*, pages 269–271. Amer. Math. Soc., Providence, RI, 2004.

[319] Joel Spencer and Géza Tóth. Crossing numbers of random graphs. *Random Structures Algorithms*, 21(3-4):347–358, 2002.

[320] S. K. Stein. Convex maps. *Proc. Amer. Math. Soc.*, 2:464–466, 1951.

[321] Ernst Steinitz. Über die Maximalzahl der Doppelpunkte bei ebenen Polygonen von gerader Seitenzahl. *Math. Z.*, 17(1):116–129, 1923.

[322] John Stillwell. *Classical topology and combinatorial group theory*, volume 72 of *Graduate Texts in Mathematics*. Springer-Verlag, New York, second edition, 1993.

[323] Samuel K. Stueckle, Barry L. Piazza, and Richard D. Ringeisen. A circular-arc characterization of certain rectilinear drawings. *J. Graph Theory*, 20(1):71–76, 1995.

[324] Andrew Suk. Disjoint edges in complete topological graphs. *Discrete Comput. Geom.*, 49(2):280–286, 2013.

[325] Andrew Suk. On the Erdős-Szekeres convex polygon problem. *ArXiv e-prints*, April 2016.

[326] Andrew Suk and Bartosz Walczak. New bounds on the maximum number of edges in k-quasi-planar graphs. *Comput. Geom.*, 50:24–33, 2015.

[327] Ondrej Sýkora and Imrich Vrt'o. On crossing numbers of hypercubes and cube connected cycles. *BIT*, 33(2):232–237, 1993.

[328] László A. Székely.. Crossing numbers and hard Erdős problems in discrete geometry. *Combin. Probab. Comput.*, 6(3):353–358, 1997.

[329] László A. Székely. Turán's brick factory problem: The status of the conjectures of zarankiewicz and hill. In Ralucca Gera, Stephen Hedetniemi, and Craig Larson, editors, *Graph Theory. Favorite Conjectures and Open Problems - 1*, pages 211–230. Springer International Publishing, 2016.

[330] Peter Guthrie Tait. Some elementary properties of closed plane curves. *Messenger of Mathematics*, 69:132–133, 1877.

[331] Roberto Tamassia, editor. *Handbook on Graph Drawing and Visualization*. Chapman and Hall/CRC, Boca Raton, FL, 2013.

[332] Carsten Thomassen. Rectilinear drawings of graphs. *J. Graph Theory*, 12(3):335–341, 1988.

[333] Carsten Thomassen.   Tutte's spring theorem.   *J. Graph Theory*, 45(4):275–280, 2004.

[334] Géza Tóth. Note on geometric graphs. *J. Combin. Theory Ser. A*, 89(1):126–132, 2000.

[335] Géza Tóth. Note on the pair-crossing number and the odd-crossing number. *Discrete Comput. Geom.*, 39(4):791–799, 2008.

[336] Géza Tóth. A better bound for the pair-crossing number. In *Thirty essays on geometric graph theory*, pages 563–567. Springer, New York, 2013.

[337] Mirosław Truszczyński. Decompositions of graphs into forests with bounded maximum degree. *Discrete Math.*, 98(3):207–222, 1991.

[338] William T. Tutte. How to draw a graph. *Proc. London Math. Soc. (3)*, 13:743–767, 1963.

[339] William T. Tutte. The non-biplanar character of the complete 9-graph. *Canad. Math. Bull.*, 6:319–330, 1963.

[340] William T. Tutte. Toward a theory of crossing numbers. *J. Combinatorial Theory*, 8:45–53, 1970.

[341] Pavel Valtr. Graph drawing with no k pairwise crossing edges. In *Graph drawing*, volume 1353 of *Lecture Notes in Comput. Sci.*, pages 205–218. Springer, Berlin, 1997.

[342] Pavel Valtr. On geometric graphs with no k pairwise parallel edges. *Discrete Comput. Geom.*, 19(3, Special Issue):461–469, 1998.

[343] Pavel Valtr. On the pair-crossing number. In *Combinatorial and computational geometry*, volume 52 of *Math. Sci. Res. Inst. Publ.*, pages 569–575. Cambridge Univ. Press, Cambridge, 2005.

[344] Oleg Verbitsky. On the obfuscation complexity of planar graphs. *Theoret. Comput. Sci.*, 396(1-3):294–300, 2008.

[345] Nikolaĭ N. Vorobjov. Estimates of real roots of a system of algebraic equations. *J. Sov. Math.*, 34:1754–1762, 1986.

[346] Jozef Širáň. Infinite families of crossing-critical graphs with a given crossing number. *Discrete Math.*, 48(1):129–132, 1984.

[347] Klaus Wagner. Bemerkungen zum Vierfarbenproblem. *Jahresber. Dtsch. Math.-Ver.*, 46:26–32, 1936.

[348] Hugh E. Warren. Lower bounds for approximation by nonlinear manifolds. *Trans. Amer. Math. Soc.*, 133:167–178, 1968.

[349] Dan E. Willard. Polygon retrieval. *SIAM J. Comput.*, 11(1):149–165, 1982.

[350] Wynand Winterbach. The crossing number of a graph in the plane. Master's thesis, University of Stellenbosch, South Africa, 2005.

[351] Douglas R. Woodall. Thrackles and deadlock. In *Combinatorial Mathematics and its Applications (Proc. Conf., Oxford, 1969)*, pages 335–347. Academic Press, London, 1971.

[352] Douglas R. Woodall. Cyclic-order graphs and Zarankiewicz's crossing-number conjecture. *J. Graph Theory*, 17(6):657–671, 1993.

[353] Mihalis Yannakakis. Embedding planar graphs in four pages. *J. Comput. System Sci.*, 38(1):36–67, 1989.

Index

Bold page numbers refer to the main occurrence of a term, typically its definition.